CONTEMPORARY MATHEMATICS

Combinatorial Methods in Topology and Algebraic Geometry

AMERICAN MATHEMATICAL SOCIETY

VOLUME 44

CONTEMPORARY MATHEMATICS

Titles in this Series

Titles in this Series

Combinatorial Methods
in Topology and
Algebraic Geometry

CONTEMPORARY
MATHEMATICS

Volume 44

Combinatorial Methods
in Topology and
Algebraic Geometry

John R. Harper and
Richard Mandelbaum, Editors

AMERICAN MATHEMATICAL SOCIETY
Providence · Rhode Island

EDITORIAL BOARD

Proceedings of a conference in honor of Arthur M. Stone held at the University of Rochester, Rochester, New York, June 29—July 2, 1982, and partially supported by the National Science Foundation Grant MCS-8116754.

This volume was prepared by the authors using the T_EX typesetting system and was transmitted to the AMS via magnetic tape. Final copy was produced on an Alphatype CRS at the AMS office. The paper used in this book is acid-free and falls within the guidelines established to ensure permanence and durability. The American Mathematical Society retains all rights except those granted to the United States Government. Copying and reprinting information can be found at the back of this volume.

1980 *Mathematics Subject Classification.* Primary 14J, 20F, 55P, 55S, 57M, 47Q, 57R, 57S.

Library of Congress Cataloging-in-Publication Data
Main entry under title:

Combinatorial methods in topology and algebraic geometry.

    (Contemporary mathematics, ISSN 0271-4132; v. 44)
    Bibliography: p.
    1. Combinatorial topology—Addresses, essays, lectures. 2. Geometry, Algebraic—Addresses, essays, lectures. 3. Combinatorial analysis—Addresses, essays, lectures. I. Harper, John R., 1941— . II. Mandelbaum, Richard, 1946— . III. Series.
QA612.C585   1985   514'.22   85-11244
ISBN 0-8218-5039-3 (alk.paper)

To Dorothy Maharam Stone and Arthur Harold Stone
in appreciation for their contribution and
service to Mathematics.

TABLE OF CONTENTS

INTRODUCTION

In recent years, combinatorial methods have re-appeared in deep and diverse ways, both in topology and algebraic geometry. In honor of Prof. Arthur H. Stone, the Department of Mathematics of the University of Rochester held a conference June 20–July 2, 1982. The purpose was to present a series of lectures surveying recent accomplishments and indicating further directions for research. These proceedings contain reports from most of the lectures that were given.

On behalf of the organizers, it is a pleasure to acknowledge the support of many people and institutions. Financial support was provided by the National Science Foundation and the Mathematics Department of the University of Rochester. We are grateful to help received from Dr. Alvin Thaler in preparing our proposal. During the conference we were assisted by Prof. A. Libgober and M. Scharlemann who organized seminars. Sally Allison, of the University of Rochester, provided us with logistical support. The departmental staff, Roberta Colon and Joan Robinson looked after myriads of details. We would especially like to thank Joan Robinson for an outstanding job of typing all the manuscripts in TEX and Prof. A. Pizer for making sure that everything connected with TEX worked out as expected.

PLENARY LECTURES

M. COHEN/W. METZLER: Combinatorial group theory and simple homotopy theory

W. TUTTE: Map coloring and differential equations

F. COHEN: Braids and homotopy theory

R. MANDELBAUM: Combinatorial group theory and Lefschetz fibrations

F. RAYMOND: Mapping class groups of Seifert manifolds

M. BARRATT: Taming Hopf invariants

B. MOISHEZON: The braid group and algebraic surfaces

M. E. RUDIN: Paracompactness

S. CAPPELL: The topology of group representations

R. KIRBY: An overview of new results in 4-dimensional topology

R. EDWARDS: Freedman's work and its consequences

J. STALLINGS: Finite graphs and free groups

J. MONTESINOS: Representing 3-manifolds as branched covers

LIST OF PARTICIPANTS

Dr. Selman Akbulut
Mathematical Institute at Berkeley
2223 Fulton St., Room 602
Berkeley, CA 94720

Dr. M. G. Barratt
Northwestern University
Evanston, IL 60201

Dr. Winfried Becker
Johann Wolfgang Goethe-Universität
Fachbereich Mathematik
6000 Frankfurt/Main
Robert-Mayer-Strasse 6-10
Germany

Dr. Richard E. Bedient
Department of Mathematics
Hamilton College
Clinton, NY 13323

Dr. Terrence P. Bisson
Canisius College
Department of Mathematics
Buffalo, NY 14208

Dr. Sylvain E. Cappell
NYU-Courant
251 Mercer St.
New York, NY 10012

Dr. Frederick R. Cohen
University of Kentucky
Department of Mathematics
Lexington, KY 40506

Dr. Marshall M. Cohen
Cornell University
Department of Mathematics
Ithaca, NY 14853

Dr. Alan H. Durfee
Holyoke Community College
Department of Mathematics
Holyoke, MA 01040

Dr. Allan L. Edmonds
Indiana University
Department of Mathematics
Bloomington, IN 47405

Dr. Robert D. Edwards
University of California at Los Angeles
Department of Mathematics
Los Angeles, CA 90024

Dr. Steven C. Ferry
University of Kentucky
Department of Mathematics
Lexington, KY 40506

Dr. Jonathan Fine
Southern Illinois University
Department of Mathematics
Carbondale, IL 62901

Dr. David Gabai
University of Pennsylvania
Department of Mathematics
Philadelphia, PA 19104

Dr. Ross Geoghegan
SUNY at Binghamton
Department of Mathematics
Binghamton, NY 13901

Dr. Charles Giffen
University of Virginia
Department of Mathematics
Charlottesville, VA 22903

Dr. Patrick M. Gilmer
Louisiana State University
Department of Mathematics
Baton Rouge, LA 70803

Dr. Norman Goldstein
University of British Columbia
Department of Mathematics
121-1984 Mathematics Rd.
Vancouver (V6T 1Y4)
British Columbia, Canada

Dr. Richard Z. Goldstein
State University of New York
Department of Mathematics
Albany, NY 12222

Dr. John Hempel
Rice University
Department of Mathematics
Houston, TX 77251

Dr. Cynthia Hog
Johann Wolfgang Goethe-Universität
Fachbereich Mathematik
6000 Frankfurt/Main
Robert-Mayer-Strasse 6-10
Germany

Dr. John E. Kalliongis
Southwest Texas State University
Department of Mathematics
San Marcos, TX 78666

Dr. Richard Kane
University of Western Ontario
Department of Mathematics
London, Ontario
Canada

Dr. Louis Kaufman
University of Illinois at Chicago Circle
Department of Mathematics,
    Statistics & Computer Science
Chicago, IL 60680

Dr. Robion C. Kirby
University of California at Berkeley
Department of Mathematics
Berkeley, CA 94720

Dr. Kyung Bai Lee
Purdue University
Department of Mathematics
W. Lafayette, IN 47907

Dr. Anatoly S. Libgober
University of Illinois at Chicago
Department of Mathematics
Box 4348
Chicago, IL 60680

Dr. W. B. Raymond Lickorish
University of Cambridge
Department of Pure Mathematics
    and Mathematical Statistics
16 Mill Lane
Cambridge (CB2 1SB), England

Dr. S. J. Lomonaco, Jr.
Inst. for Defense Analyses, STD
1801 N. Beauregard St.
Alexandria, VA 22311

Dr. Martin Lustig
Johann Wolfgang Goethe-Universität
Fachbereich Mathematik
6000 Frankfurt/Main
Robert-Mayer-Strasse 6-10
Germany

Dr. Darryl McCullough
University of Oklahoma
Department of Mathematics
Norman, OK 73019

Dr. Wolfgang Metzler
Johann Wolfgang Goethe-Universität
Fachbereich Mathematik
6000 Frankfurt/Main
Robert-Mayer-Strasse 6-10
Germany

Dr. Michael Mihalik
Vanderbilt University
Department of Mathematics
Nashville, TN 37240

Dr. Kenneth C. Millett
University of California
    at Santa Barbara
Department of Mathematics
Santa Barbara, CA 93106

Dr. Boris Moishezon
Columbia University
Department of Mathematics
New York, NY 10027

Dr. José M. Montesinos
University of Zaragoza
Fac de Ciencias
Zaragoza, Spain

Dr. Carlos J. Moreno
University of Illinois
Department of Mathematics
Urbana, IL 61801

Dr. Kazumi Nakano
State University of New York
Department of Mathematics
Brockport, NY 14420

Dr. Joseph Neisendorfer
Ohio State University
Department of Mathematics
Columbus, OH 43210

Dr. Hae Soo Oh
Louisiana State University
Department of Mathematics
Baton Rouge, LA 70803

Dr. Donal O'Shea
Mt. Holyoke College
Clapp Lab
South Hadley, MA 01075

Dr. Richard Randall
University of Iowa
Department of Mathematics
Iowa City, IA 52240

Dr. Frank A. Raymond
University of Michigan
Department of Mathematics
Ann Arbor, MI 48109

Dr. Leonard R. Rubin
University of Oklahoma
Department of Mathematics
Norman, OK 73019

Dr. Martin Scharlemann
University of California
    at Santa Barbara
Department of Mathematics
Santa Barbara, CA 93106

Dr. Alan Siegel
Department of Mathematics
Mt. Holyoke College
South Hadley, MA 01075

Dr. Allen Sieradski
University of Oregon
Department of Mathematics
Eugene, OR 97403

Dr. John R. Stallings
University of California at Berkeley
Department of Mathematics
Berkeley, CA 94720

Dr. Mark Steinberger
Northern Illinois University
Department of Mathematics
DeKalb, IL 60115

Dr. Bruce Trace
Department of Mathematics
University of Alabama
Tuscaloosa, AL 35486

Dr. Carol Tretkoff
Brooklyn College
Department of Computer Science
Bedford Avenue and Avenue H
Brooklyn, NY 11210

Dr. Marvin Tretkoff
Stevens Institute of Technology
Department of Mathematics
Castle Point Station
Hoboken, NJ 07030

Dr. Thomas W. Tucker
Colgate University
Department of Mathematics
Hamilton, NY 13346

Dr. Edward C. Turner
State University of New York
Department of Mathematics
Albany, NY 12222

Dr. W. T. Tutte
University of Waterloo
Department of Mathematics
Waterloo (N2L 3G1)
Ontario, Canada

Dr. Samuel A. Weinberger
Princeton University
Department of Mathematics
Princeton, NJ 08544

Dr. James E. West
Cornell University
Department of Mathematics
Ithaca, NY 14853

Dr. George W. Whitehead
Massachusetts Institute of Technology
Department of Mathematics
Cambridge, MA 02139

Dr. John Wood
University of Illinois at Chicago Circle
Department of Mathematics
Chicago, IL 60680

TOPOLOGY AND COMBINATORIAL GROUP THEORY

Contemporary Mathematics
Volume **44**, 1985

COLLAPSES OF $K \times I$ AND GROUP PRESENTATIONS

M. Cohen*, W. Metzler and K. Sauermann

§1. INTRODUCTION. In his paper "On the duncehat" [Ze₁], E.C. Zeeman made the following conjecture. (See §2 for notation.)

(Z)
$\qquad$ If X^2 is a compact, contractible polyhedron
$\qquad$ then $(X^2 \times I) \searrow *$.(i.e.,"$X^2$ is 1-collapsible").

This clearly implies the truth of the conjecture

(AC) $\quad$ If X^2 is a compact, contractible polyhedron then $X^2 \overset{3}{\searrow} *$.

On the other hand (AC) implies a weakened form of (Z), namely [K-M] that there is a polyhedron Y^2 such that $X^2 \nearrow Y^2$ and $(Y^2 \times I) \searrow *$.

These conjectures are of interest because of their well known connections to the 3 and 4 dimensional PL Poincaré conjectures ([Ze₁], [G-R], [A-C]) and — of particular interest here — because of the questions concerning group presentations to which they have been tied: (AC) has been completely characterized ([Wr₂], [M]) in terms of elementary presentation ("Andrews-Curtis") moves. Zeeman's conjecture in dimensions greater than 2 was first disproved using presentations of certain non-trivial group extensions ([C₂], [Ro]). Lickorish [L] and [Wr₁] implicate particular group presentations in the failure of $X^2 \times I$ to collapse to vertical lines $\{p\} \times I$.

In this paper we introduce the notion of *prismatic collapsibility* of a complex K^2 and show that this is inextricably tied to whether the based homotopy classes of the attaching maps for the 2-cells of K^2 form a basis-up-to-conjugation of $\pi_1(K^1)$. These considerations have already led [C-M-Z] to a criterion** for the recognition of bases in the free group on two generators. Our results (most of which were announced in [C-M-Z]) are as follows:

1980 Mathematics Subject Classification: 20F05, 57M05, 57M20, 57Q10, 57R60..

*Partially supported by an NSF Research Grant.

**discovered independently from another viewpoint by Kaneto [Ka]. See also Osborne-Zieschang [O-Z] and [Cohn].

THEOREM. *If K^1 and L^1 are based finite connected PLCW 1-complexes (non-degenerate, so that $K^1 \neq *$) and $f : \pi_1(L^1, *) \to \pi_1(K^1, *)$ is an isomorphism then there is a PL embedding $\alpha : (L^1, *) \to (K^1 \times I, * \times I)$ such that $(K^1 \times I) \searrow \alpha(L^1)$ and $(p \circ \alpha)_* = f$, where $p : K^1 \times I \to K^1$ is the natural projection and $(p \circ \alpha)_*$ is the induced map of fundamental groups.*

As a consequence of Theorem 1 we get

THEOREM 2. *If K^1 is a finite connected 1-complex and $\{w_1, \ldots, w_n\}$ is a basis of the free group $\pi_1(K^1, *)$ then there are attaching maps $\phi_i : (\partial D^2, 1) \to (K^1, *)$ $(1 \leq i \leq n)$ such that $[\phi_i] = w_i$ and such that the resulting 2-complex $K^2 = K^1 \cup e_1^2 \cup \ldots \cup e_n^2$ is prismatically 1-collapsible.*

To say that K^2 is *prismatically 1-collapsible* means, roughly, that there is a subpolyhedron $Y \subset |K^2 \times I|$ such that (a) $|K^2 \times I| \searrow Y \searrow *$ and (b) Y is the union of $|K^1 \times I|$ (= a set of "cylinders" or "prisms") with a set of "sheets" $S_i \subset \bar{e}_i^2 \times I (1 \leq i \leq n)$. For example, Zeeman showed [Ze₁] that the duncehat is prismatically 1-collapsible. See (2.6) for a precise definition.

Note that $[\phi_i]$ in Theorem 2 is a *based* homotopy class. Note also that if the basis $w_1, \ldots, w_n$ is given as a set of words in $F(a_1, \ldots, a_n)$ and K^2 is the standard complex with one vertex realizing these words then the conclusion of Theorem 2 holds for cyclic permutations of the w_i. For we can realize such permutations geometrically without changing the homeomorphism class of K^2.

When $n = 2$, and K^1 is a wedge product of circles, Theorem 2 can be strengthened: Define a set of words $\{w_1, \ldots, w_n\} \subset F(a_1, \ldots, a_n)$ to be a *basis-up-to-conjugation* if there are words $g_1, \ldots, g_n$ such that $\{g_i w_i g_i^{-1} | 1 \leq i \leq n\}$ is a basis of $F(a_1, \ldots, a_n)$.

THEOREM 3. *If $\{w_1, w_2\}$ is a basis-up-to-conjugation of $F(a_1, a_2)$ and if K^1 is the wedge product of two circles then there are attaching maps $\phi_i : (\partial D^2, 1) \to (K^1, *)$ $(i = 1, 2)$ such that $[\phi_i] = w_i$ and such that the resulting 2-complex $K^1 \cup e_1^2 \cup e_2^2$ is prismatically 1-collapsible.*

We do not know whether the analogue of Theorem 3 holds when $n > 2$. [This is conceivable. For example, it is a trivial generalization of Zeeman's prismatic collapse of duncehat $\times I$ that the 2-complex built naturally from the words $w_1 = a_2 a_1 a_2^{-1}, \ldots, w_{n-1} = a_n a_{n-1} a_n^{-1}, w_n = a_1 a_n a_1^{-1}$ is prismatically 1-collapsible. See (2.8).] But we do know that the converse of Theorem 3 holds for all n:

THEOREM 4. *Suppose $K^2 = K^1 \cup e_1^2 \cup \ldots \cup e_n^2$ is prismatically 1-collapsible. Let T be a tree (possibly a vertex, possibly a maximal tree, but in any case a subcomplex) of K^1. Suppose that $e_i^2 (1 \leq i \leq n)$ has attaching map $\phi_i : (\partial D^2, 1) \to (K^1, T)$ which represents $w_i = [\phi_i] \epsilon \pi_1(K^1, T)$. Then $\{w_1, \ldots, w_n\}$ is a basis-up-to-conjugation of $\pi_1(K^1, T)$.*

In particular, when $n = 2$ and $K^0 = \{*\}$, Theorems 3 and 4 give a geometric characterization of bases-up-to-conjugation in $F(a_1, a_2) = \pi_1(S^1 \vee S^1, *)$: *The* words w_1, w_2 form a basis-up-to-conjugation if and only if they represent based homotopy classes in which attaching maps may be chosen so as to yield a prismatically 1-collapsible complex. (Note that the attaching maps may realize these words in an unreduced form and may contain many little wiggles).

As a corollary to Theorem 4 we can deduce the following result of A. Zimmermann [Zi].

THEOREM 5. If K^2 is the standard PLCW 2-complex with one vertex v determined by (possibly non-reduced) words w_i in a_j, a_j^{-1} $(1 \leq i, j \leq n)$ and if K^2 is prismatically 1-collapsible then the link of v in K^2 contains an articulation point.

An articulation point is a point x (not necessarily a vertex) of a graph G such that $G - \{x\}$ has more components than G. In fact, Zimmermann's results are more general than Theorem 5. See (5.6) and [Zi].

Organization of the paper: In §2 we give definitions, notation and some examples. §3 deals with 1-complexes; we prove Theorem 1 and discuss extensions of it. All the results of this section come from K. Sauermann's thesis [S]. In §4 we give a necessary algebraic condition (basis-up-to-conjugation) for a 2-complex to be collapsible. Further, we give a necessary and sufficient geometric criterion (the singular disk property) for collapsibility of polyhedra. §5 deals with prismatic collapsibility and includes the proofs of Theorems 2-5. Finally, in §6 we discuss the use of our methods in understanding (non-prismatic) 1-collapsibility. We consider the conjectures (Z) and (AC) in terms of "cylinders with sheets and holes". Our basic viewpoint in doing this is that the flexible view of (Z) which we have adopted ("there exists an attaching map in the based homotopy class yielding a 1-collapsible complex") makes (Z) a geometric means of studying (AC). The paper closes with a discussion of the singular disk property as a tool for disproving the Zeeman conjecture.

Note from the first author: I would like to record here that, at the outset [C-C], my involvement in the subject matter of this paper was due to the enthusiasm and insight of my friend George Cooke. His death was a deep personal loss and a loss to Mathematics.

§2. DEFINITIONS, NOTATION AND EXAMPLES

We work in the category of compact polyhedra and PL maps. A superscript (as in X^n) indicates the dimension of a polyhedron. I^n is the standard n-cube and $D^n \cong I^n$ is the unit ball in Euclidean space with PL structure determined by some fixed homeomorphism to I^n. We write $1 = (1,0) \epsilon \partial D^2$.

We call Y a *spine* of X and write $X \searrow Y$ if X collapses piecewise linearly to Y. We write $K \searrow^s L$ or $K \searrow^{cw} L$ if the simplicial (respectively PLCW) complex K collapses simplicially (or CW) to its subcomplex L. We write $X \overset{n}{\nearrow\hspace{-0.9em}\searrow} Y$ if there is a sequence of PL expansions and collapses $X = X_0 \to X_1 \to \ldots X_q = Y$ with $\dim X_i \leq n$ for all i. See [H], [R-S] or [Ze2] for background concerning the PL category and PL collapsing. See below for PLCW complexes.

PLCW complexes

It is extremely useful to be able to work with CW complexes, even in the PL category. Thus we make the following definition.

(2.1) *A PLCW complex is a CW complex K such that the underlying space $X = |K|$ has a PL structure for which*

  (1) *The closures of all cells e_i and (thus) of all skeleta $K^n (n \geq 0)$ are subpolyhedra of X.*

  (2) *Each n-cell e_i has a PL attaching map $\phi_i : \partial D^n \to K^{n-1}$ such that the subpolyhedron $e_i \cup K^{n-1}$ is PL homeomorphic, rel K^{n-1}, to $D^n \underset{\partial D^n}{\bigcup} C(\phi_i)$, where $C(\phi_i)$ is the PL mapping cylinder of ϕ_i (see [C_1, §9, 10], [A]).*

REMARK 1: These are just Whitehead's "membrane complexes" ([Wh_2, §6], [Wh_3, §3]) reformulated in terms of our current understanding of the PL category and CW complexes.

REMARK 2: We shall have occasion to speak of *characteristic maps in PLCW complexes*. Technically, some care is required in doing so because the usual CW characteristic map $\Phi_i : D^n \to \bar{e}_i$ corresponding to the PL attaching map $\phi_i = \Phi_i|\partial D^n$ will ordinarily not be PL. (For if Φ_i were PL and $\Phi_i|\partial D^n$ squeezed a line segment to a point then $\Phi_i|(\text{Int } D^n)$ would not be one-one). We will repress the technicalities and speak as though we are dealing with ordinary CW characteristic maps, but we wish, at least here at the outset, to state what is really involved:

A characteristic map $\Phi_i : D^n \to \bar{e}_i$ is any map gotten as follows. Triangulate ∂D^n and K^{n-1} so that the attaching map given by (2.1) is simplicial. Let $D_0^n \subset D^n$ be a smaller concentric n-disk. Extend the triangulation of ∂D^n to a triangulation of $Cl(D^n - D_0^n)$ by triangulating this collar as the simplicial mapping cylinder $C(1_{\partial D^n})$ (as defined in [C_1, §4]). Then triangulate D_0^n as the cone on ∂D_0^n and triangulate D^n as $D_0^n \cup C(1_{\partial D^n})$. Let Φ_i be the composition

$$D^n = D_0^n \cup C(1_{\partial D^n}) \to D_0^n \cup C(\phi_i) \overset{\cong}{\to} \bar{e}_i$$

where the first map is the standard extension of ϕ_i (i.e., the join of $1_{D_0^n}$ with ϕ_i, as in [C_1, §3,4]) and the second map is given by (2.1) above. This PLCW characteristic map *is* PL and is quite as good for our purposes as the usual topological characteristic map. (See [Wh_4, §10], [C_1, §9–10], [A]) for details.) More significantly we shall use this fact:

(2.2) *If Δ is a subpolyhedron of $D^2 \times I$ such that*

$$D^2 \times I \searrow (\partial D^2 \times I) \cup \Delta$$

then

$$(\Phi_i \times 1)(D^2 \times I) \searrow (\Phi_i \times 1)(\partial D^2 \times I) \cup (\Phi_i \times 1)(\Delta).$$

REMARK 3: The usual inductive construction of CW complexes applies to PLCW complexes. (See the references in Remark 2, especially [Wh_4, Lemma 2]). Thus if K^{n-1} is an $(n-1)$ - dimensional PLCW complex and $\{\phi_i : \partial D_i^n \to$

$K^{n-1}\}$ is a finite[†] family of PL maps then the identification space

$$K^n = K^{n-1} \cup \bigcup_i (D_i^n \cup C(\phi_i))$$

naturally becomes a PLCW complex.

REMARK 4: An ordinary CW collapse $K \searrow^{cw} L$ of PLCW complexes induces a PL collapse $|K| \searrow |L|$. See [Wh₂, Lemma 6].

(2.3) NOTATIONAL CONVENTION: We shall assume throughout this paper that

   a) $\Phi_i : D^2 \to K^2$ is a characteristic map for the 2-cell e_i.

   b) $\phi_i = \Phi_i | \partial D^2$ is the attaching map for e_i.

   c) $F_i = \Phi_i \times 1 : D^2 \times I \to K^2 \times I$.

Prismatic collapsibility

A key idea in this paper will be that of a prismatic collapse. We define this in the PL and PLCW categories:

(2.4) A *triangulation* (J, K) of $(X^2 \times I, X^2 \times 0)$ is said to be *prismatic* if the natural projection $X \times I \to X \times 0$ is simplicial as a map from J to K.

(See the proof of (6.1) to see how one obtains a prismatic triangulation from an arbitrary one.)

(2.5) A *polyhedron* X^2 is *prismatically 1-collapsible* if there is a prismatic triangulation (J, K) of $(X \times I, X \times 0)$ such that: if $\sigma_1, \dots, \sigma_t$ are the 2-simplexes of K^2 then there is a set $\sigma_1^*, \dots, \sigma_t^*$ of 2-simplexes of J for which

   (i) σ_i^* projects isomorphically onto σ_i under the natural projection $X \times I \to X \times 0$.

   (ii) $X \times I = |J| \searrow |J_1 \cup \cup_i \sigma_i^*| \searrow *$ (PL collapses), where $|J_1| = |K^1 \times I|$.

(2.6) A *PLCW complex* K^2 is *prismatically 1-collapsible* if for each 2-cell e_i we may choose a characteristic map $\Phi_i : D^2 \to K^2$ and a properly embedded 2-disk $\Delta_i^2 \subset (D^2 \times I)$ such that there are PL collapses

$$|K^2 \times I| \searrow |K^1 \times I| \cup \bigcup_i F_i(\Delta_i^2) \searrow *$$

In these circumstances we refer to $K^1 \times I$ as *the cylinders* and to the $F_i(\Delta_i^2)$ as *the sheets*. (Remember (2.3)! $F_i = (\Phi_i \times 1)$.). *Properly embedded* means that $\Delta_i^2 \cap \partial(D^2 \times I) = \Delta_i^2 \cap (\partial D^2 \times I) = \partial \Delta_i^2$.

REMARK 5: Notice that the inclusion $\partial \Delta_i^2 \subset (\partial D^2 \times I)$ must be a homotopy equivalence, since otherwise $\partial \Delta_i^2$ would be null-homotopic in $\partial D^2 \times I$ and $F_i(\Delta_i^2)$ would paradoxically determine a non-zero 2-cycle in a collapsible 2-complex. Conversely, given a) a PL circle $\Sigma^1 \subset (\partial D^2 \times I)$, with the inclusion a homotopy equivalence and b) a characteristic map $\Phi_i : D^2 \to K^2$ for a 2-cell e_i, a sheet is determined. For Σ^1 bounds a proper PL 2-disk Δ^2 which, by the 3-dimensional Schönflies theorem, separates $D^2 \times I$ into two 3-balls. One easily proves that $D^2 \times I \searrow (\partial D^2 \times I) \cup \Delta^2$, and thus, by (2.2) that

$$F_i(D^2 \times I) \searrow F_i((\partial D^2 \times I) \cup \Delta^2).$$

[†]locally finite when the discussion is not restricted to compact polyhedra.

As an example of prismatic collapsing, we generalize Zeeman's 1-collapse of the duncehat. This example is a special case of Theorem 3 and will be used in its proof.

(2.7) *Let $P = (a_1, a_2 : R_1, R_2)$ be a presentation of the trivial group, where $R_i = g_i a_i g_i^{-1} (i = 1, 2)$ for some (not necessarily reduced) words $g_i \in F(a_1, a_2)$. Let $K_P = S_1^1 \vee S_2^1 \cup e_1^2 \cup e_2^2$ be the naturally associated PLCW complex with 2-cells whose attaching maps ϕ_i monotonically trace out the relators R_i. Then K_P is prismatically 1-collapsible.*

To see this, choose proper 2-disks Δ_i^2 in $D_i^2 \times I$ with $\partial \Delta_i^2 \subset (\partial D_i^2 \times I)$, as drawn in Figure 1.

Here we view $\partial D_i^2 \times I$ as a rectangle whose vertical sides are identified and which is to be monotonically wrapped around the cylinders $(S_1^1 \vee S_2^1) \times I$ in a height-preserving fashion according to the word R_i.

Clearly $\partial \Delta_i^2 \subset (\partial D_i^2 \times I)$ is a homotopy equivalence; so by the preceding discussion $|K_P \times I| \searrow ($ cylinders $) \cup ($sheets$)$. The cylinders are pictured in Figure 2 (where the 3 vertical lines are identified).

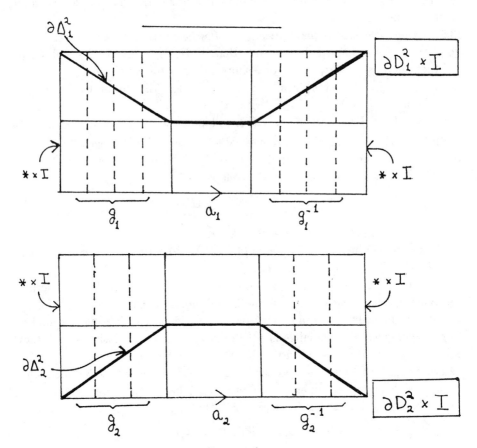

FIGURE 1

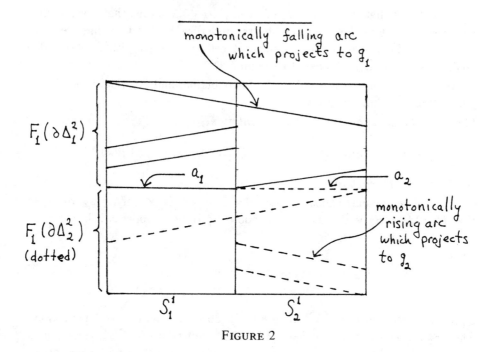

FIGURE 2

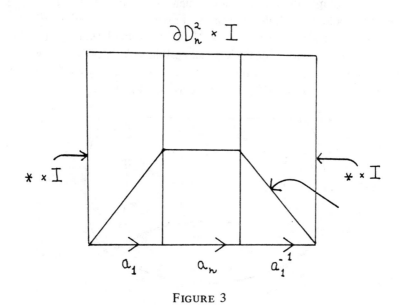

FIGURE 3

It is easy to see that $(S_1^1 \vee S_2^1) \times I \searrow F_1(\partial\Delta_1^2) \cup F_2(\partial\Delta_2^2)$, (i.e., that the cylinders collapse to the boundaries of the sheets) and that $F_i(\Delta_i^2)$ then has free face $S_i^1 \times \frac{1}{2} (i = 1, 2)$. Collapsing away the two 2-cells we are left with a contractible 1-complex which then collapses to a point. $\square$

Another example (mentioned in the Introduction, and a special case of [Wa, Thm. 4] or [B, Thm. 6]) is given by

(2.8) *The standard complex for the presentation*

$$P = (a_1, \ldots, a_n : a_2 a_1 a_2^{-1}, a_3 a_2 a_3^{-1}, \ldots, a_n a_{n-1} a_n^{-1}, a_1 a_n a_1^{-1})$$

is prismatically 1-collapsible.

In this case one may choose sheets $\Delta_i^2 = D_i^2 \times \frac{1}{2} \subset D_i^2 \times I$, if $1 \le i \le n-1$, so that $F_i(\partial\Delta_i^2) = (S_i^1 \vee S_{i+1}^1) \times \frac{1}{2}$, and choose Δ_n^2 to be a proper 2-disk in $D_n^2 \times I$ with $\partial\Delta_n^2$ as pictured in Figure 3.
Then

$$K_P \times I \searrow \text{cylinders} \cup \text{sheets}$$
$$\searrow \text{sheets} = F_1(\Delta_1^2) \cup \ldots \cup F_n(\Delta_n^2).$$

$S_1^1 \times \frac{1}{2}$ is a free face of $F_1(\Delta_1^2)$ in $\cup_i F_i(\Delta_i^2)$. So $F_1(\Delta_1^2)$ may be collapsed away. Then $S_2^1 \times \frac{1}{2}$ is a free face of $F_2(\Delta_2^2)$ in what's left (since $\partial F_2(\Delta_2^2)$ corresponds to $a_3 a_2 a_3^{-1}$ and had only been incident with $F_1(\Delta_1^2)$ which is no longer there). We collapse this away and continue in this manner to prove $K_P \times I \searrow *$. $\square$

(2.9) As an example of a 2-complex which is *not prismatically 1-collapsible*, consider the standard complex K^2 of the presentation $(a, b : ab, a^2b^3)$. The link graph of the vertex v of K^2 (Figure 4) has no articulation point. Thus, by Thm. 5, $K^2 \times I$ is not prismatically collapsible. (Alternatively, we could use Theorem 4 and any criterion which shows that $w_1 = ab$, $w_2 = a^2b^3$ do not form a basis-up-to-conjugation of $F(a, b)$; compare [C-M-Z].)

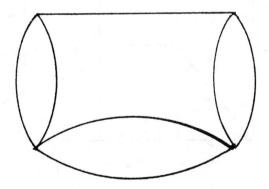

FIGURE 4

Nevertheless, Example (2.9) is well known to be 1-collapsible [Ze$_2$, p. 344]. In (6.2) we give a collapse of $K^2 \times I$ (which we learned from W. Becker) to illustrate our analysis of *prismatic 1-collapsibility with holes and sheets*.

§3. ONE DIMENSIONAL COMPLEXES

The purpose of this section is to prove Theorem 1 of the Introduction. First we prove a stronger result in the special case where K and L are wedge products of circles.

(3.1) *Suppose that* $f : \pi_1(L, e^0) \to \pi_1(K, e^0)$ *is an isomorphism, where* $L = K = S_1^1 \vee \ldots \vee S_n^1$ *with wedge point* e^0. *Then there is an embedding* $\alpha : (L \times I, e^0 \times I) \to (K \times I, e^0 \times I)$ *such that*

(i) *The composition* $L \xrightarrow{\times 0} L \times I \xrightarrow{\alpha} K \times I \xrightarrow{\text{proj.}} K$ *induces the isomorphism* f *on* π_1

(ii) $K \times I \searrow \alpha (L \times I)$.

(iii) *There are neighborhoods* U_i *of* e^0 *in* $S_i^1 (1 \le i \le n)$ *such that* $\alpha(U_i \times I) = U_i \times [\frac{1}{2} - \epsilon, \frac{1}{2} + \epsilon]$ *for some* $\epsilon > 0$ *and* $\alpha(U_i \times \frac{1}{2}) = U_i \times \frac{1}{2}$.

REMARK: Our proof will give embeddings of $L \times I$ with large numbers of little oscillations ("detours") in the vicinity of $e^0 \times I$. These are necessary. We will give an example (3.2) in which there does not exist an embedding (to say nothing of collapsibility) $\alpha : L \to K \times I$, with $\alpha(e^0) = (e^0, \frac{1}{2})$, for which the curves $\alpha(S_i^1)$ project down to loops which monotonically read off the words $f(b_i)$. There is too much congestion around the base point. The idea of the proof of (3.1) is to give each band $\alpha(S_i^1 \times I)$ its own runway – $U_i \times [\frac{1}{2} - \epsilon, \frac{1}{2} + \epsilon]$ – on which to depart and arrive at the base point without running into the others, except where they are joined along $\alpha(e^0 \times I)$.

PROOF OF (3.1). Let $\{a_1, \ldots, a_n\}$ and $\{b_1, \ldots, b_n\}$ be the bases determined by the oriented circles $S_1^1, \ldots, S_n^1$ of $\pi_1(K) \equiv \pi_1(K, e^0)$ and $\pi_1(L) \equiv \pi_1(L, e^0)$ respectively. Nielsen's Theorem [N] asserts that we can get from the sequence of basis elements $(a_1, \ldots, a_n)$ to the sequence of basis elements $(f(b_1), \ldots, f(b_n))$ of $\pi_1(K)$ by the following moves: Once a sequence $(R_1, \ldots, R_n)$ has been achieved, build a new sequence by changing exactly one of the words (say R_i) by

(a) $R_i \to R_i^{-1}$ or

(b) $R_i \to R_i R_j$ for some $j \ne i$.

To mimic the algebra geometrically, we shall use the embeddings ρ_i and $h_{ij}(j \ne i)$ of $L \times I$ into $L \times I$ given by

(a)
$$\rho_i | S_j^1 = \text{ identity if } i \ne j$$
$$\rho_i(x, y, t) = (x, -y, t) \text{ if } (x, y) \in S_i^1 \subset \mathbb{R}^2 .$$

(b)
$$h_{ij}(x, y, t) = (x, y, \ 0.4 + (0.2)t) \text{ if } (x, y) \in S_i^1$$
$$h_{ij} : (S_i^1 \times I) \to (S_i^1 \vee S_j^1) \times I \text{ as in Figure 5.}$$

The inductive construction of the embedding $\alpha : L \times I \to K \times I$ can now be described as follows:

Let $\alpha_0 = $ identity. Suppose inductively an embedding $\alpha_k : L \times I \to K \times I$ has been constructed which satisfies (i), (ii), and (iii) of (3.1) with respect to the isomorphism $f_k : \pi_1(L) \to \pi_1(K)$ with $f_k(a_1) = R_1, \ldots, f_k(a_n) = R_n$. Let f_{k+1} differ from f_k in that (a)$f_{k+1}(a_i) = R_i^{-1}$ or (b)$f_{k+1}(a_i) = R_i R_j$. Define α_{k+1} accordingly as (a) $\alpha_{k+1} = \alpha_k \circ \rho_i$ or (b) $\alpha_{k+1} = \alpha_k \circ h_{ij}$. It is clear then that $\alpha_{k+1}(e^0 \times I) \subset e^0 \times I$ and that α_{k+1} also satisfies (i), (iii). Moreover, one easily checks that $L \times I \searrow h_{ij}(L \times I)$. (Follow the circled numbers in Figure 5). Therefore

$$K \times I \searrow \alpha_k(L \times I) \searrow \alpha_k h_{ij}(L \times I) = \alpha_{k+1}(L \times I).$$

Thus (ii) is also carried by the induction. Finally, if N Nielsen moves are required, so that $f = f_N$, then $\alpha = \alpha_N$ satisfies the requirements of our theorem. $\square$

NOTE: In general the embedded wedge product $\alpha(L)$ projects to non-reduced words (even if the local detours in the above proof are ignored) because the Nielsen method involves cancellation in words of the form $R_i R_j$. The occurrence of $a_i a_i^{-1}$ in a word to which $\alpha(L)$ projects is viewed as a "large detour". However, the following example concerns a basis which can be achieved by Nielsen transfor-

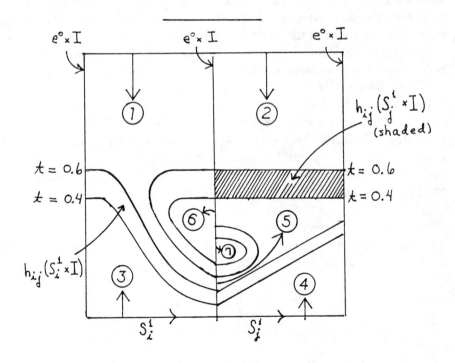

FIGURE 5

mations without cancellation, so that the embedding of (3.1) can be constructed here according to a sequence of Nielsen moves which produces no large detours.

(3.2) *Let* $L = K = $ *the wedge product of five circles. The isomorphism* $\pi_1 L = F(b_1, \ldots, b_5) \to \pi_1 K = F(a_1, \ldots, a_5)$ *given by*

$$b_1 \to w_1 = a_1 a_4^{-1}$$
$$b_2 \to w_2 = a_2 a_4^{-1} a_5 a_4$$
$$b_3 \to w_3 = a_3 a_4^{-1} a_5^{-1} a_4$$
$$b_4 \to w_4 = a_4$$
$$b_5 \to w_5 = a_5$$

cannot be realized by any embedding $\alpha : (L, e^0) \to (K \times I, (e^0, \frac{1}{2}))$ *which has neither large nor small detours.*

PROOF. Suppose, on the contrary, an embedding were given in which each circle $\alpha(S_i^1)$ (also called w_i) projects down to trace out the reduced word w_i, rotating monotonically clockwise or counterclockwise each time it enters a cylinder. Without loss of generality w_4 and w_5 give the embedded circles at height $\frac{1}{2}$. (See Figure 6).

The reader may now reach a contradiction by verifying each of the following assertions:

a) No segment of w_1, w_2 or w_3 can begin at a point above height 1/2 and end at a point below $t = 1/2$ or vice versa. (An allowable embedding of w_1 is drawn dashed in Figure 6). Thus each of w_1, w_2, w_3 lies totally in the upper half of the cylinders or totally in the lower half of the cylinders.

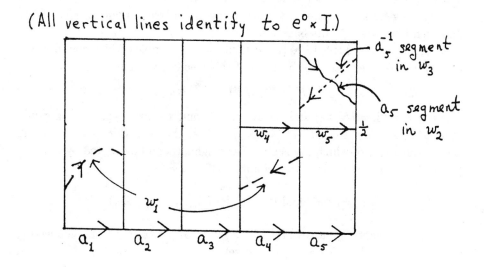

FIGURE 6

b) w_1 and w_2 cannot both lie in the upper or both lie in the lower half of the cylinders. (For the final segments a_4 and a_4^{-1} would have to cross as they return to the base point).

c) w_1 and w_3 cannot both lie in the same half of the cylinders (same reason as b)).

d) w_2 and w_3 cannot both lie in the same half of the cylinders. (For example, to lie in the upper half the beginnings of the a_4 segments of w_2 and w_3 must lie below the ends of the two a_4^{-1} segments; i.e., the ends of the a_5 and a_5^{-1} segments must both lie below the beginnings of the a_5 and a_5^{-1} segments. This forces the a_5 and a_5^{-1} segments to intersect, as in Figure 6). □

We record without proof (see [S], page 45) the fact that small detours may be avoided if the unit interval I is replaced by a triod (= a cone on three points):

(3.3) *If $K = L = S_1^1 \vee \ldots \vee S_n^1$ and $f : \pi_1(L, e^0) \to \pi_1(K, e^0)$ is an isomorphism then there is an embedding $\alpha : (L, e^0) \to (K \times T, e^0 \times T)$, where T is a triod, such that*

(i) *The composition $L \xrightarrow{\alpha} K \times T \xrightarrow{proj.} K$ induces the isomorphism f on π_1.*

(ii) $K \times T \searrow \alpha(L \times T)$

(iii) *The path $(proj) \circ \alpha(S_1^1)$ monotonically traces out the not necessarily reduced word $[f(S_i^1)] \epsilon \pi_1(K^1, e^0) \equiv F(a_1, \ldots, a_n)$.* □

PROOF OF THEOREM 1. Suppose that K and L are arbitrary non-degenerate 1-complexes and $f : \pi_1(L, y_0) \to \pi_1(K, x_0)$ is an isomorphism. We must find an embedding

$\beta : (L, y_0) \to (K \times I, x_0 \times I)$ which realizes f and for which $K \times I \searrow \beta(L)$. When L is a tree, the proof is left to the reader. (L embeds in I^2 and hence in $K \times I$). In general, the plan is to squeeze maximal trees to points, apply (3.1), and then blow these points back up.

Choose a maximal tree in K (in particular a CW complex which contains K^0) and let T_K be a regular neighborhood of this tree in $|K|$. Then

$$|K| = T_K \cup K_1 \cup \ldots \cup K_n$$

where $n = \text{rank } \pi_1(K, x_0)$ and the K_i are disjoint PL arcs representing generators of $\pi_1(K, T_K)$. Let $\partial K_i = \{k_{i1}, k_{i2}\}$ and let the arcs which connect K_i to the maximal tree of which T_K is a regular neighborhood be denoted by $A_{2i-1} = [a_{i1}, k_{i1}]$ and $A_{2i} = [a_{i2}, k_{i2}]$. (See Figure 7). Thus

$$|K| = (\text{ maximal tree }) \cup \bigcup_{i=1}^{n} (A_{2i-1} \cup K_i \cup A_{2i}).$$

Similarly we define T_L, arcs L_i ($1 \le i \le n$) with $\partial L_i = \{\ell_{i1}, \ell_{i2}\}$ and connecting arcs $B_{2i-1} = [b_{i1}, \ell_{i1}]$, $B_{2i} = [b_{i2}, \ell_{i2}]$ to get

$$|L| = T_L \cup L_1 \cup \ldots \cup L_n$$

$$= (\text{ maximal tree }) \cup \bigcup_{i=1}^{n} (B_{2i-1} \cup L_i \cup B_{2i}).$$

Squeezing the maximal tree of K to a point we obtain a 1-complex $\overline{K} = K/(\text{ max. tree })$ with the single vertex $\overline{x}_0 = \{ \text{ max. tree } \} \, \epsilon \, \overline{K}$. Similarly we obtain $\overline{L}$ with unique vertex $\overline{y}_0 \, \epsilon \, \overline{L}$. (For any subset X of K or L we let $\overline{X}$ denote its image in $\overline{K}$ or $\overline{L}$). We identify L_i, K_i with their homeomorphic images $\overline{L}_i, \overline{K}_i$ and we identify f with its induced isomorphism $\pi_1(\overline{L}, \overline{y}_0) \to \pi_1(\overline{K}, \overline{x}_0)$. Let $\alpha : (\overline{L}, \overline{y}_0) \to (\overline{K} \times I, (\overline{x}_0, \frac{1}{2}))$ be an embedding realizing f, as given by (3.1) with $\overline{L} = \overline{L} \times \frac{1}{2}$. From (3.1)(iii) we may assume (switching the names of A_{2i-1} and A_{2i} if necessary) that $\alpha(\overline{B}_j) = \overline{A}_j \times \frac{1}{2}$, $1 \le j \le 2n$. Also, from the proof of (3.1) we may assume that the vertical axis $\{\overline{x}_0\} \times I$ is always crossed transversely; i.e., if $\overline{y}_0 \ne z \, \epsilon \, \overline{L}$ and $\alpha(z) = (\overline{x}_0, t_z)$ – so $z \epsilon$ Int $(\overline{L}_i)$ for some i – then there is a closed interval $N(z)$ about z, which goes homeomorphically into $\overline{K} \times t_z$, crossing from the ith cylinder to the pth cylinder, for some $i, p (1 \le i, p \le n)$ with $\alpha(Bdy\, N(z)) = \{k_{ij}, k_{pq}\} \times t_z$. Finally, we assume, for these crossing points $z \ne \overline{y}_0$, that $t_z < \frac{1}{3}$ or $t_z > \frac{2}{3}$.

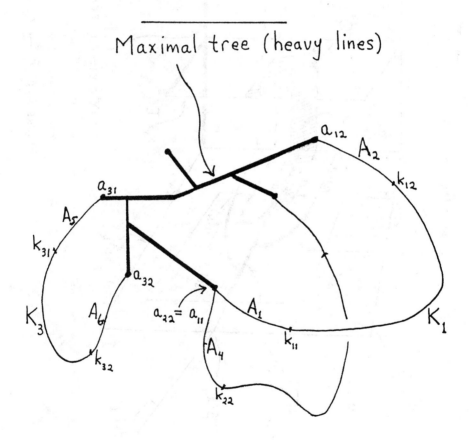

FIGURE 7

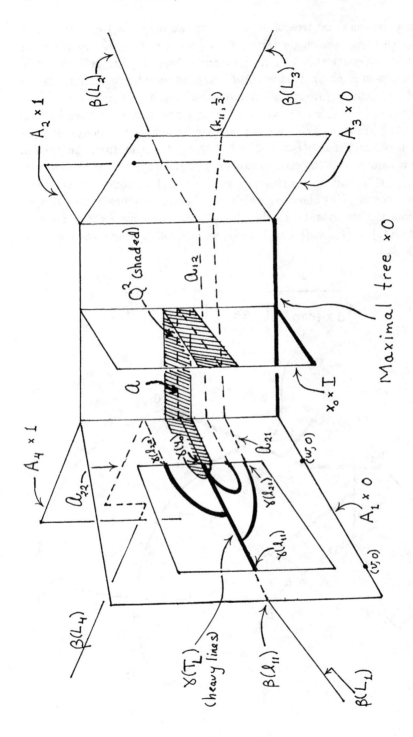

FIGURE 8. $\beta : T_L \longrightarrow T_K \times L$

The desired embedding $\beta : L \to K \times I$ is now defined in a series of steps:

(1) β agrees with α on $\alpha^{-1}(\bigcup_{i=1}^{n} K_i \times I) = (\bigcup_{i=1}^{n} L_i) - \bigcup_z \overset{\circ}{N}(z)$.

(2) If $z \neq \bar{y}_0$, $\alpha(z) = (\bar{x}_0, t_z)$ and $\alpha(Bdy\, N(z)) = \{k_{ij}, k_{pq}\} \times t_z$ then β takes $N(z)$ homeomorphically to the unique arc in $T_K \times t_z$ connecting (k_{ij}, t_z) with (k_{pq}, t_z).

(3) To define β on T_L, see Figure 8 and proceed as follows:

    a) Let γ be an embedding of T_L into a rectangle $[v, w] \times [\frac{1}{3}, \frac{2}{3}]$ in the interior of $A_1 \times I$, so that $\gamma(y_0) = (w, \frac{1}{2})$, $\gamma(\ell_{11}) = (v, \frac{1}{2})$ and $\gamma(\ell_{ij}) \in w \times [\frac{1}{3}, \frac{2}{3}]$ if $(i, j) \neq (1, 1)$.

    b) Run a horizontal arc $\mathcal{A}_{11}$ in $T_K \times \frac{1}{2}$ from $\gamma(\ell_{11})$ to $(k_{11}, \frac{1}{2})$ and, if $(i, j) \neq (1, 1)$, run an arc $\mathcal{A}_{ij}$ in $T_K \times I$ from $\gamma(\ell_{ij})$ to $(k_{ij}, \frac{1}{2})$ such that $\mathcal{A}_{ij}$ is the union of a horizontal arc in $(A_1 \cup$ max. tree $) \times I$ with a line segment in $A_{2i-1} \times I$ or $A_{2i} \times I$. (See the dotted arcs in Figure 8). Then γ is isotopic, by pulling the free faces $\gamma(\ell_{ij})$ along the arcs $\mathcal{A}_{ij}$, to an embedding $\beta_1 :$ $T_L \to T_K \times I$ with $\beta_1(T_L) = \gamma(T_L) \cup \mathcal{A}_{11} \cup \mathcal{A}_{12} \cup \ldots \cup \mathcal{A}_{n2}$.

    c) Run a horizontal arc $\mathcal{A}$ in $T_K \times \frac{1}{2}$ from $\gamma(y_0) = \beta_1(y_0)$ to $(x_0, \frac{1}{2})$. For small $\epsilon > 0$ consider the 2-cell $Q^2 = N_\epsilon(\gamma(y_0)) \cup [\mathcal{A} \times (\frac{1}{2} - \epsilon, \frac{1}{2} + \epsilon)]$ (shaded in Figure 8). We isotop the cone $\beta_1(T_L) \cap Q^2$ inside Q^2, rel $\beta_1(T_L) \cap \partial Q^2$, so that $\beta_1(y_0)$ is dragged over to $(x_0, \frac{1}{2})$. (See Figure 9.) The final result is an embedding $\beta : T_L \to T_K \times I$ with $\beta = \beta_1$ on $\beta_1^{-1}[\beta_1(T_L - Q^2)]$ and $\beta(y_0) = (x_0, \frac{1}{2})$.

Together (1), (2), (3) give an embedding β of (L, y_0) into $(K \times I, (x_0, \frac{1}{2}))$. It is clear, since $\beta(T_L) \subset T_K \times I$ and $\beta = \alpha$ on $\alpha^{-1}(K - T_K)$, that (proj.) $\circ \beta$ induces the given isomorphism $f : \pi_1(L, y_0) \to \pi_1(K, x_0)$. We must show that $K \times I \searrow \beta(L)$.

Let $p : K \times I \to \overline{K} \times I$ be the quotient map. Since $\overline{K} \times I \searrow \alpha(\overline{L})$, we see at once that

$$\overline{K} \times I \searrow \alpha(\overline{L}) \cup (N(\bar{x}_0) \times [\frac{1}{3}, \frac{2}{3}])$$

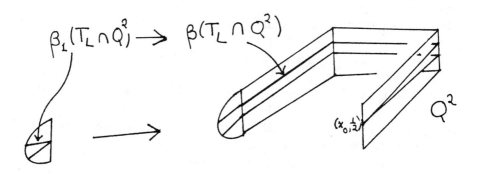

FIGURE 9

where $N(\bar{x}_0) = \bigcup_{i=1}^n \bar{A}_i$ is a regular neighborhood of $\bar{x}_0$ in $\bar{K}$. Thus

$$K \times I \searrow p^{-1}(\alpha(\bar{L}) \cup (N(\bar{x}_0) \times [\tfrac{1}{3}, \tfrac{2}{3}]))$$

$$= \alpha[\bar{L} - \bigcup\{N(z) | \alpha(z) = (\bar{x}_0, t_z)\}]$$

$$\cup \bigcup\{T_K \times t_z | \alpha(z) = (\bar{x}_0, t_z)\} \cup (T_K \times [\tfrac{1}{3}, \tfrac{2}{3}]).$$

This collapse occurs on general principles, since $p : K \times I \to \bar{K} \times I$ is a map with collapsible point inverses. (Alternatively, one can explicitly mimic the collapse in $\bar{K} \times I$ by a collapse in $K \times I$, using the fact that T_K is collapsible as in [S]). Further, since β_1 agrees with α on $\bar{L} - \bigcup\{N(z) | \alpha(z) = (\bar{x}, t_z)\}$ and since $T_K \times t_z$ collapses to the arc $\beta_1(N(z))$, the above polyhedron further collapses to yield

$$K \times I \searrow \beta_1(L - T_L) \cup (T_K \times [\tfrac{1}{3}, \tfrac{2}{3}])$$

$$\searrow \beta_1(L - T_L) \cup [v, w] \times [\tfrac{1}{3}, \tfrac{2}{3}] \cup \bigcup_{i,j} \mathcal{A}_{ij} \cup \mathcal{A}$$

$$\searrow \beta_1(L) \cup \mathcal{A} .$$

But then

$$K \times I \searrow \beta_1(L) \cup Q^2 \searrow \beta(L).$$

as required. □

§4. COLLAPSIBLE 2-COMPLEXES

In this section we give some useful criteria concerning collapsibility of 2-complexes. The first is well-known and is stated without proof for the sake of completeness.

(4.1)
   (a) If $X^2 \searrow A \supset B$ and $X^2 \searrow B$ then $A \searrow B$.
   (b) If $X^2 \supset A \supset B$, where X^2 is a pseudomanifold[††] and $H_*(A, B; \mathbb{Z}_2) = 0$, then $A \searrow B$. □

(4.2) If $S^1 \subset X \searrow *$ then S^1 bounds a singular PL 2-disk in X which is collapsible and which is a spine of X.

There is no restriction on the dimension of X in (4.2). We will be mainly interested in the case $X = |K^2 \times I|$ as discussed in §6.

PROOF OF (4.2). We collect some facts on trails of collapses: (4.2) will be a special case.

Let S be a (finite) sequence of elementary simplicial collapses of a simplicial complex K to a subcomplex L. If $M \subset K$ is a subcomplex, then $T_S(M)$, *the trail of M with respect to S,* is intuitively the material which is overrun as $|M|$ is deformed in $|K|$ by S. A simple definition is given by:

$T_S(M)$ is the smallest subcomplex of K satisfying
   a) $M \subset T_S(M)$,
   b) if a simplex σ^{n+1} collapses in S from the face σ^n and if $\sigma^n \in T_S(M)$ then $\sigma^{n+1} \in T_S(M)$.

[††]i.e., in any triangulation, each 1-simplex is a face of at most two 2-simplexes.

This definition is equivalent to the inductive one given by Zeeman ([Ze₂], Ch. 7, p.31). Note that

    c) the first simplex of $T_S(M)$ which is removed by S has its free face in M.

By rearranging the sequence of collapses, Zeeman proves ([Ze₂], Ch. 7, Lemma 45)

(1) $$K \searrow L \cup T_S(M) \searrow L.$$

Another fact about trails is that the elementary collapses in S which remove simplexes of $T_S(M)$ give rise to

(2)   a PL epimorphism $f : |M \times I| \to |T_S(M)|$ which takes $|M \times 0|$ to $|M|$ by the natural identification.

The combination of (1) and (2) yields

(3) $$|K| \searrow |L| \cup f(|M \times I|) \searrow |L| .$$

In particular, when $|L| = *$, we get

    (1') $K \searrow T_S(M) \searrow *$;

    (2') there exists a PL epimorphism $f : |cM| \to |T_S(M)|$ which maps the base $|M|$ of the cone $|cM|$ identically to itself;

    (3') $|K| \searrow f(|cM|) \searrow *$.

    (4.2) is now a special case of (2') and (3'): Let (K, M) be a triangulation of (X, S^1) such that K is simplicially collapsible. Thus $|cM|$ is a 2-disk and (2') provides us with the desired singular PL 2-disk spanned by S^1; (3') gives the fact that this singular disk is collapsible and is a spine of X. $\square$

ADDENDUM: By c) above, the singular disk constructed has a free face in S^1.

    The next lemma is useful in relating the hypothesis of PL collapsibility to a given $PLCW$ structure.

(4.3) *If the PLCW complex $K^2 = K^1 \cup e_1^2 \cup \ldots \cup e_n^2$ is PL collapsible then there is a subdivision K_1 of its 1-skeleton K^1 such that the PLCW complex $K_2 = K_1 \cup e_1^2 \cup \ldots \cup e_n^2$ is CW collapsible.*

    PROOF. Let K_* be a simplicially collapsible simplicial subdivision of K^2. (Actually any CW collapsible CW subdivision of K^2 will do). Let $K_1 = (K^1)_*$ and let the order of the 2-dimensional simplicial collapses be $(\sigma_1, \tau_1), \ldots, (\sigma_p, \tau_p)$, where τ_i is the relevant free face of the 2-simplex σ_i. For each i $(1 \leq i \leq p)$ let $e_{k_i}^2$ be the unique 2-cell of K^2 with $\mathring{\sigma}_i \subset e_{k_i}^2$. Since τ_1 is a free face of σ_1 in K_* it is certainly a free face of $e_{k_1}^2$ in K_2. Perform the CW collapse $K_2 \searrow K_2 - (e_{k_1}^2 \cup \mathring{\tau}_1)$. Next, let σ_j be the first 2-simplex with $e_{k_j}^2 \neq e_{k_1}^2$. Since $\mathring{\sigma}_1 \cup \ldots \cup \mathring{\sigma}_{j-1} \subset e_{k_1}^2$ and τ_j is a free face of σ_j in $K_* - (\mathring{\sigma}_1 \cup \ldots \cup \mathring{\sigma}_{j-1})$, it is certainly a free face of $e_{k_j}^2$ in $K_2 - (e_{k_1}^2 \cup \tau_1^0)$. Perform the CW collapse $K_2 - (e_{k_1}^2 \cup \mathring{\tau}_1) \searrow K_2 - (e_{k_1}^2 \cup \mathring{\tau}_1) - (e_{k_j}^2 \cup \mathring{\tau}_j)$. We continue in this manner to collapse away all n of the 2-cells of K_2

and n of the 1-cells $\overset{\circ}{\tau}_i$. Thus $K_2 \searrow$ (a subcomplex of its 1-skeleton). Since K_2 is contractible this 1-complex is a tree and thus has a CW collapse to a point.   □

(4.4) *Let K^2 be a PL collapsible PLCW complex with base point $e^0 \epsilon K^0$. Let $e_1, \ldots, e_q$ be some of the (open) 2-cells of K^2, with attaching maps $\phi_i : \partial D^2 \to K^1 (1 \leq i \leq q)$. Let $L = K^2 - e_1 - \ldots - e_q$. Then*

   (a) *there   are   paths α_i : $(I, 0, 1)$ $\to$ $(K^1, e^0, \phi_i(1))$ such   that $\{[\alpha_i * \phi_i * \overline{\alpha}_i] | 1 \leq i \leq q\}$ is a basis of $\pi_1(L, e^0)$.*

   (b) *if $\phi_i(1) = e^0 (1 \leq i \leq q)$ then $\{[\phi_i] | 1 \leq i \leq q\}$ is a basis of $\pi_1(L, e^0)$.*

PROOF.   We may assume (because of (4.3)) that $K^2 \searrow^{cw} e^0$ and that $\phi_i(1) \epsilon K^0 (1 \leq i \leq q)$. Let $B_i = \Phi_i(\frac{1}{2}D^2)$, $x_i = \Phi_i(\frac{1}{2}, 0)$, and $a_i = \Phi_i\{(t, 0) | \frac{1}{2} \leq t \leq 1\}$. We may perform a PL collapse of $|K^2|$ which stepwise mimics the 2-dimensional CW collapse of K^2, except that whenever one of our specified 2-cells e_i is to be collapsed away via the free face $\overset{\circ}{\tau}_i$, we instead collapse away all of $\overset{\circ}{\tau}_i \cup e_i$ except $a_i \cup B_i$. (See Figure 10).

The result is a PL collapse $|K^2| \searrow T \cup a \cup B$, where T is a tree in $|K^1|$ containing $|K^0|$, $B = B_1 \cup \ldots \cup B_q$ and $a = a_1 \cup \ldots \cup a_q$. Then

$$|K^2| - (\text{ Int } B) \searrow (T \cup a) \cup \partial B.$$

The right hand term is a tree with disjoint loops attached. Choosing arcs $\beta_i : (I, 0, 1) \to (T \cup a, e^0, x_i)$ and homeomorphisms $h_i : (\partial D^2, 1) \to (\partial B_i, x_i)$ with $h_i(x) = \Phi_i(\frac{1}{2}x)$ for all $x \epsilon \partial D^2$, we clearly get a basis $\{[\beta_i * h_i * \overline{\beta}_i] | 1 \leq i \leq q\}$ of $\pi_1(T \cup a \cup \partial B, e^0)$ and thus of $\pi_1(|K^2| - \text{ Int } B, e^0)$.

Finally to get a basis of $\pi_1(L, e_0) = \pi_1(K^2 - e_1 - \ldots - e_q, e^0)$, use the radial deformation retraction of $D^2 - \frac{1}{2}\overset{\circ}{D}{}^2$ onto ∂D^2 to induce deformation retractions $R_i : \overline{e}_i - (\text{ Int } B_i) \to \phi_i(\partial D^2)$. Combining the R_i, we get a deformation retraction

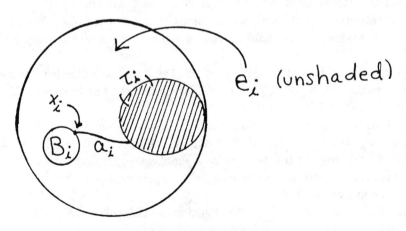

FIGURE 10

$R : |K^2| - \text{Int}\, B \to |L|$. Set $\alpha_i = R \circ \beta_i$. Then the induced isomorphism R_* takes basis elements $[\beta_i * h_i * \overline{\beta}_i]$ onto basis elements $[\alpha_i * \phi_i * \overline{\alpha}_i]$ of $\pi_1(L, e_0)$. This proves (a). If $\phi_i(1) = e^0$ then β_i could clearly be chosen above so that α_i is the constant map at e^0. Thus (b) follows. $\square$

An alternative formulation of (4.4) which is useful in practical situations is given by

(4.5) *Suppose that K^2 is CW collapsible and T is a maximal tree in K^1 such that $K^2 \searrow^{cw} T$. Let $\pi_1(K^1)$ be viewed as the free group generated by the oriented cells of $K^1 - T$ in the usual way. Assume that the attaching maps ϕ_i for the 2-cells may be read off as words w_i in these generators. Let v_i be any cyclic permutation of w_i. Then $v_1, \ldots, v_n$ is a basis of $\pi_1(K^1)$.*

PROOF. Squeeze T to a point and apply (4.4b). $\square$

§5. PRISMATIC COLLAPSIBILITY

In this section we prove Theorems 2–5 and also a theorem (5.5) relating the different notions of prismatic collapsibility.

(5.1) PROOF OF THEOREM 2. Let L^1 denote the wedge product of n circles and identify $\pi_1(L^1, *)$ with the free group $F(a_1, \ldots, a_n)$ in the usual way. Then the given basis $\{w_1, \ldots, w_n\}$ of $\pi_1(K^1, *)$ determines an isomorphism $f : \pi_1(L^1, *) \to \pi_1(K^1, *)$ with $f(a_i) = w_i$. Let $\alpha : (L^1, *) \to (K^1 \times I, * \times I)$ be an embedding given by Theorem 1. The restriction of α to the ith circle of L^1 yields an embedding $\alpha_i = (\phi_i, t_i) : \partial D^2 \to K^1 \times I$ with PL maps $\phi_i : (\partial D^2, 1) \to (K^1, *)$ and $t_i : \partial D^2 \to I$. These ϕ_i satisfy

$$(1) \qquad\qquad [\phi_i] = w_i$$

and are taken as attaching maps in defining $K^2 = K^1 \cup e_1^2 \cup \ldots \cup e_n^2$.

The points $(x, t_i(x)) \in \partial D^2 \times I$ form the boundary of a properly embedded 2-disk $\Delta_i^2 \subset (D^2 \times I)$. By §2, Remark 5,

$$(2) \qquad\qquad K^2 \times I \searrow (K^1 \times I) \cup \bigcup_i F_i(\Delta_i^2).$$

The choice of the ϕ_i and Δ_i^2 implies that $\bigcup_i F_i(\Delta_i^2)$ is a wedge product of 2-disks intersecting $K^1 \times I$ in $\alpha(L^1)$. Since α is an embedding given by Theorem 1 we have

$$(3) \qquad\qquad (K^1 \times I) \cup \bigcup_i F_i(\Delta_i^2) \searrow \bigcup_i F_i(\Delta_i^2) \searrow *.$$

(1), (2), and (3) together imply the assertion. $\square$

(5.2) PROOF OF THEOREM 3. We are given a basis-up-to-conjugation $\{w_1, w_2\}$ of $F(a_1, a_2) = \pi_1(K^1, *)$, where $K^1 = S^1 \vee S^1$, and we must again exhibit attaching maps $\phi_i : (\partial D^2, 1) \to (K^1, *)$ with $[\phi_i] = w_i$ such that the resultant K^2 is prismatically 1-collapsible. In a slight embellishment of the preceding argument we shall construct appropriate maps $\gamma_i = (\phi_i, t_i) : \partial D^2 \to K^1 \times I (i = 1, 2)$ with $\phi_1(1) = \phi_2(1) = *$ (but not necessarily $t_1(1) = t_2(1)$) and choose the ϕ_i as our attaching maps.

For $i = 1, 2$, write $w_i = g_i v_i g_i^{-1}$ where $\{v_1, v_2\}$ is a basis of $F(a_1, a_2)$. Express $g_i \epsilon F(a_1, a_2)$ as a word $g_i = \bar{g}_i(v_1, v_2)$ in v_1, v_2. Let $L^1 = S_1^1 \vee S_2^1$, with the circles determining generators b_1, b_2 of $\pi_1(L^1, *) = F(b_1, b_2)$. Set $h_i = \bar{g}_i(b_1, b_2) \epsilon F(b_1, b_2)$. Let β_1 and β_2 be maps of $(\partial D^2, 1)$ into $(L^1 \times I, * \times I)$, as in the proof of (2.7), whose images wind in from above and below and contain $S_1^1 \times \{\frac{1}{2}\}$ and $S_2^1 \times \{\frac{1}{2}\}$ respectively, whose projections to L^1 spell out $h_1 b_1 h_1^{-1}$ and $h_2 b_2 h_2^{-1}$, and for which $(L^1 \times I) \searrow (\operatorname{im} \beta_1 \cup \operatorname{im} \beta_2)$.

Let the embedding $\alpha : (L^1 \times I, * \times I) \to (K^1 \times I, * \times I)$ realize the fundamental group isomorphism $b_i \to v_i$ $(i = 1, 2)$ as in (3.1). Set $\gamma_i = (\alpha \circ \beta_i)$. Then $\gamma_i : (\partial D^2, 1) \to (K^1 \times I, * \times I)$ represents $w_i = g_i v_i g_i^{-1}$ and $(K^1 \times I) \searrow \alpha(L^1 \times I) \searrow (\operatorname{Im} \gamma_1 \cup \operatorname{Im} \gamma_2)$. If now $\gamma_i = (\phi_i, t_i)$ and we build $K^2 = K^1 \bigcup_{\phi_1} e_1^2 \bigcup_{\phi_2} e_2^2$ then

$$K^2 \times I \searrow (K^1 \times I) \cup (\text{ sheets attached by } \gamma_1, \gamma_2)$$
$$\searrow (\text{ sheets attached by } \gamma_1, \gamma_2) \searrow *$$

The first collapse occurs (see §2, Remark 5) because the inclusion $\{(x, t_i(x)) | x \epsilon \partial D^2\} \subset \partial D^2 \times I$ is a homotopy equivalence, and $\gamma_i(x) = F_i(x, t_i(x))$. The second collapse occurs because $(K^1 \times I) \searrow (\operatorname{Im} \gamma_1) \cup (\operatorname{Im} \gamma_2)$, and the last collapse occurs because the γ_i sheet $(i = 1, 2)$ has a free face, namely $\alpha(S_i^1 \times \frac{1}{2})$. □

(5.3) PROOF OF THEOREM 4. $\pi_1(K^1, T)$ denotes the set of homotopy classes of maps of pairs $(S^1, 1) \to (K^1, T)$. If f, g are two such maps we do not require that $f(1) = g(1)$. Nevertheless $\pi_1(K^1, T)$ becomes a group via the homotopy equivalence of pairs, $(K^1, T) \sim (K^1/T = *, *)$.

We are given a prismatic collapse

$$|K^2 \times I| \searrow |K^1 \times I| \cup \bigcup_i F_i(\Delta_i^2) \searrow * ,$$

where $F_i = \Phi_i \times 1$, and we must show that the elements $w_i = [\phi_i]$ (which may be thought of as "edge path words" in $K^1 - T$ if T is a maximal tree) form a basis-up-to-conjugation of $\pi_1(K^1, T)$.

Assign the middle term displayed above (the cylinders ∪ sheets) the structure of a CW complex J^2 in which the sheets $S_i = F_i(\operatorname{Int} \Delta_i^2)$ are some of the 2-cells and where J^2 induces a subdivision of the product complex $K^1 \times I$. Since the inclusion $\partial \Delta_i^2 \to \partial D^2 \times I$ is a homotopy equivalence, we may choose a homeomorphism $g_i : \partial D^2 \to \partial \Delta_i^2$ such that $g_i(1) \epsilon \{1\} \times I$ and $(\text{proj}) \circ g_i \simeq 1_{\partial D^2}$ (rel $\{1\}$). Now take $\psi_i = F_i \circ g_i$ as attaching map for S_i. Note that $[p\psi_i] = [\phi_i] \epsilon \pi_1(K^1, T)$, where $p = \text{proj.}: K^1 \times I \to K^1$, because $F_i|(\partial D^2 \times I) = \phi_i \times 1$.

Let e^0 be a vertex of $T \times 0$, and hence of $J^0 \cap (T \times 0)$. By (4.4) there are paths $\alpha_i : (I, 0, 1) \to (J^1, e^0, \psi_i(1))$ such that $\{[\alpha_i * \psi_i * \bar{\alpha}_i] | 1 \le i \le n\}$ is a basis of $\pi_1(J^2 - S_1 - \ldots - S_n, e^0) = \pi_1(|K^1 \times I|, e^0)$. Projecting to K^1 we conclude that $\{[(p\alpha_i) * (p\psi_i) * p\bar{\alpha}_i] | 1 \le i \le n\}$ is a basis of $\pi_1(K^1, e_0)$. But the inclusion induces an isomorphism $\pi_1(K^1, e_0) \to \pi_1(K^1, T)$; moreover

$$p\alpha_i(0) = e^0 \epsilon T, \quad p\alpha_i(1) = p\psi_i(1) = pF_i g_i(1) = \phi_i(1) \epsilon T .$$

Thus $p\alpha_i$ determines an element $[p\alpha_i] \in \pi_1(K^1, T)$ and $\{[p\alpha_i][\phi_i][p\alpha_i]^{-1} \mid 1 \leq i \leq n\}$ is a basis of $\pi_1(K^1, T)$. Hence $\{[\phi_1], \ldots, [\phi_n]\}$ is a basis-up-to-conjugation of $\pi_1(K^1, T)$. $\square$

(5.4) PROOF OF THEOREM 5. Assuming prismatic collapsibility of the standard complex $K \times I$, we must show that the link of the unique vertex has an articulation point. By Theorem 4, the w_i determine a basis up to conjugation of $F(a_1, \ldots, a_n)$. Although they are not necessarily reduced, the w_i thus constitute a *simple set of cyclic words* in the sense of [Wh₁]. The Whitehead lemma [Wh₁, p.51] now gives the conclusion, because the link of v in $|K|$ is exactly the graph considered by Whitehead for a set of cyclic words. $\square$

Next we give an implication connecting the different notions of prismatic collapsibility in §2:

(5.5) *If K^2 is a PLCW-complex and $X = |K^2|$ is prismatically 1-collapsible in the sense of* (2.5), *then K^2 is prismatically 1-collapsible in the sense of* (2.6).

PROOF. Prismatic collapsibility in the sense of (2.5) is clearly preserved under prismatic simplicial subdivision (2.4). Thus we may assume without loss of generality that there is a simplicial subdivision $(K')^2$ of K^2 and a prismatic triangulation (J, K'^2) of $(X \times I, X \times 0)$ such that

(1) the attaching maps for the 2-cells e_i^2 of K^2 are simplicial maps $\phi_i : \partial D_i^2 \to K'^2$ with respect to some triangulation of ∂D_i^2.

(2) $J \searrow^s (J_1 \bigcup_i \sigma_i^*) \searrow^s *$, where $|J_1| = |K'^1 \times I|$.

Each ϕ_i maps at least one 1-simplex nondegenerately, because otherwise the corresponding $\bar{e}_i^2$ would be a 2-sphere contradicting the contractibility of $|K|^2$. Without loss of generality we may assume that ϕ_i maps every 1-simplex nondegenerately.

Now consider the 1-simplexes τ_j^1 of ∂D_i^2. By the nondegeneracy of ϕ_i on 1-simplexes we may extend the triangulation of ∂D_i^2 to a triangulation of D_i^2 and choose Φ_i such that the 2-simplex of D_i^2 which meets ∂D_i^2 in a given 1-simplex τ_j^1 is mapped by Φ_i simplicially onto a 2-simplex $\sigma_j^2 \subset \bar{e}_i^2$ with $\sigma_j^2 \in K'^2$. The corresponding σ_j^{*2} of definition (2.5) meets $\phi_i(\tau_j^1) \times I$ in an interval. We take the $(\phi_i \times 1)$-preimage of this interval in $\tau_j^1 \times I$, and together with connecting lines in (0-skeleton of $\partial D_i^2) \times I$ these preimages constitute a polygonal simple closed curve in $\partial D_i^2 \times I$. (See Figure 11). This curve bounds a properly embedded 2-disk $\Delta_i^2 \subset D_i^2 \times I$. We define $\bar{e}_i^{*2}$ to be $F_i(\Delta_i^2)$ and clearly have $K^2 \times I \searrow (K^1 \times I) \cup (\cup e_i^{*2})$. It remains to show that $(K^1 \times I) \cup (\cup e_i^{*2}) \searrow *$.

Let S be the sequence of simplicial collapses $J_1 \cup (\cup \sigma_k^{*2}) \searrow *$ in (2) above. The curve $F_i(\partial \Delta_i^2)$, which by construction is the boundary ∂e_i^{*2} of e_i^{*2}, is part of $|J_1|$. It bounds a singular PL 2-disk d_i in $|J_1 \cup (\cup \sigma_k^{*2})| \cap (\bar{e}_i^2 \times I)$, which consists of all σ_k^{*2} with $\sigma_k^2 \subset \bar{e}_i^2$ and certain regions in the $\sigma_\ell^1 \times I$, [where $\sigma_\ell^1 \subset e_i^2$, $\sigma_\ell^1 \in K'^1$,] which connect the σ_k^{*2}. By inspection of d_i or by homotopy considerations we see that the first collapse of S that removes part of d_i must enter d_i from ∂e_i^{*2}.

We denote by S' the subsequence of S consisting of the 2-dimensional collapses of the subdivided $K^1 \times I$ along with the first collapses entering d_i for each i. If

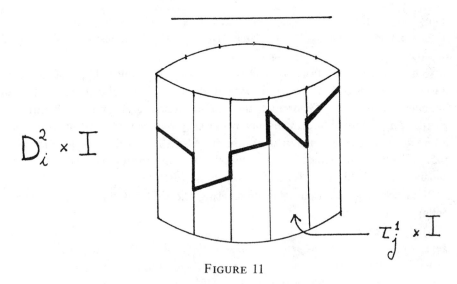

$$D_i^2 \times I$$

$$\tau_j^1 \times I$$

FIGURE 11

in S' we replace every such first collapse entering a d_i by a total collapse of the corresponding $\bar{e}_i *^2$ from the same face, we get a collapse of $(K^1 \times I) \cup (\cup e_i *^2)$ onto a 1-dimensional polyhedron. [In general, S' will no longer be a possible collapse of $J_1 \cup (\cup_k \sigma_k^{*^2})$, as some parts of the d_i might be left, which block other elements of S'. But still our conclusion holds, as $e_i^{*^2}$ is removed all at once. See (4.3) for the whole argument.] As this 1-dimensional polyhedron still has the homotopy type of $|K^2|$, it must be a tree and thus can be collapsed to a point. $\square$

Because of (5.5) we may replace the condition of CW prismatic collapsibility (2.6) in Theorem 5 by PL prismatic collapsibility (2.5) and still conclude that the link of the single vertex of the *original cell structure* K^2 has an articulation point. (See [C-M-Z], 4.1).

In his thesis [Zi], A. Zimmermann derives an even stronger criterion by direct geometric methods (without use of Theorem 4 and Whitehead's lemma). We cite without proof:

(5.6) *If K^2 is a PLCW complex and $|K^2|$ is prismatically 1-collapsible then every 2-dimensional subcomplex $L^2 \subset K^2$ contains a vertex v such that the link of v with respect to L has an articulation point.*

That L^2 is 2-dimensional, must be taken literally: L^2 has to contain 2-cells. But K^2 and L^2 may coincide. Note that K^2 in (5.6) may have more than one vertex, and there are no restrictions on the attaching maps of the 2-cells.

§6. THE GENERAL PROBLEM OF COLLAPSING $X^2 \times I$

In this section we discuss the general problem of collapsing $X^2 \times I$ when X^2 is contractible. We will broaden the preceding discussion of collapsing via cylinders and sheets to one of collapsing via "cylinders-with-holes and (several) sheets".

At the end we will conclude with some comments about possible algebraic and geometric obstructions to 1-collapsibility.

We first make the observation that any collapse $X^2 \times I \searrow *$ can be achieved through simplicial cylinders-with-holes and sheets:

(6.1) *If $X^2 \times I \searrow *$ then there is a prismatic simplicial subdivision (J, K) of $(X^2 \times I, X^2 \times 0)$ such that, if $|J_1| = |K^1 \times I|$ and if $\sigma_1, \ldots, \sigma_q$ are the 2-simplexes of K, then there are 2-simplexes $\sigma_{ij}^* \ (1 \le i \le q, 1 \le j \le n_i)$ and $\tau_k \ (1 \le k \le p)$ of J (called "sheets" and "holes" respectively) for which*

   (i) *σ_{ij}^* projects isomorphically onto σ_i under the natural projection $X^2 \times I \to X^2 \times 0$,*
   (ii) *τ_k is a vertical 2-simplex(i.e., $\tau_k \in J_1$),*
   (iii) *$J \searrow [J_1 \cup \bigcup_{i,j} \sigma_{ij}^* - \bigcup_k \operatorname{Int} \tau_k] \searrow *$.*

NOTE: For Euler characteristic reasons, $n_1 + \ldots + n_q = q + p$; thus the number of sheets equals the number of 2-simplexes of K plus the number of holes.

PROOF OF (6.1). Choose a triangulation (L, L_0) of $(X \times I, X \times 0)$ such that $L \searrow *$. Let (J, K) be a subdivision of (L, L_0) such that the PL projection $X \times I \to X \times 0$ becomes simplicial. By [Ch] there is a sequence of elementary simplicial collapses $J \searrow *$. Do the 3-dimensional collapses of the sequence (in the order in which they occur) and stop. Then J will have collapsed to a collapsible complex consisting of a set of simplicial prisms-with-holes and sheets. $\square$

We cannot prove the analogue of (6.1) in the PLCW situation, but in practice PLCW complexes K^2 have often been shown to be *1-collapsible via cylinders-with holes and CW sheets.*(In short: "1-collapsible with holes and sheets".) This means that

$$|K^2 \times I| \searrow [|K^1 \times I| \cup \bigcup_{i,j} F_i(\Delta ij) - \bigcup_k (\operatorname{Int} B_k)] \searrow * ,$$

where

   a) the Δ_{ij} are properly embedded 2-cells in $D^2 \times I$
   b) as usual, $F_i = \Phi_i \times 1$ where the Φ_i are characteristic maps of the 2-cells of K^2.
   c) the open cells $F_i(\operatorname{Int} \Delta_{ij})$ are pairwise disjoint
   d) the B_k are pairwise disjoint embedded 2-disks in $|K^1 \times I| - |K^0 \times I| - \bigcup_{i,j} F_i(\Delta_{ij})$.

See [Ze₁], [L₂], [Wa], [B], [W-W] for examples. It is notable that [K-M] does not give this type of collapse.

As an illustration of how this works and how this leads to information about group presentations we return to the example of (2.9): (The collapse here, which we learned from W. Becker, is well-known, but the connection to bases of blown-up free groups is new).

(6.2) *Let K^2 be the standard complex of the presentation $(a, b : ab, a^2 b^3)$. Then $|K^2 \times I|$ is collapsible via cylinders-with-holes and sheets in a manner which leads to the basis $\{ab, a(ca)b^3, a(ca)ba^{-1}\}$ of $F(a, b, c)$.*

PROOF. In the cylinder $S_a^1 \times I$ of $(K^1 \times I) = (S_a^1 \vee S_b^1) \times I$ let B_1 be a small oriented round disk half way up (see Figure 12) connected by an arc α to the base point.

If $c = [\alpha * \partial B_1 * \overline{\alpha}]$ then $\{a, b, c\}$ is a basis of

$$\pi_1(|K^1 \times I| - (\operatorname{Int} B_1), *) \equiv F(a, b, c).$$

Let $\phi_1 : \partial D^2 \to K^1$ monotonically sweep out the word ab, and in $D^2 \times I$ let Δ_{11} and Δ_{12} be the two slanted disks pictured in Figure 13 so that F_1 maps $\partial D^2 \times I$ to $K^1 \times I$ with the images of $\partial \Delta_{11}$ and $\partial \Delta_{12}$ as pictured in Figure 14. Let $\phi_2 : \partial D^2 \to K^1$ monotonically sweep out the word $a^2 b^3$ and let Δ_{21} be the two cell of $D^2 \times I$ pictured in Figure 15. Notice that $F_2(\partial \Delta_{21}) = F_1(\partial \Delta_{11} \cup \partial \Delta_{12})$ $\cup (S_b^1 \times \{1\})$. (See the heavy path in Figure 16.)

Let $L^2 = |K^1 \times I| - (\operatorname{Int} B_1)$ (the cylinders with a hole) and, attaching 3 sheets, let $J^2 = L^2 \cup F_1(\Delta_{11}) \cup F_1(\Delta_{12}) \cup F_2(\Delta_{21})$. Then we claim that

$$|K^2 \times I| \searrow |J^2| \searrow *$$

The first collapse comes (using 2.2)) from the 3-dimensional collapses

$$F_2(D^2 \times I) \searrow |K^1 \times I| \cup F_2(\Delta_{21}) \quad \text{and}$$
$$F_1(D^2 \times I) \searrow |K^1 \times I| \cup F_1(R) \quad (R = \text{ region between } \Delta_{11} \text{ and } \Delta_{12})$$
$$\searrow L^2 \cup F_1(\Delta_{11} \cup \Delta_{12}).$$

The latter complex collapses by 2-dimensional moves to the union of its sheets because $|L_2| \searrow F_1(\partial \Delta_{11} \cup \partial \Delta_{12}) \cup F_2(\partial \Delta_{21})$, as pictured in Figure 16. This leaves us with the complex of Figure 17 which collapses to a point.

By Theorem (4.4) the attaching maps $F_i|\partial \Delta_{ij}$ for the sheets, in conjunction with paths lying in any maximal tree T which remains after all 2-collapses in the collapse of J^2 are completed, should determine a basis of $\pi_1(L^2, *)$. Choose $T = T_1 \cup T_2$ where T_1, T_2 are the left halves of $F_1(\partial \Delta_{12})$ and $F_1(\partial \Delta_{11})$ in Figure 14 or Figure 17. Since $F_1(\partial \Delta_{12})$ and $F_2(\partial \Delta_{21})$ go through the base point we may choose the constant path as the path in T connecting them to the base point. We connect $F_1(\partial \Delta_{11})$ to $*$ by the path T_1. Then we have (see Figure 14),

$$[F_1(\partial \Delta_{12})] = ab, \quad [F_2(\partial \Delta_{21})] = a(ca)b^3, \quad \text{and}$$
$$[T_1 * F_1(\partial \Delta_{11}) * \overline{T}_1] = a(ca)b \, a^{-1} \, \epsilon \, \pi_1(L^2, *) = F(a, b, c,). \quad \square$$

The *general situation* is well illustrated by (6.2): Suppose K^2 is the standard complex determined by the presentation $P = (a_1, \ldots, a_n : R_1, \ldots, R_n)$. If $B_1, \ldots, B_m$ are PL pairwise disjoint 2-balls ("holes") in $|K^1 \times I| - |K^0 \times I|$ then the oriented boundaries ∂B_i, along with arcs to the base point $* = (e^0, 0)$, determine a basis $\{a_1, \ldots, a_n, b_1, \ldots, b_m\}$ of the fundamental group of the cylinders-with-holes. (See Figure 18).

Suppose further that K^2 is 1-collapsible with sheets $F_i(\Delta_{ij})$ and holes $B_1, \ldots, B_m$ (so that necessarily $F_i(\partial \Delta_{ij}) \cap B_k = \varnothing$ for all i, j, k). Then the

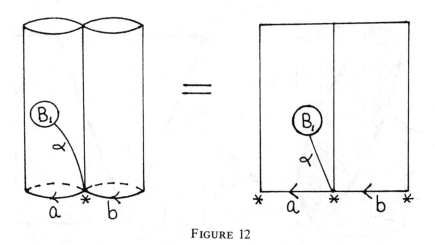

FIGURE 12

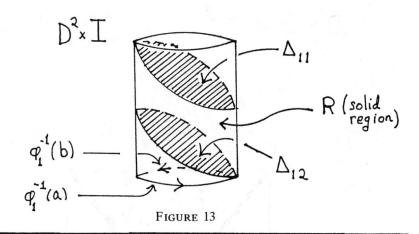

FIGURE 13

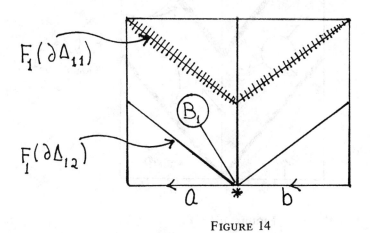

FIGURE 14

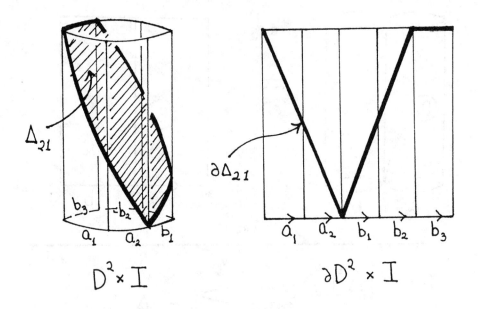

FIGURE 15

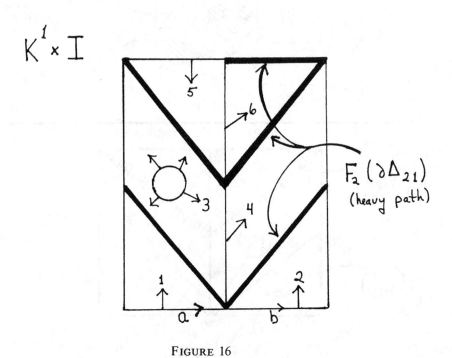

FIGURE 16

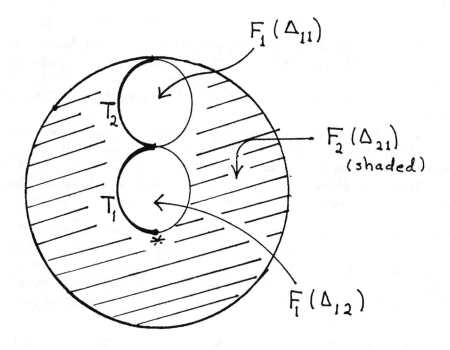

FIGURE 17

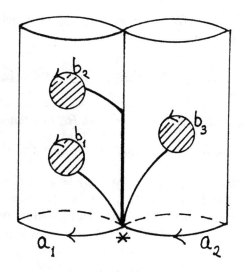

FIGURE 18

sheets $F_i(\Delta_{ij})$ (playing the role of $\bar{e}_1, \ldots, \bar{e}_q$ in (4.4)) determine a basis R_{ij} of π_1 (cylinders-with-holes). Specifically

$$R_{ij} = [\alpha_{ij} * F_i | \partial \Delta_{ij} * \bar{\alpha}_{ij}] \in F(a_1, \ldots, a_n, b_1, \ldots, b_m)$$

Here $\partial \Delta_{ij}$ is an oriented simple closed curve in $\partial D^2 \times I$ which has its base pont z_{ij} on $\{1\} \times I$ and which projects vertically by an orientation preserving homotopy equivalence to $\partial D^2 \times 0$. The α_i are paths from $*$ to $F_i(z_{ij}) \in e^0 \times I$ which lie in a tree in the cylinders-with-holes to which (cylinder-with-holes) $\cup$ (sheets) collapses.

What does vertical projection $|K^1 \times I| - \bigcup_k \operatorname{Int} B_k \to |K^1 \times 0|$ do to this basis of the fundamental group? On the one hand it is clear that the induced map is just the natural projection $F(a_1, \ldots, a_n, b_1, \ldots, b_m) \to F(a_1, \ldots, a_n)$. On the other hand it is clear, since $F_i = (\Phi_i \times 1)$, that $[(\text{proj}) \circ (F_i | \partial \Delta_{ij})] = R_i$. Thus $R_{ij} \to w_{ij} R_i w_{ij}^{-1}$ where $w_{ij} = [(\text{proj}) \circ \alpha_{ij}]$. We summarize this as follows:

(6.3) *If the PLCW complex K^2 naturally determined by the presentation $P = (a_1, \ldots, a_n : R_1, \ldots, R_n)$ is 1-collapsible with holes and sheets then a basis $\{R_{ij} | 1 \leq i \leq n, 1 \leq j \leq q_i\}$ of an expanded free group $F(a_1, \ldots, a_n, b_1, \ldots, b_m)$ is determined for which the natural projection $F(a_1, \ldots, a_n, b_1, \ldots, b_m) \to F(a_1, \ldots, a_n)$ takes R_{ij} to an element of the form $w_{ij} R_i w_{ij}^{-1}$.*

The fact that a set of normal generators $R_1, \ldots, R_n$ of $F(a_1, \ldots, a_n)$ can be "blown up" to become an actual basis of a larger free group might at first be surprising and seem to indicate that we have an obstruction to the existence of a CW prismatic collapse with holes and sheets. Algebraically this is not a valid reaction:

(6.4) *If $P = (a_1, \ldots, a_n : R_1, \ldots, R_n)$ is a presentation of the trivial group then there exists another such presentation*

$$Q = (a_1, \ldots, a_n, b_1, \ldots, b_m : \{R_{ij} | 1 \leq i \leq n, 1 \leq j \leq q_i\})$$

such that $\{R_{ij}\}$ is a basis of $F(a_1, \ldots, a_n, b_1, \ldots, b_m)$ and the natural projection $F(a_1, \ldots, a_n, b_1, \ldots, b_m) \to F(a_1, \ldots, a_n)$ takes each R_{ij} to a conjugate of R_i.

REMARK: (6.4) seems to be somehow dual to Theorem 2 of [Ra].

PROOF. Let $\{w_{ij} R_i w_{ij}^{-1} | 1 \leq i \leq n, 1 \leq j \leq q_i\}$ be a set of conjugates of the relators R_i which generate $F(a_1, \ldots, a_n)$. We may apply a sequence $S_1, S_2, \ldots, S_t$ of Nielsen moves which transform these conjugates (given in some order) to the sequence $(a_1, \ldots, a_n, 1, 1, \ldots, 1)$. If there are m 1's occurring apply the inverse sequence $S_t^{-1}, \ldots, S_1^{-1}$ of moves to the sequence of symbols $(a_1, \ldots, a_n, b_1, \ldots, b_m)$. The result will be a basis R_{ij} of $F(a_1, \ldots, a_n, b_1, \ldots, b_m)$ which projects to $w_{ij} R_i w_{ij}^{-1}$ upon setting each b_i equal to 1. $\square$

Speculation concerning possible strategies and obstructions

I. (6.4) leads us to a notion of how the Andrews-Curtis conjecture might be proved: Start with a presentation $P = (a_1, \ldots, a_n : R_1, \ldots, R_n)$ of the trivial group. Choose conjugates $w_{ij} R_i w_{ij}^{-1}$ which generate $F(a_1, \ldots, a_n)$ and use these to blow P up to a presentation $Q = (a_1, \ldots, a_n, b_1, \ldots, b_m : R_{ij})$ as in (6.4). Place m-holes $B_1, \ldots, B_m$ somewhere in the n cylinders $(S_1^1 \vee \ldots \vee S_n^1) \times I$ and distribute $n + m$ closed curves $F_i(\partial \Delta_{ij})$ in the cylinder-with-holes so that these curves represent the words R_{ij} and project vertically to curves representing the words R_i. Now build a 2-complex K_P modelled on the presentation P (perhaps with detours in the attaching maps as in the proof of Theorem 2) so that $(K_P \times I) \searrow X^2 = $ (cylinders-with-holes) $\cup$ (sheets). If all this is possible we are done, for we may collapse the cylinders with holes to a wedge product W^1 of circles representing the generators $a_1, \ldots, a_n, b_1, \ldots, b_m$. This induces a simple homotopy equivalence $X^2 \searrow^{3} (W^1 \cup $ sheets $)$ where the sheets are attached by maps representing the basis elements R_{ij} of $F(a_1, \ldots, a_n, b_1, \ldots, b_m)$. We homotop these attaching maps a bit to get a 2-complex W^2 with attaching maps as in the proof of Theorem 2. Then

$$(W^1 \cup \text{ sheets } \nearrow^{3} W^2 \nearrow (W^2 \times I) \searrow *.$$

If this "argument" runs into difficulties, one can replace I by the cone on finitely many points (as in 3.3), or indeed, by any tree.

II. The preceding dream of a proof is fraught with difficulties. The choice of a blown-up basis $\{R_{ij}\}$ is not unique and once we choose one there is the problem of choosing appropriate holes and attaching-curves for the sheets. Indeed, starting with a particular K_P, it might be proved a counterexample to the Zeeman conjecture in the (possibly) limited sense that $K_P \times I$ does not collapse to a point via CW cylinders-with-holes and sheets because the presentation P allows no way to place the holes and sheets so as to satisfy the conclusion of (6.3). It is highly probable that the Zeeman conjecture is false even for some complexes K^2 for which $K^2 \searrow^{3} *$.

III. Finally, if (6.3) does not lead to an algebraic obstruction to Zeeman's conjecture, (4.2) may give an accessible geometric obstruction:

If we choose a simple closed curve S^1 in $K^2 \times I$ (or even $K^2 \times 0$) the existence of a "good" singular disk bounded by this curve—one which is collapsible *and* to which $|K^2 \times I|$ collapses—is necessary and sufficient for $|K^2 \times I|$ to be collapsible. There are numerous examples in which just one of these requirements seems to lead to a disproof of the Zeeman conjecture: for instance, in some cases every singular disk which we have been able to visualize either doesn't have a free face in S^1 or is not simply connected. One of our goals is to make a rigorous proof out of these observations, taking into account all possible singular disks.

Alternatively, the proof of (4.2) suggests that it may be useful to focus on more than the trail of one curve; for instance we might concentrate on the trail of the 1-skeleton $K^1 \times 0 \subset K^2 \times 0$.

REFERENCES

[A] E. Akin, *Transverse cellular mappings of polyhedra*, Trans. Amer. Math. Soc. **169** (1972), 401–438. [MR 48, #5088].

[A-C] J. J. Andrews and M. L. Curtis, *Free groups and handlebodies*, Proc. Amer. Math. Soc., **16** (1965), 192–195. [MR 30, #3454].

[B] W. Becker, *Untersuchungen zur Zeeman-Vermutung*, Diplom-Arbeit, Frankfurt/Main 1978.

[Ch] D. R. J. Chillingworth, *Collapsing three-dimensional convex polyhedra*, Proc. Camb. Phil. Soc., **63** (1967), 353–357. [MR 35, #995]. Correction, Math. Proc. Camb. Phil. Soc., **88** (1980), [MR 81g: #5800].

[C_1] M. Cohen, *Simplicial structures and transverse cellularity*, Ann. of Math., *(2)* **85** (1967), [MR 35, #1037].

[C_2] M. Cohen, *Whitehead torsion, group extensions and Zeeman's conjecture in high dimensions*, Topology **16** (1977), 79–88. [MR 55, #9105].

[C-C] M. Cohen and G. Cooke, *Collapsing $|K^1 \times I|$ onto $|L^1|$*, unpublished manuscript.

[C-M-Z] M. Cohen, W. Metzler and A. Zimmermann, *What does a basis of $F(a,b)$ look like?* Math. Ann. **257**, (1981), 435–445. [MR 82m: 20 027].

[Cohn] H. Cohn, *Markoff forms and primitive words*, Math. Ann. **196** (1972), 8-22, [MR 45, #6899].

[G-R] D. Gillman and D. Rolfsen, *The Zeeman conjecture for standard spines is equivalent to the Poincaré conjecture*, Topology **22** (1983), 315–323.

[H] J. F. P. Hudson, *Piecewise linear topology*, W. A. Benjamin, Inc., New York (1968), [MR 40, #2094].

[Ka] T. Kaneto, *On presentations of the fundamental group of the 3-sphere associated with genus two Heegard diagrams*, Kokyuroku of R.I.M.S., Kyoto University 369 (1979), p. 144–163 (Japanese). [English language preprint available.]

[K-M] R. Kreher and W. Metzler, *Simpliziale Transformationen von Polyedern und die Zeeman-Vermutung*, Topology **22**, (1983), 19–26.

[L_1] W.B.R. Lickorish, *On collapsing $X^2 \times I$*, Topology of Manifolds, ed., by J. C. Cantrell and C. H. Edwards, pp. 157–160, Markham Publishing Co., Chicago, 1970, [MR 42, #6826].

[L_2] W.B.R. Lickorish, *An improbable collapse*, Topology **12**, (1973), 5–8 [MR 47, #1067].

[M] W. Metzler, *Äquivalenzklassen von Gruppenbeschreibungen, Identitäten und einfacher Homotopietyp in niederen Dimensionen*, in: Homological Group Theory, Proc. Durham Symp. (1977), ed. C.T.C. Wall, London Math. Soc. Lecture Note Series **36** (1979), 291–326, [MR 81h: #57001].

[N] J. Nielsen, *Die Isomorphismengruppe der freien Gruppen*, Math. Ann. **91** (1924), 169–209.

[O-Z] R. P. Osborne and H. Zieschang, *Primitives in the free group on two generators*, Invent. Math. **63**, (1981), 17–24.

[Ra]   E.S. Rapaport, *Groups of order 1*, Proc. A.M.S. **15** (1964), 828–833. [MR 29, #4788].

[Ro]   O. S. Rothaus, *On the non-triviality of some group extensions given by generators and relations*, Ann. Math. *(2)* **106** (1977), 599–612 [MR 57, #7612].

[R-S]   C. P. Rourke and B.J. Sanderson, *Introduction to piecewise linear topology*, Ergebnisse der Mathematik und ihrer Grenzgebiete Band 69, Springer-Verlag New York (1972), [MR 50, #3236].

[S]   K. Sauermann, *Einige Phänomene des einfachen Homotopietyps in niederen Dimensionen*, Dissertation, Frankfurt/Main (1979). [Copies may be obtained by writing to: Fachbereich Mathematik, Robert-Mayer-Strasse 6-10, D-6000 Frankfurt, Federal Republic of Germany.]

[Wa]   L. W. Wajda, *One-collapsing the general dunce hat*, Ph.D. Dissertation, SUNY at Buffalo, (1978).

[W-W]   D. E. Webster and L. W. Wajda, *On Zeeman's Conjecture [The collapsibility of $K(2, q, r, s) \times I]$*, Manuscript, 1978.

[Wh$_1$]   J.H.C. Whitehead, *On certain sets of elements in a free group*, Proc. London Math. Soc. *(2)* **41** (1936), 48–56 (Math. Works of J.H.C. Whitehead, Vol. II, 69–77).

[Wh$_2$]   J.H.C. Whitehead, *Simplicial spaces, nuclei and m-groups*, Proc. London Math. Soc. **45** (1939), 243–327, (Math. Works of J.H.C. Whitehead, Vol. II, 99–183).

[Wh$_3$]   J.H.C. Whitehead, *On incidence matrices, nuclei and homotopy types*, Ann. of Math. *(2)* **42** (1941), 1197–1239 [MR 3–142] (Math. Works of J.H.C. Whitehead, Vol. II, 259–301).

[Wh$_4$]   J.H.C. Whitehead, *Combinatorial Homotopy I.*, Bull. Amer. Math. Soc. **55** (1949), 213–245 [MR 11–48]. (Math. Works of J.H.C. Whitehead, Vol. III, 85–117).

[Wr$_1$]   P. Wright, *Collapsing $K \times I$ to vertical segments*, Proc. Camb. Phil. Soc., **69** (1971), 71–74. [MR 42, #8475].

[Wr$_2$]   P. Wright, *Group presentations and formal deformations*, Trans. Amer. Math. Soc., **208** (1975), 161–169. [MR 52, #1710].

[Ze$_1$]   E. C. Zeeman, *On the dunce hat*, Topology **2** (1964), 341–358. [MR 27, #6275].

[Ze$_2$]   E. C. Zeeman, *Seminar on Combinatorial Topology*, Institut des Hautes Etudes Scienfitiques, 1963 (mimeographed).

[Z$_1$]   A. Zimmermann, *Eine spezielle Klasse kollabierbarer Komplexe $K^2 \times I$*. Thesis, Frankfurt/Main, 1978. [Copies may be obtained by writing to Fachbereich Mathematik, Robert-Mayer-Strasse 6-10, D-6000 Frankfurt, Federal Republic of Germany.]

CORNELL UNIVERSITY AND THE UNIVERSITY OF FRANKFURT

Contemporary Mathematics
Volume 44, 1985

ON THE ANDREWS-CURTIS-CONJECTURE AND RELATED PROBLEMS*

Wolfgang Metzler

§1. Introduction.

Two compact, connected polyhedra K^n, L^n, which have the same simple homotopy type, may be transformed into each other by a finite sequence of formal deformations not exceeding the dimension $n + 1$ ($K^n \overset{n+1}{\curvearrowright} L^n$), *provided* $n \neq 2$, s. Wall [33]. In the case $n = 2$ only

$$(1) \qquad\qquad K^2 \curvearrowright L^2 \Rightarrow K^2 \overset{4}{\curvearrowright} L^2$$

is known.

$K^2 \overset{3}{\curvearrowright} L^2$ holds if and only if presentations $\mathbf{P} = \{a_1, \ldots, a_k; R_1, \ldots, R_\ell\}$ and $\mathbf{Q} = \{b_1, \ldots, b_m; S_1, \ldots, S_n\}$ of $\pi_1(K)$ resp. $\pi_1(L)$, read off in the usual way from K^2 and L^2, can be transformed into each other by a finite sequence of the following elementary presentation moves:

(I)   a) *free transformations among the defining relators;*

       b) *conjugation of a relator*: $R_j \to w R_j w^{-1}$;

(II)   *free transformations of the generators*;

(III)   *prolongation, i.e. introduction of a new generator a and a new*

       relator $R = a$, and the inverse transformation (if possible).

This connection between simple homotopy type and combinatorial group theory is due to P. Wright [35], s. Kreher-Metzler [20], Theorem 2 for an n-dimensional and simplicial version of the essential step, converting a formal PL deformation $K^n \overset{n+1}{\curvearrowright} L^n$ into another one, in which every $(n + 1)$-simplex is collapsed immediately after its introduction (*transient moves*).

Following Rapaport [30], we speak of *Q-transformations* and *Q-equivalent presentations*, if only transformations of type I are involved. Transformations of type I + II generate *Q*-equivalences*. *Q**-equivalences*, or simply: *presentation equivalences* are generated by elementary transformations of type I + II + III.

*dedicated to Winfried Becker, Marshall M. Cohen, Cynthia Hog, Günther Huck and Martin Lustig.

Thus P. Wright's theorem may be formulated as follows:

(2) $K^2 \overset{3}{\nearrow} L^2 \Leftrightarrow \mathbf{P}$ and $\mathbf{Q}$ are Q^{**}-equivalent for presentations $\mathbf{P}$ of $\pi_1(K)$,
       $\mathbf{Q}$ of $\pi_1(L)$, given by arbitrary cell decompositions of K^2 resp. L^2.

In [2] Andrews and Curtis made the conjecture

(AC)   that *any balanced presentation of* $\pi = 1$ *is* Q^{**}-*trivial*, i.e., Q^{**}-equivalent to $\{a; a\}$ or the empty presentation, or equivalently, that *for any contractible compact* K^2, $K^2 \overset{3}{\nearrow} *$ *holds.*

As they point out ([2], Theorem 2), the truth of (AC) would imply that any 5-dimensional regular neighbourhood of $K^2 \simeq *$ (in a PL 5-manifold) is a PL ball.

But there exist several notorious examples, for which no Q^{**}-trivialization is known, although these balanced presentations can be shown to define the trivial group:

(α) $\{a, b, c; b^{-2}c^{-1}bc, c^{-2}a^{-1}ca, a^{-2}b^{-1}ab\}$, s. [30];
(β) $\{a, b; a^3ba^{-2}b^{-1}, b^3ab^{-2}a^{-1}\}$, s. [12], p.41;
(γ) $\{a, b; b^5a^{-4}, aba(bab)^{-1}\}$, s.[1].

They serve as possible candidates to disprove (AC).

We mention the following facts and observations about these examples:

(i)   Note the symmetry of (α) and (β); these presentations give rise to *relative presentations* in the sense of [24], §4, with $\mathbb{Z}_3$ resp. $\mathbb{Z}_2$ as operator group.

(ii)   Each of the relators in (β) defines a one-relator, non-hopfian group, s. [22], p.197.

(iii) If in (β) *one* relator is replaced by an arbitrary other one, yielding a perfect group π, then still $\pi = 1$ holds. Moreover, the exponents 2 and 3 in the remaining relator may be substituted by n and $n + 1$ (Miller and Schupp [25]).

(iv)   Although (γ) is not known to be Q^{**}-trivial, the corresponding 2-complex has a topological 5-ball as a regular neighborhood, S. Akbulut and Kirby [1].* (By [37], Proposition 1.4 there is a unique 5-dimensional PL regular neighborhood for any contractible K^2.)

(AC) may be generalized to the question,

(AC′)   whether in (1) the dimension bound can be improved from 4 to 3,

s. Wall [33]. At present this generalization is an open problem too.

The aim of this survey is to give an account on some recent partial results. They concern relations to (other) problems of combinatorial group theory, (3- and 4-dimensional) PL-topology and geometry. Some of these relations are characterizations of $(AC^{(\prime)})$, some point out possible strategies of their solution. At least they reveal the importance of $(AC^{(\prime)})$ as a topic, where algebra and geometry blend together in a much richer manner than could be expected from

---

*added in print: Freedman's work implies this fact for an arbitrary $K^2 \simeq *$: Embed K^2 PL in S^5 with regular neighborhood N. Then ∂N is a homotopy 4 sphere and thus homeomorphic to S^4 (Freedman). By Morton Brown's Schoenflies theorem now N is a topological 5-ball. (This argument was pointed out to me by W. B. R. Lickorish.)

the original motivation of 5-dimensional regular neighborhoods. In particular, algebraic aspects cannot be considered "blindfolded", and topological aspects cannot do without combinatorial group theory. This interaction was basic some decades ago, s. [23], and its stimulation was one of the purposes of the Rochester conference.

Another purpose of this paper is to encourage cooperation in a field, in which working alone may cause weariness. Friends and students have created a different atmosphere for rewarding joint work. The dedication is to those of them, who participated in the conference.

§2 Aspects concerning Combinatorial Group Theory.

We first describe *two characterizations of Q-equivalence in terms of auto-morphism groups of free groups.* For this purpose, we need the following notations: Let F^n be the free group with free generators $a_1, \ldots, a_n$ and G an arbitrary group. If $\varphi \in \operatorname{Aut} F^n$ is given by $\varphi(a_i) = w_i(a_1, \ldots, a_n)$ with words $w_i \in F^n$, then we assign to each "vector" $(g_1, \ldots, g_n) \in \underbrace{G \times \ldots \times G}_{n \text{ factors}}$ the element $\varphi(g_1, \ldots, g_n) = (w_1(g_i), \ldots, w_n(g_i))$. Thus we obtain

(4)   an operation of Aut F^n on the set $G \times \ldots \times G$ fulfilling the rule of composition $\psi(\varphi(g_1, \ldots, g_n)) = (\psi \circ \varphi)(g_1, \ldots, g_n))$,

where $\psi \circ \varphi$ is Nielsen's traditional notation for $F^n \overset{\psi}{\to} F^n \overset{\varphi}{\to} F^n$.

A *Q-transformation* applied to a presentation $\mathbf{P} = \{a_1, \ldots, a_m; R_1, \ldots, R_n\}$ may be defined by these means as

(5) $\mathbf{P} \to \varphi(\mathbf{P}), \quad \varphi \in W,$

where $\mathbf{P}$ is interpreted as a vector with $m + n$ components in $G = F(a_1, \ldots, a_m)$. $W \subset \operatorname{Aut} F^{m+n} = \operatorname{Aut} F(a_1, \ldots, a_m, b_1, \ldots, b_n)$, i.e. *the group of (m, n) Q-trans-formations,* is generated by

(6) a) free transformations among the b_j;

   b) conjugation of a b_j by a word $w \in F(a_i)$.

In addition we will consider the larger group V with $W \subset V \subset \operatorname{Aut} F^{m+n}$, which is generated by

(7) a) free transformations among the b_j;

   b) multiplication of a b_j from left or right by an a_i-word.

Because of the relative Nielsen Theorem [24], Thm. 7, see also Denk [13] for a general relative Nielsen reduction method, W may be defined alternatively as

(8)   the subgroup W of V consisting of those automorphisms $\varphi \in V$, which stabilize the presentation $\{a_i; 1, \ldots, 1\}$.

Note that this stabilization with respect to the operation (4) is different from the usual one in automorphism groups of free groups, as it concerns composition of components instead of substitution of (free) "variables".

To every presentation $\mathbf{P} = \{a_1, \ldots, a_m; R_1, \ldots R_n\}$ we assign the element

(9) $\sigma_{\mathbf{P}} \in V$ given by the automorphism of $F(a_i, b_j)$, which fixes every a_i and maps b_j to $R_j(a_i) \cdot b_j$.

Then

(10) $\sigma_{\mathbf{P}}(\{a_i; 1, \ldots, 1\}) = \mathbf{P}$

is immediate.

THEOREM 1.   *Two presentations* $\mathbf{P} = \{a_1, \ldots, a_m; R_1, \ldots, R_n\}$ *and*
$\mathbf{Q} = \{a_1, \ldots, a_m; S_1, \ldots, S_n\}$ *are Q-equivalent if and only if for some* $\varphi, \psi \in W$

(11) $\sigma_{\mathbf{Q}} = \varphi \circ \sigma_{\mathbf{P}} \circ \psi$ *holds* .

PROOF.

a) If $\varphi(\mathbf{P}) = \mathbf{Q}$ for some $\varphi \in W$, then $\{a_i; 1, \ldots, 1\} \overset{\sigma_{\mathbf{P}}}{\to} \{a_i; R_j\} \overset{\varphi}{\to} \{a_i; S_j\} \overset{\sigma_{\mathbf{Q}}^{-1}}{\to}$
$\{a_i; 1, \ldots, 1\}$. Because of (8) this implies $\sigma_{\mathbf{Q}}^{-1} \circ \varphi \circ \sigma_{\mathbf{P}} \in W$, and with $\psi^{-1} = \sigma_{\mathbf{Q}}^{-1} \circ \varphi \circ \sigma_{\mathbf{P}}$ we get (11).

b) If (11) is fulfilled for some $\varphi, \psi \in W$, then $\varphi \circ \sigma_{\mathbf{P}} = \sigma_{\mathbf{Q}} \circ \psi^{-1}$, hence

$$(\varphi \circ \sigma_{\mathbf{P}})(\{a_i; 1, \ldots, 1\}) \overset{(10)}{=\!=} \varphi(\mathbf{P}) =$$
$$(\sigma_{\mathbf{Q}} \circ \psi^{-1})(\{a_i; 1, \ldots, 1\}) \overset{(8)}{=\!=} \sigma_{\mathbf{Q}}(\{a_i; 1, \ldots, 1\}) \overset{(10)}{=\!=} \mathbf{Q}. \quad \square$$

Denote by $S(\mathbf{P})$ the *stabilizer of* $\mathbf{P}$ *in* W with respect to the operation (4). Then we get:

THEOREM 2.   *Two presentations* $\mathbf{P} = \{a_1, \ldots, a_m; R_1, \ldots, R_n\}$ *and*
$\mathbf{Q} = \{a_1, \ldots, a_m; S_1, \ldots, S_n\}$, *the relators* R_j, S_j *of which are not proper powers,*
are Q-equivalent if and only if $S(\mathbf{P})$ *and* $S(\mathbf{Q})$ *are conjugate subgroups of* W.

PROOF.

a) If $\varphi(\mathbf{P}) = \mathbf{Q}$, then $\varphi \circ S(\mathbf{P}) \circ \varphi^{-1} = S(\mathbf{Q})$ is obvious; and the argument doesn't need the restriction on the defining relators.

b) Suppose first $S(\mathbf{P}) = S(\mathbf{Q})$, then in particular, $S(\mathbf{P})$ and $S(\mathbf{Q})$ contain the same conjugation elements (6) b) of W. By the restriction on the defining relators these are conjugations of a b_j with all powers R_j^k resp. S_j^ℓ. Hence $R_j = S_j^{\pm 1}, j = 1, \ldots, n$ and $\mathbf{P}, \mathbf{Q}$ are shown to be Q-equivalent.

In the general case $\varphi \circ S(\mathbf{P}) \circ \varphi^{-1} = S(\mathbf{Q}), \varphi \in W$, we consider in addition $\varphi(\mathbf{P})$, which by a) has the same stabilizer as $\mathbf{Q}$. Then the preceding argument yields that $\varphi(\mathbf{P})$ and $\mathbf{Q}$ are Q-equivalent, and because of $\varphi \in W$ the same holds for $\mathbf{P}$ and $\mathbf{Q}$. $\quad \square$

The *double coset* characterization of Theorem 1 has an analogy in *Heegard theory* of 3-manifolds, s. J. Birman [3], [4] and R. Craggs [11]. Moreover, Craggs [11] has generalized Theorem 1 in an axiomatic approach, which covers the case of handlebody decompositions of 3-manifolds as well as Q^{**}-equivalences. A reason for this analogy is the fact that $(Q^{**}-)$ presentation classes are 3-manifold invariants (Reidemeister-Singer), s. [24], (5), (6) and §3 (18) below for a short proof. One passes from 3-manifolds to presentation classes by "forgetting that certain free transformations live on a surface". According to this pattern, the Stallings-Jaco-splitting has an equivalent for Q-classes of presentations of $\pi = 1$:

If $\mathbf{P} = \{b_1, \ldots, b_m; R_1, \ldots, R_n\}$ is an (arbitrary) finite presentation, we consider the set of elements

(11)
$$\begin{aligned}
\mathbf{a}_i &= (a_i, b_i) \in F(a_1, \ldots, a_m) \times F(b_1, \ldots, b_m), \\
\mathbf{b}_j &= (1, R_j) \in F(a_1, \ldots, a_m) \times F(b_1, \ldots, b_m), \\
i &= 1, \ldots, m, \ j = 1, \ldots, n.
\end{aligned}$$

It is easy to see that

(12) $\mathbf{P}$ is a presentation of $\pi = 1$ if and only if the $\mathbf{a}_i$ and $\mathbf{b}_j$ together generate $F(a_i) \times F(b_i)$, compare [22], p. 193 f..

Q-transformations of $\mathbf{P}$ give rise to Nielsen transformations of the $\mathbf{a}_i, \mathbf{b}_j$, namely: free transformations among the $\mathbf{b}_j$, conjugations of a $\mathbf{b}_j$ by $\mathbf{a}_i^{\pm 1}$ and products of these transformations. Hence we obtain in particular:

THEOREM 3. *A necessary condition for the Q-triviality[1) of $\mathbf{P}$ is that $m = n$ and the $\mathbf{a}_i, \mathbf{b}_i$ are Nielsen-equivalent to the generators $(a_i, b_i), (1, b_i)$ of $F(a_i) \times F(b_i), i = 1, \ldots, n$, i.e. $\mathbf{a}_i, \mathbf{b}_i$ can be lifted to a basis of $F(a_i) * F(b_i)$. Moreover, the lift of $\mathbf{a}_i$ may be prescribed to be $a_i \cdot b_i, i = 1, \ldots, n$.*

Splitting homomorphisms $F^{2n} \to F^n \times F^n$, are obtained by assigning to the generators of F^{2n} the images $\mathbf{a}_i, \mathbf{b}_i$.

As in the case of 3-manifold splitting, Theorem 3 in fact is a characterization (if the $a_i \cdot b_i$ together with appropriate lifts of the $\mathbf{b}_i, i = 1, \ldots, n$ constitute a basis of $F(a_i) * F(b_i)$, then $\mathbf{P}$ is Q-trivial), s. Craggs [9],[10], Metzler [24], Theorem 4. These papers also cover Q^*- and Q^{**}-equivalences. Note that Q^*-triviality implies Q-triviality ([24], (22)).

In order to minimize the overlap with [24], we have given an argument for Theorem 3, which does not involve *Peiffer identities*. But it should be mentioned that these have turned out to be a key phenomenon for the study of homotopy problems of 2-complexes, s. Brown-Huebschmann [5], Sieradski [31], [32].

Theorem 1 and Theorem 3 are the starting point of various ideas to solve (AC) or (AC'). Some of these concern matrix representation of endomorphisms and automorphisms of free groups; a report of this "work in progress" is given in [21].

Theorem 3 gives rise to a *geometric idea*, which we indicate briefly: F^n acts on hyperbolic n-space and thus $F^n \times F^n$ acts on the product space. The standard generators give rise to tesselations of the product space, in which "opposite $(2n-1)$-faces" of a fundamental domain correspond to a shift by one of the generators. With some care, involving a (CW-)argument analogous to the ordinary *shearing* of euclidean cubes, one can see that the existence of such *pseudocube tesselations* is preserved inductively under Nielsen transformations of generators. Combining with Theorem 3, the *nonexistence of tesselations for the $\mathbf{a}_i$, $\mathbf{b}_i$ of some balanced presentation $\mathbf{P}$ of $\pi = 1$ would exhibit a nontrivial Q-class.*

_____

[1)]$\mathbf{P}$ is called *Q-trivial,* if $\mathbf{P}$ is Q-equivalent to a presentation with $m = n$, in which the $i-th$ defining relator equals the $i-th$ generator.

This approach probably needs complicated metric techniques, for instance determining the volume of (hypothetical) fundamental domains.

A similar idea is due to John Ratcliffe.

Any strategy to solve $(AC^{(1)})$ must beware of the attempt to obtain an algorithm, which decides, whether two arbitrary presentations $\mathbf{P}$ and $\mathbf{Q}$ are $Q^{(*)(*)}$-equivalent or not:

The statements:

(a) $\{a^1,\ldots,a_m; R_1,\ldots,R_n\}$ is a presentation of $\pi = 1$;

(b) $\mathbf{P} = \{a_1,\ldots,a_m; R_1,\ldots,R_n, \underbrace{1,\ldots,1}_{m \text{ additional 1's}} \}$

is Q-equivalent to

$\mathbf{Q} = \{a_1,\ldots,a_m; \underbrace{1,\ldots,1}_{n \text{ additional 1's}} ,a_1,\ldots,a_m\};$

(c) $\mathbf{P}$ is Q^*-equivalent to $\mathbf{Q}$ for $\mathbf{P}$, $\mathbf{Q}$ of (b);

(d) $\mathbf{P}$ is $\mathbf{Q}^{**}$-equivalent to $\mathbf{Q}$ for $\mathbf{P}$, $\mathbf{Q}$ of (b);

clearly fulfill $(a) \Rightarrow (b) \Rightarrow (c) \Rightarrow (d) \Rightarrow (a)$. Thus *the general Q-, Q^*-, and Q^{**}-problems are unsolvable because of the unsolvability of the triviality problem (a)*.

Presentation classes (i.e.: Q^{**}-classes) Φ and Ψ may be added by taking "disjoint" representatives $\{a_i; R_j\}$, $\{b_k; S_l\}$ and passing to the presentation $\{a_i, b_k; R_j, S_l\}$. The sum $\Phi + \Psi$ corresponds to a) the free product of the groups, which are presented, b) the one-point union of 2-complexes, c) the connected sum of 3-manifolds, s. [9], section 2 and [24], (8). The class Φ_o of Q^{**}-trivial presentations (of $\pi = 1$) clearly is the neutral element of the *semigroup of presentation classes.* There are various open problems concerning this semigroup, s. [24], (8). We close this section by giving an argument, which reduces the question of the *existence of a prime "factorization"* for any given Φ to the original (AC)-problem: If $\Phi = \Phi_1 + \ldots + \Phi_t$ is a decomposition realized by presentations $\mathbf{P} = \{a_1,\ldots,a_m; R_1,\ldots,R_n\}$, $\mathbf{P}_i = \{a_{i1},\ldots,a_{im_i}; R_{i1},\ldots,R_{in_i}\}$ with groups π, π_i, and if π can be generated by ℓ elements, then because of the Grushko-Neumann theorem

(13)   at most ℓ of the π_i are different from the trivial group.

The differences $m - n, m_i - n_i$ are invariants of Φ, resp. Φ_i, which fulfill

$$(14) \qquad\qquad m - n = \sum_{i=1}^{t}(m_i - n_i),$$

and, if $\mathbf{P}_i$ presents the trivial group, we clearly have

$$(15) \qquad\qquad m_i \leq n_i \,.$$

The free products of those π_i, which are different from 1, may be viewed as a subgroup of π. This implies $k \geq \sum_{\pi_i \neq 1}(m_i - n_i)$, where k is the rank of the free abelian part of the abelianized π. By (14) and (15) the right hand side of this inequality is equal to $(m-n)+\sum_{\substack{\pi_i=1\\n_i>m_i}}(n_i-m_i)$, the second summand of which is an upper bound for the number of Φ_i with $\pi_i=1$, $m_i-n_i\neq 0$. Hence

(16) at most $k-(m-n)$ of the Φ_i fulfill $\pi_i=1, m_i-n_i\neq 0$.

(13) and (16) together imply that, *if (AC) is true, an inductive "factorization" of a given Φ must terminate after finitely many steps.*Moreover, (13) and (16) yield the existence of a prime factorization for the coarser "presentation classes", for which the list of elementary Q^{**}-transformations is enlarged by addition and deletion (if possible) of a balanced presentation of $\pi=1$. But the original classes certainly are of greater relevance, and (13) and (16) suggest to look in particular for prime Φ's, which belong to balanced presentations of $\pi=1$.

§3. Aspects concerning PL-Topology.

a) Every compact, connected polyhedron K^2 determines a presentation class $\Phi(K^2)$. Even to a 3-dimensional compact polyhedron K^3 we may assign a presentation class $\Phi(K^3)$, if K^3 has a 2-spine K^2: define $\Phi(K^3)=\Phi(K^2)$.

(17) $\Phi(K^3)$ is well defined,

for, if $K^3\searrow K^2$ and $K^3\searrow L^2$, then $\Phi(K^2)=\Phi(L^2)$ by (2).

In particular, every compact, connected 3-manifold M^3 with nonempty boundary determines a presentation class $\Phi(M^3)$. If $\partial M=\emptyset$, we first remove the interior of an (arbitrary) PL 3-ball and define $\Phi(M^3)$ to be the presentation class of the complement.

Because of the Hauptvermutung for 3-manifolds

(18) $\Phi(M^3)$ is a topological invariant.

Clearly, $\Phi(S^3)$ is the trivial presentation class Φ_o. Thus a counterexample to (AC), which can be realized by a 3-manifold, would disprove the 3-dimensional Poincaré conjecture. On the other hand, a balanced presentation of $\pi=1$, which cannot be Q^{**}-transformed to a 2-spine of a 3-manifold, provides a counterexample to (AC).

The question, *whether a given presentation class is a 3-manifold class,*in my opinion is highly relevant. It is located, "between" the question of *imbeddibility of an individual K^2* (which can be destroyed easily by (local) Q^{**}-moves applied to K^2 (s. the transition from Fig. 3 to Fig. 2 below)) and the coarser one, *whether a given group π is the fundamental group of a 3-manifold.*

In what follows, I will give a summary of some results of P. Wright and F. Quinn concerning this question. It may be of interest, as [29] is unpublished so far. By [34]

(19)   every compact, connected K^2 is Q^{**}-equivalent to a *closed fake surface P^2,*

that is, a compact connected 2-polyhedron, in which the star of each point is one of the PL types shown in Figure 1.

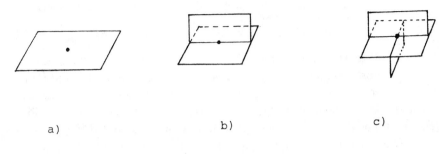

a)   b)   c)

FIGURE 1

The *intrinsic 2-skeleton* consists of the points of type a); b) and c) define the *intrinsic* $1-$ resp. $0-skeleton$. Furthermore, we require that the components of these skeleta define a cell structure. This can be achieved within the presentation class, compare [36], Proposition 1.

P^2 can be locally "thickened" to a 3-manifold. But this local thickening cannot always be done coherently to yield an orientable thickening of P^2 at the closure P^1 of the intrinsic 1-skeleton: Comparing start and return along certain closed curves of P^1 there might arise a contradiction, as the three adjacent sheets could be permuted by an odd permutation. In this case, one has to expand to a nonorientable 1-handle. It is easy to see that this yields the construction of a unique thickening of P^2 at P^1; however, an attempt to extend it over the 2-cells now may be futile: A circle concentric to the boundary of each 2-cell is thickened already to either α) an annulus $S^1 \times I$ or β) a Möbius band, and in the latter case, the corresponding disc cannot be thickened in a compatible way.

P. Wright has shown that nevertheless

(20) P^2 has a 2-fold branched covering, branched over a finite subset of P^2, which embeds in an orientable 3-manifold, s. [36], Theorem 1.

The branching locus consists of one point in each open 2-cell of P^2, which is of type β).

Instead of passing to branched coverings, F. Quinn in [29] drops the requirement that the thickening of P^2 should be locally euclidean at all points: Start with a thickening of P^2 at P^1 (orientable or nonorientable) as described ahead of (20). In the case α) we thicken the rest of the 2-cell by expanding to $D^2 \times I$; in the case β) we expand to the cone on the (real) projective plane, the "concentric circle" above becoming an essential curve in the base of the cone.

Thus we obtain a *singular 3-manifold* $\check{M}^3$, which we define to be a compact connected polyhedron, in which the link of every point is D^2 (boundary point), S^2 (interior point) or projective 2-space (singular point), and which has a nonempty set of boundary points $\partial \check{M}^3$.

Remarks: 1) F. Quinn doesn't require the properties "compact, connected" and "$\partial M \neq \varnothing$" by definition. 2) Our assumption of compactness implies that the cardinality of the set of singular points is finite, because the star of a single point doesn't contain any other singular point. 3) Each branch point in the construction of P. Wright becomes a singular point in F. Quinn's thickening.

An analogous argument as for ordinary 3-manifolds yields that each $\overset{\vee}{M}{}^3$ has a 2-dimensional spine. Hence, because of (17), $\Phi(\overset{\vee}{M}{}^3)$ is defined. By the thickening construction we have moreover:

(21)   Every presentation class Φ can be realized as a $\Phi(\overset{\vee}{M}{}^3)$.

We now describe the three *elementary surgery moves* of Quinn for singular 3-manifolds:

(I) Add a cone on projective 2-space to $\partial \overset{\vee}{M}{}^3$ along a 2-disc in their boundary.

(II) If $\overset{\vee}{M}{}^3$ is locally a thickening of a 2-disc D^2 without singular points, then replace this piece by a thickening of a 2-disc with two essential[2a] singular points in its interior,[2b] s. (22).

(III)   If $\overset{\vee}{M}{}^3$ is locally a thickening of

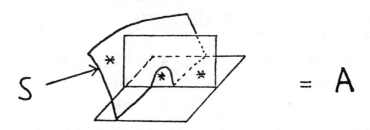

FIGURE 2

with given singular points[2c] , then replace it by a thickening of the local model with given singular points [2c], s. (22). See Figure 3.

In (I) the resulting singular manifold $\overset{\vee}{M}{}^3_1$ collapses to $\overset{\vee}{M}{}^3$, hence $\Phi(\overset{\vee}{M}{}^3_1) = \Phi(\overset{\vee}{M}{}^3)$.

In (II) we have

(22) $$\overset{\vee}{M}{}^3 \searrow \overset{\vee}{N}{}^3 \cup D^2 \nearrow \overset{\vee}{M}{}^3_1,$$

[2a] "Essential" means that the link of the singular point with respect to the disc is a generator of π_1 of the projective plane, which is the link with respect to $\overset{\vee}{M}{}^3$. We will assume the analogous situation in (III).

[2b] $(D^2 \times I, S^1 \times I) \sim (cP \#_\partial cP, S^1 \times I)$ (with P the projective plane) is the precise replacement.

[2c] The intersection of the "sheet" S with the $T \times I$-part is contained in the boundary of S. In Fig. 3, all singular points are in the horizontal parts of $T \times I$, in Fig. 2 none. [d] Quinn denotes these moves and their inverses by the number of singular points which are involved in the local situation before and after, for example (I) is a $(0,1)$ move, its inverse a $(1,0)$ move.

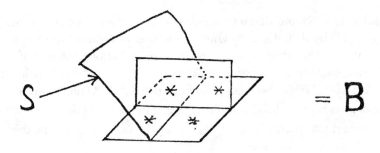

$$S \longrightarrow \qquad\qquad\qquad = B$$

FIGURE 3

where $D^2 \cap \check{N}^3 = \partial D^2 \subset \partial \check{N}^3$ with an annulus as regular neighborhood of ∂D^2 in $\partial \check{N}^3$. (22) implies $\Phi(\check{M}_1^3) = \Phi(\check{M}^3)$ too.

In (III) the local situation is described more precisely by

$$(23) \qquad \check{M}^3 \searrow \check{N}^3 \cup A \stackrel{3}{\nearrow} \check{N}^3 \cup B \nearrow \check{M}_1^3 \ (\text{rel. } \check{N}^3),$$

where $A \cap \check{N}^3 = \{ \text{ free faces of } A \} \subset \partial N$, having a torus with two holes as a regular neighborhood of $A \cap \check{N}^3$ in $\partial \check{N}^3$, and the same properties hold for B instead of A. Once more we get $\Phi(\check{M}_1^3) \stackrel{(23)}{=\!=\!=} \Phi(\check{M}^3)$.

By the preceding discussion we have proved:

(24)   If $\check{M}^3$ is transformed into $\check{M}_1^3$ by a finite sequence of elementary moves and their inverses, then $\Phi(\check{M}^3) = \Phi(\check{M}_1^3)$ holds.

The key result of Quinn now is that the converse is also true:

(25)   If $\Phi(\check{M}^3) = \Phi(\check{M}_1^3)$ holds for singular 3-manifolds, then $\check{M}^3$ can be transformed into $\check{M}_1^3$ by a finite sequence of elementary moves and their inverses, [29].

Compare [17] for a detailed proof.

Together with (21) this result implies *that there is a bijection between presentation classes and "deformation classes" of singular 3-manifolds. The elementary surgery moves in the latter play the role of the elementary Q^{**}-transformations.*

This bijection may not only be valuable for the (AC')-problem throughout, but in particular for the question of *3-manifold presentation classes,* which is now translated into one: *Which singular 3-manifolds can be deformed into 3-manifolds (without singular points)?*

The study of this question is a common project of C. Hog, M. Lustig and the author. A preliminary remark:

If one removes from $\check{M}^3$ open regular neighborhoods around each singular point, then an ordinary 3-manifold $M^3 \subset \check{M}^3$ remains, the boundary of which is the disjoint union of ∂M^3 and a projective plane for each singular point. We call the latter part of ∂M^3 the *singularity boundary.* By consideration of the elementary moves it is easy to see that

(26)   the Euler characteristic of ∂M^3 is an invariant of the deformation class of $\check{M}^3$, see [17], p. 59.

But there seem to exist much richer invariant connections between the peripheral system of $\partial \check{M}^3$ and that of the singularity boundary, which we want to understand better. So far we have climbed only some steps in exploiting the hierarchy a) homology, b) Reidemeister's homotopy-chains, c) π_1, d) isotopy of curves with respect to this situation.

b) A further connection between (AC) and the 3-dim. Poincaré conjecture is given by the Zeeman conjecture:

(Z)   If K^2 is a compact, contractible polyhedron, then $K^2 \times I \searrow *$.

This conjecture clearly implies (AC), for we would have $K^2 \nearrow K^2 \times I \searrow *$. But (Z) also implies the 3-dim. Poincaré conjecture (Zeeman, [38]).

Although the Zeeman conjecture seems the most likely of the three to be false, there exist two (different) weakened forms of it, each of which has turned out to be in fact equivalent to one of the others:

(27)   If $\Phi(K^2) = \Phi(L^2)$, then K^2 may be expanded to K_1^2 ($K^2 \nearrow K_1^2$), such that $K_1^2 \times I \searrow L^2$ holds (s. Kreher-Metzler ([20], Theorem 1) and the appendix of [24]).

This yields even an equivalence to (AC') (of a generalized weakened form of (Z)).

On the other hand, Ikeda [19] as well as Gillman and Rolfsen [14] have drawn attention to the fact that it is not necessary to prove (Z) in full generality in order to obtain the 3-Poincaré conjecture, as every compact, connected 3-manifold M with $\partial M \neq \varnothing$ collapses to a closed fake surface.[3)] Such a closed fake surface is called a *standard spine* of M^3. Gillman and Rolfsen's main result now is:

(28)   If K^2 is a standard spine of D^3 then $K^2 \times I \searrow *$.

Together with the converse implication proven in [38], this yields: *The Zeeman conjecture for standard spines of compact contractible 3-manifolds is equivalent to the 3-dim. Poincaré conjecture, s. [14].*

We mention further the paper of Cohen, Metzler and Sauermann [8] in this volume, where several common aspects of (AC) and (Z) are treated, in particular the discussion of its last section §6.

c) Whereas *in dimension 3* the *existence* of 3-manifold regular neighborhoods for appropriate complexes in a given presentation class is a crucial problem, it is a well known phenomenon that *4-dim.* (manifold) regular neighborhoods of a 2-complex in general are *not unique*, s. Neuzil [27]. The complexity arises from linking problems.

By a standard argument based on the original idea of Andrews and Curtis [2], the double of a 4-dim. regular neighborhood $N^4(K^2)$ is PL-homeomorphic to

---

[3)]In [14] it is not assumed that the intrinsic skeleta define a cell structure.

S^4 if $\Phi(K^2) = \Phi_o$. A counterexample to (AC) "thus" may give rise to an exotic PL-structure on S^4, (s. [24], Theorem 6).

We list some (partial) results of G. Huck [18] which throw some light on the situation. For this purpose, we confine ourselves to *standard complexes* $K_{\mathbf{P}}$ modeled on a presentation $\mathbf{P}$. The relators of $\mathbf{P} = \{a_1, \ldots, a_m; R_1, \ldots, R_n\}$ are given as (nonreduced) words in the a_i, a_i^{-1}. Such a presentation is called *perfect*, if $m = n$ and each R_i reduces to a_i in the abelianized $F(a_i)$. $K_{\mathbf{P}}$ is called a *generalized dunce hat*, if moreover the R_i freely reduce to a_i. Huck's results are the following:

(29)
If $\mathbf{P}$ is a perfect presentation, then $K_{\mathbf{P}}$ may be PL embedded in S^4 with a simply connected complement. Moreover, the embedding can be chosen such that the closed complement Q of the reg. neighborhood of K^2 has a spine, which belongs to the trivial presentation class. Thus, in particular, $Q \times I$ is a PL 5-ball.

(30)
The 4-dim. PL manifolds, which occur as regular neighborhoods of $K_{\mathbf{P}}$ with $\Phi(\mathbf{P}) = \Phi_o$, occur already as regular neighborhoods of generalized dunce hats.

(31)
If P is in the trivial presentation class, then it may be changed by α) prolongations and β) insertions of $a_i^{\epsilon} a_i^{-\epsilon}$, $\epsilon = \pm 1$ into the relators to $\mathbf{Q}$, such that $K_{\mathbf{Q}}$ has a 4-ball as a regular neighborhood.

These observations give rise to various ideas. For instance, (30) may be a hint to look for invariants of 4-dimensional regular neighborhoods of generalized dunce hats. Then, if some regular neighborhood of a contractible $K_{\mathbf{P}}$ does not fulfill the criteria, (AC) would be disproved.

d) Because P. Wright's transient move lemma [35] as well as an analogue to (27) is valid in dimension n (s. [20]), one should try to obtain a proof of (AC') by establishing

$$K^n \searrow L^n \Rightarrow K^n \overset{n+1}{\searrow} L^n$$

using purely simplicial arguments, which do not involve the Hurewicz-theorem as Wall's proof in [33].

The following is a sketch of a strategy into this direction: If the simplicial complex K^n is transformed into the simplicial complex L^n by a finite sequence of transient simplicial moves (s. §1), then this process yields a simplicial mapping $f : K'^n \to L^n$, where K'^n is a stellar subdivision of K^n rel. K^{n-1}. (See Figure 4.)

Taking into account the corresponding inverse transformation, one can see that the f-labelled subdivided n-simplexes of K^n may be "added" in a sequence of elementary steps, in which the results reduce monotonously[4] to the n-simplexes of L^n by *fin cancellations*.

---

[4] with respect to the total number of n-patches in the subdivided n-simplexes.

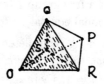

FIGURE 4

An elementary step ($n = 2$): The hatched front n-simplex $\sigma^n = OQR$ is mapped via an elementary stellar subdivision rel σ^n. The point S, starring σ^n, is mapped to P. Let PQR be the new n-simplex.

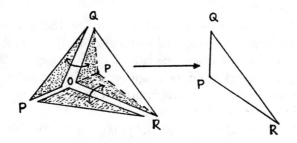

FIGURE 5

Addition and fin cancellation ($\leftarrow\!\!\!\!\rightarrow$), as it occurs in a single transient move ($n = 2$) and its inverse, compare Figure 4.

This process is analogous to a sequence of elementary Nielsen reductions converting an arbitrary basis of a free group $F(a_i)$ into the standard basis, with fin cancellation as an n-dimensional equivalent to a_i, a_i^{-1}-cancellation.

The above observation easily can be completed to a *characterization of* $K^n \overset{n+1}{\nwarrow} L^n$ *in terms of cancellation properties of simplicial maps.*

But as opposed to the case $n = 1$, there exist non-simple homotopy equivalences in dimensions $n \geq 2$, and thus the existence of a monotonous reduction process cannot be a mere consequence of the existence of homotopy inverse maps f and g. Moreover, if $n \geq 2$, there is no unique reduced form for fin cancellation; and the characterization for polyhedra K^n, L^n involves expansions of dimensions $\leq n$ and simplicial structures, which are unknown a priori. Because of the unsolvability of the general Q^{**}-problem it is clear that an "n-dimensional Nielsen-reduction algorithm" doesn't exist in our situation.

Nevertheless, a simple homotopy equivalence $K^n \overset{n+2}{\nwarrow} L^n$ gives rise to simplicial maps, which are candidates to be improved until they fulfill the required cancellation property[5] It is a topic of further research, to understand in this context the exceptional case $n = 2$, the only one, in which the possibility of this improvement is questionable.

e) Finally I want to give a personal summary on the odds of the $(AC^{(')})$-problem: The strategies involving combinatorial group theory probably are more useful to disprove (AC) and are presently developing beyond [21]; a) of this section is an open strategy and d) more apt for a positive decision. The latter in particular may be responsible for prolongations (via subdivisions), which cannot be understood in purely algebraic terms.

If I had to bet, I would bet against (AC) (Jan. 19, 1983).

BIBLIOGRAPHY

[1] Akbulut, S. and Kirby, R.: *Homotopy 4-spheres and a correction to "An exotic involution of S^4".* Preprint (1981).

[2] Andrews, J. J. and Curtis, M. L.: *Free groups and handlebodies. Proc. Amer. Math. Soc.* **16**, 192–195 (1965).

[3] Birman, J. S.: *Poincaré's conjecture and the homeotopy group of a closed, orientable 2-manifold. J. Austr. Math. Soc.* **17**, 214–221 (1975).

[4] Birman, J. S.: *On the equivalence of Heegard splittings of closed, orientable 3-manifolds.* In: Neuwirth, L.P., ed.: Knots, Groups and 3-Manifolds. Papers dedicated to the Memory of R. H. Fox, *Ann. of Math. Studies* **84**, 137–164 (1975).

---

[5] The concept of higher-dimensional group theory (R. Brown [6]) may be a helpful tool.

[5] Brown, R. and Huebschmann, J.: *Identities among relations.* In: Brown, R., Thickstun, T. L., eds.: Low-Dimensional Topology. Proc. Bangor (1979), I, *London Math. Soc. Lecture Notes Series* **48**, 153–202 (1982).

[6] Brown, R.: *Higher-dimensional group theory*, ibidem, 215–240.

[7] Casler, B. G.: *An imbedding theorem for connected 3-manifolds with boundary.* Proc. Amer. Math. Soc. **16**, 559–566 (1965).

[8] Cohen, M. M., Metzler, W. and Sauermann, K.: *Collapses of $K \times I$ and group presentations.* This volume.

[9] Craggs, R.: *Free Heegard Diagrams and extended Nielsen transformations, I.* Mich. Math. J. **26**, 161–186 (1979).

[10] Craggs, R.: *Free Heegard Diagrams and extended Nielsen transformations, II.* Ill. Journal of Math. **23**, 101–127 (1979).

[11] Craggs, R.: Letter to the author. (Sept. 1980).

[12] Crowell, R. H. and Fox, R. H.: *Introduction to knot theory.* Ginn and Co., New York, Toronto, London (1963).

[13] Denk, G.: *Kombinatorische Gruppentheorie mit Operatoren: Die relativen Sätze von Nielsen und Grushko-Neumann.* Thesis, Frankfurt am Main (1983).

[14] Gillman, D. and Rolfsen, D.: *The Zeeman conjecture for standard spines is equivalent to the Poincaré conjecture.* To appear in: Topology

[15] Gillman, D. and Rolfsen, D.: *Manifolds and their special spines.* To appear in: Lomonaco, S. J. (Jr.), ed.: Low Dimensional Topology, A.M.S. Contemporary Mathematics.

[16] Gillman, D. and Laszlo, P.: *A computer search for contractible 3-manifolds.* To appear in: Topology and its Applications.

[17] Hog, C.: *Pseudoflächen und singuläre 3-Mannigfaltigkeiten.* Staatsexamensarbeit, Frankfurt am Main (1983).

[18] Huck, G.: Private communication (1982).

[19] Ikeda, H.: *Acyclic fake surfaces.* Topology **10**, 9–36 (1971).

[20] Kreher, R. and Metzler, W.: *Simpliziale Transformationen von Polyedern und die Zeeman-Vermutung.* Topology **22**, 19–26 (1983).

[21] Lustig, M. and Metzler, W.: *Integral Representations of* Aut F^n *and Presentation Classes of Groups.* This volume.

[22] Lyndon, R. C. and Schupp, P. E.: *Combinatorial Group Theory.* Springer-Verlag, Berlin, Heidelberg, New York (1977).

[23] Magnus, W.: Max Dehn. The Math. Intelligencer 1, 132–143 (1978).

[24] Metzler, W.: *Äquivalenzklassen von Gruppenbeschreibungen, Identitäten und einfacher Homotopietyp in niederen Dimensionen.* In: C.T.C. Wall, ed.: Homological Group Theory. *Proc. Durham Symp., London Math. Soc. Lecture Notes Series* **36**, 291–326 (1979).

[25] Miller, C. F. and Schupp, P. E.: Letter to M. M. Cohen (Oct. 1979).

[26] Neuwirth, L.: *An algorithm for the construction of 3-manifolds from 2-complexes.* Proc. Camb. Phil. Soc. **64**, 603–613 (1968).

[27] Neuzil, J. P.: *Embedding the dunce hat in S^4*. Topology **12**, 411-415 (1973).

[28] Osborne, R.P. and Stevens, R.S.: *Group presentations corresponding to spines of 3-manifolds*, I–IV. I: Amer. J. of Math. **96**, 454–471 (1974), II–IV: preprints.

[29] Quinn, F.: *Presentations and 2-complexes, fake surfaces and singular 3-manifolds*. Parts 2–4 of a preprint (1981).

[30] Rapaport, E. S.: *Groups of order 1, some properties of presentations*. Acta Math. **121**, 127–150 (1968).

[31] Sieradski, A. J.: *Framed Links for Peiffer Identities*. Math. Z. **175**, 125–137 (1980).

[32] Sieradski, A. J.: *A combinatorial interpretation of the third integral Homology group of a group*. Preprint (1982).

[33] Wall, C. T. C.: *Formal deformations*, Proc. London Math. Soc. *(3)* **16**, 342–352 (1966).

[34] Wright, P.: *Formal 3-deformations of 2-polyhedra*. Proc. Amer. Math. Soc. **37**, 305–308 (1973).

[35] Wright, P.: *Group presentations and formal deformations*. Trans. Amer. Math. Soc. **208**, 161–169 (1975).

[36] Wright, P.: *Covering 2-dimensional polyhedra by 3-manifold spines*. Topology **16**, 435–439, (1977).

[37] Young, S. F.: *Contractible 2-complexes*. Manuscript, University of Cambridge, Christ's College (1976).

[38] Zeeman, E. C.: *On the dunce hat*. Topology **2**, 341–358 (1964).

FACHBEREICH MATHEMATIK, FRANKFURT AM MAIN

Contemporary Mathematics
Volume 44, 1985

INTEGRAL REPRESENTATIONS OF Aut F^n
AND PRESENTATION CLASSES OF GROUPS

MARTIN LUSTIG AND WOLFGANG METZLER

§ 1. Introduction.

$Q^{(*)(*)}$-classes of group presentations can be characterized in terms of endomorphisms and automorphisms of free groups of finite rank (s.[17],[18]). Thus we became interested in methods to effectively apply these characterizations. One device might be to work with J. Nielsen's or J. McCool's *presentations* of Aut F^n, another one, which we have concentrated on so far, is the use of *matrix representations*. Some representations, which deserve interest of their own, are exhibited here, along with a discussion, how to apply them to our original problem.

In order to distinguish presentation classes of a group π, even an abelian one, the most familiar representation

$$(1) \qquad\qquad \rho_1 : \operatorname{Aut} F^n \to GL(n, \mathbb{Z}),$$

obtained by the abelianization of F^n, is of no help. We need representations with a smaller kernel than

$$(2) \qquad\qquad K^n = \operatorname{Ker} \rho_1.$$

It would be desirable to have a faithful, finite dimensional representation of Aut F^n at hand, but for $n \geq 2$ no such representation is known, and recent work of M. Burrow [4] gives support to the conjecture of W. Magnus (s.[12]) that no one exists.

Our first attempt to obtain smaller kernels deals with a concept, which has been pursued by Magnus and his students since about ten years ago. They have taken up methods of R. Fricke and H. Vogt ($\sim$ 1890) concerning trace invariants of 2×2 matrices, which lead to a representation of Aut F^n by polynomial substitutions, s. R. Horowitz [7], [8], A. Whittemore [20] and W. Magnus [13], [14]. More precisely, these *Fricke-Characters* give rise to an algebraic variety ξ_n, on which Aut F^n operates with a natural fixed point W. By the induced operation on the tangent space $\operatorname{Tan}_W \xi_n$, we obtain an integral representation of $\operatorname{Aut} F^n$, which is a generalization of Theorem 3.3. in Magnus [13], where the case $n = 3$ is treated.

As traces are invariant under conjugation of matrices, inner automorphisms are annihilated so far. But a slight modification of this construction yields a representation

$$(3) \qquad \rho_2 : \operatorname{Aut} F^n \to GL(\frac{n^2 + n}{2}, \mathbb{Z}),$$

the kernel of which turns out to be

$$(4) \qquad \operatorname{Ker} \rho_2 = [\operatorname{Ker} \rho_1, \operatorname{Ker} \rho_1] = [K^n, K^n],$$

s. Theorem 1 and Theorem 2 below.

By consideration of the finitely many generators of K^n, given by Magnus in [9] (listed in (26)), it is easy to see that ρ_2 can be factored as follows:

$$(5) \qquad \rho_2 : \operatorname{Aut} F^{(n)} \to \operatorname{Aut} F^{(n)}/F_3^{(n)} \overset{\sigma_2}{\to} GL(\frac{n^2 + n}{2}, \mathbb{Z}),$$

where σ_2 is faithful and the first map is induced by the projection $F \to F/F_3$, with F_3 being the third term in the lower central series

$$(6) \qquad F = F_1 \triangleright F_2 \triangleright F_3, \dots, F_{m+1} = [F_1, F_m].$$

Certainly it is worthwhile to regain known distinctions of presentation classes of non-trivial groups π (comp. [5],[16]) by application of ρ_2 to the characterizations [17], [18]. Moreover, one should try to generalize this approach to a purely algebraic method, which is applicable to a larger class of presentations, alternatively to homotopy-theoretic methods. We have postponed this idea, because we are mainly interested in the Andrews-Curtis-Conjecture for contractible 2-complexes, for which π is the trivial group. Even in the more general case of a presentation of a *perfect* group π, W. Browning [3] has shown, how to push a possible Andrews-Curtis-invariant "towards infinity" with respect to higher commutator subgroups of F^n, and thus a fortiori with respect to the lower central series (compare Theorem 5 and its proof).

This result suggests to look for a family of representations

$$(7) \qquad \rho_m : \operatorname{Aut} F \to \operatorname{Aut} F/F_{m+1} \overset{\sigma_m}{\to} GL(N, \mathbb{Z}),$$

where the first map is induced by the projection $F \to F/F_m$ and the second one is faithful. On every individual level m, Browning's theorem implies that ρ_m cannot exhibit nontrivial presentation classes, but there might be an obstruction against a "simultaneous solution", i.e. a "limit solution" of the characterizing equations and criteria of [17], [18], as m tends to infinity, s. Theorem 6 below.

Representations of type (7) exist by general theorems, s. Merzlyakov [15]. In our second approach (§ 3), we concentrate on a particular nice construction, which arises from Magnus' embedding of F^n in a power series ring $A(\mathbb{Z}, n)$ over noncommuting variables [10], see also [11], chapter 5. The ρ_m are obtained by truncating the power series ring at degree $m + 1$. This implies that the corresponding matrices are "telescoped into each other" with growing m, yielding the faithful representations of Aut F^n given by $A(\mathbb{Z}, n)$ as their limit, s. (29), (30) and Theorem 3. These representations are obvious from Magnus' ideas in [10] and Andreadakis [1], but they don't seem to be well known and have not been sufficiently utilized so far. Our concluding paragraph describes them in the context of the motivating problem of this paper.

Before we start with details, we want to thank R. Bieri, D. Johnson and W. Magnus for helpful communication. We are indebted to G. McHardy, who has helped us to avoid the worst errors of English style.

§ 2. A representation by means of traces of $SL(2, \mathbb{C})$-Matrices.

We denote by F^n the free group of rank n with standard basis $a_1, \ldots, a_n$. A homomorphism of F^n into $SL(2, \mathbb{C})$ is determined by an arbitrary selection of images $X_i = \begin{pmatrix} z_{i1}, z_{i2} \\ z_{i3}, z_{i4} \end{pmatrix}$ assigned to the basis elements a_i. Thus we get a bijection (8)

$$\mathrm{Hom}(F^n, SL(2, \mathbb{C})) \to \{(z_{i1}, z_{i2}, z_{i3}, z_{i4}) \in \mathbb{C}^{4n} | z_{i1}z_{i4} - z_{i2}z_{i3} = 1, i = 1, \ldots, n\}.$$

The algebraic variety on the right-hand side will be denoted by χ_n henceforth. Aut F^n operates on F^n and hence on $\mathrm{Hom}(F^n, SL(2, \mathbb{C}))$,

$$\varphi : (x : F^n \to SL(2, \mathbb{C})) \mapsto (x \bullet \varphi : F^n \to SL(2, \mathbb{C})),$$

where the order of composition is reversed. The maps of the corresponding operation on χ_n, denoted by φ^χ, are polynomial transformations, i.e. morphisms of algebraic varieties. The values $X_i = \begin{pmatrix} 1 & 0 \\ 0 & 1 \end{pmatrix}$ define a point $V \in \chi_n$, which clearly is fixed under Aut F^n. One might now be tempted to consider the representation of Aut F^n given by the induced operation on the tangent space of χ_n at V, but the kernel equals K^n of (2).

More information can be preserved, if we pass to the traces of the matrices X_i and their products. Using the formula

$$(9) \qquad tr(A \bullet B) = tr(A) \bullet tr(B) - tr(A \bullet B^{-1}), A, B \in SL(2, \mathbb{C}),$$

Horowitz has proved in [7] that the traces of all products of the X_i, X_i^{-1} can be obtained as polynomial expressions in finitely many of these traces: for each $w \in F^n$ there exists a polynomial $p_w^* \in \mathbb{Z}[x_{i_1 i_2 \ldots i_k}], 1 \le i_1 < \ldots < i_k \le n$, such that for any representation $x : F^n \to SL(2, \mathbb{C}), x(a_i) = X_i$, the following holds:

$$(10) \qquad tr(x(w)) = p_w^*(tr(X_{i_1} \bullet \ldots \bullet X_{i_k})).$$

In [6] Helling presents a formula, which expresses $tr(A_1 \bullet A_2 \bullet A_3 \bullet A_4)$ as a polynomial with rational coefficients in the traces $tr(A_{i_1} \bullet \ldots \bullet A_{i_k})$, $1 \le i_1 < \ldots < i_k \le 4, k \le 3$. Thus we may eliminate inductively certain variables in p_w^* until we get

$$(11) \qquad p_w \in \mathbb{Q}[(x_i)_{1 \le i \le n}, (y_{ij})_{1 \le i < j \le n}, (z_{ijn})_{1 \le i < j < k \le n}]$$

with the property analogous to (10). The polynomials p_w^* and p_w are not uniquely determined; p_w may vary within the ideal

(12) $I_n = \{p \in \mathbb{C}[x_i, y_{ij}, z_{ijk}] \mid p(tr(X_i), tr(X_i, X_j), tr(X_i X_j X_k)) = 0$
$$\forall x : F^n \to SL(2, \mathbb{C}), x(a_i) = X_i\},$$

without losing property (10). (The indices i, ij, ijk are restricted as in (11) from now on.)

Assigning to each point $x \in \chi_n$ the element $(tr(X_i), tr(X_i X_j), tr(X_i X_j X_k)) \in \mathbb{C}^{n+\binom{n}{2}+\binom{n}{3}}$ we get an algebraic morphism $t : \chi_n \to \mathbb{C}^{n+\binom{n}{2}+\binom{n}{3}}$. I_n is the ideal, which consists exactly of those polynomials, which vanish on $t(\chi_n)$. Let ξ_n denote the Nullstellengebilde of I_n, fulfilling $\xi_n \supset t(\chi_n)$. For $n \geq 4$ this may well be a proper inclusion, see (15) and the discussion given subsequently. Nevertheless, the operation of $\operatorname{Aut} F^n$ on χ_n can be transferred by t to an operation of $\operatorname{Aut} F^n$ on ξ_n, given for $\varphi : a_i \mapsto w_i$ by

(13) $\varphi\xi : (x_i, y_{ij}, z_{ijk}) \mapsto$
          $(P_{w_r}(x_i, y_{ij}, z_{ijk}), P_{w_r \bullet w_s}(x_i, y_{ij}, z_{ijk}), P_{w_r \bullet w_s \bullet w_t}(x_i, y_{ij}, z_{ijk})),$

where the indices r, rs, rst vary as i, ij, ijk.

$W = (2, \ldots, 2) = t(V)$ evidently is a fixed point under this operation. The induced mappings of the tangent space at W, the differential

(14) $d_W \varphi^\xi : \operatorname{Tan}_W \xi_n \to \operatorname{Tan}_W \xi_n$

give rise to representations $\operatorname{Aut} F^n \to GL(N, \mathbb{C})$, where N is the dimension of $\operatorname{Tan}_W \xi^n$. In order to compute N, an explicit finite list of generators of I_n would be helpful. Horowitz [7] shows that

(15) $I_1 = I_2 = (0),$
              $I_3 = (z^2 - pz + Q)$

with $P = x_1 y_{23} + x_2 y_{13} + x_3 y_{12} - x_1 x_2 x_3$
and $Q = x_1^2 + x_2^2 + x_3^2 + y_{12}^2 + y_{13}^2 + y_{23}^2$
       $- x_1 x_2 y_{12} - x_1 x_3 y_{13} - x_2 x_3 y_{23} + y_{12} y_{23} y_{13} - 4.$

Unfortunately, for $n \geq 4$ no finite generator-system for I_n is known. For $n = 4$, the best known result is Whittemore's long list [20] of generators of an ideal $J \subset I_4$, which is invariant under $\operatorname{Aut} F^4$. This holds in spite of the fact that under certain regularity conditions independent variables can be chosen, and the remaining ones can be expressed in these variables by explicit algebraic formulae, s. Magnus [13], [14].

But we are able to compute the (dimension of the) tangent space of ξ_n at the (singular) point W; all derivations of elements of I_n at this point vanish because of:

THEOREM 1. $\operatorname{Tan}_W \xi_n = \mathbb{C}^{n+\binom{n}{2}+\binom{n}{3}}.$

PROOF. Consider the set of projections

$$\pi_{opq} : F^n \to F^3, a_o \mapsto a_1, a_p \mapsto a_2, a_q \mapsto a_3, o < p < q, a_\ell \mapsto 1, \ell \neq o, p, q.$$

They induce embeddings

$$\pi_{opq}^\xi : \xi_3 \to \xi_n, (x_i, y_{ij}, z_{ijk}) \mapsto$$
$$(p_{\pi(a_r)}(x_i, y_{ij}, z_{ijk}), p_{\pi(a_r \bullet a_s)}(x_i, y_{ij}, z_{ijk}), p_{\pi(a_r \bullet a_s \bullet a_z)}(x_i, y_{ij}, z_{ijk}))$$

and corresponding linear maps

$$d_W \pi_{opq}^\xi : \mathrm{Tan}_W \xi_3 \to \mathrm{Tan}_W \xi_n \subset \mathbb{C}^{n + \binom{n}{2} + \binom{n}{3}}.$$

Since the gradient of the generating polynomial of I_3 at W is zero, we get

$$\mathrm{Tan}_W \xi_3 = \mathbb{C}^7 .$$

The matrices of $d_W \pi_{opq}^\xi$, written in the natural coordinates x_i, y_{ij}, z_{ijk} have the form

$$M(\pi_{opq}) =$$

(16)

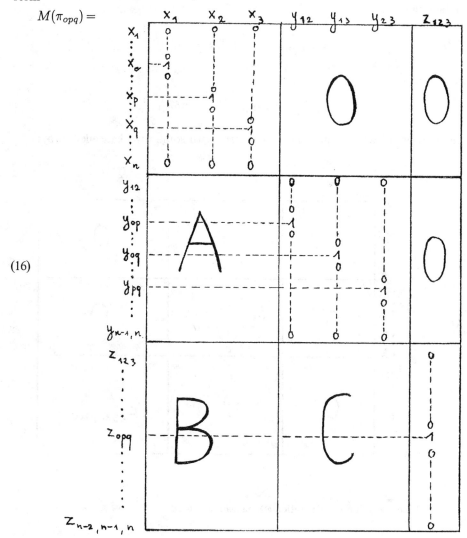

So it is immediate that the union of all $d_W \pi^\xi_{opq}$ generates the whole space $\mathbb{C}^{n+\binom{n}{2}+\binom{n}{3}}$. $\square$

For any homomorphism $\varphi : F^m \to F^n$ the analogous notation to (16) for the matrix will be maintained; thus elements of Tan ξ are column-vectors and the derivatives of the polynomials describing $d_W \varphi^\xi$ are rows of $M(\varphi)$. With the convention that the composition of $F^1 \xrightarrow{\lambda} F^m \xrightarrow{\varphi} F^n$ is denoted by $\lambda \circ \varphi$, which is the traditional notation of of Nielsen [19] and Magnus-Karrass-Solitar [11], we get*

$$(17) \qquad\qquad M(\lambda \circ \varphi) = M(\lambda) \bullet M(\varphi).$$

A similar method as in the proof of Theorem 1 will be used to obtain further information about the matrices $M(\varphi)$. For every choice $1 \leq i_1 < \ldots < i_k \leq n$ the embedding

$$(18) \qquad\qquad \iota_{i_1 \ldots i_k} : F^k \to F^n, a_s \mapsto a_{i_s}$$

induces maps

$$d_W \iota^\xi_{i_1 \ldots i_k} : \mathrm{Tan}_W \, \xi_n \to \mathrm{Tan}_W \, \xi_k,$$

the matrices of which equal the unit matrix with some rows cancelled away:
$$M(\iota_{i_1 \ldots i_k}) =$$

(19)

*Sometimes we will use the notation $\varphi\lambda$ instead of $\lambda \circ \varphi$.

where

$$a_{r,u} = \begin{cases} 1\ldots & u = i_r \\ 0\ldots & \text{otherwise}, \end{cases}$$

$$b_{rs,uv} = \begin{cases} 1\ldots & (u,v) = (i_r, i_s) \\ 0\ldots & \text{otherwise}, \end{cases}$$

$$b_{rst,uvw} = \begin{cases} 1\ldots & (u,v,w) = (i_r, i_s, i_t) \\ 0\ldots & \text{otherwise}. \end{cases}$$

For any $\varphi \in \operatorname{Aut} F^n$ that stabilizes $Gr(a_{i_1}, \ldots, a_{i_k})$ and every $a_t \notin \{a_{i_1} \ldots, a_{i_k}\}$ there exists a uniquely determined automorphism $\hat{\varphi} \in \operatorname{Aut} F^k$ with

$$\hat{\varphi} \circ \iota_{i_1 \ldots i_k} = \iota_{i_1 \ldots i_k} \circ \varphi,$$

and therefore

$$(20) \qquad M(\hat{\varphi}) \bullet M(\iota_{i_1 \ldots i_k}) = M(\iota_{i_1 \ldots i_k}) \bullet M(\varphi).$$

Because of (19), $M(\iota_{i_1 \ldots i_k}) \bullet M(\varphi)$ reproduces certain rows of $M(\varphi)$; thus sufficiently many of the $M(\hat{\varphi})$ determine $M(\varphi)$.

As a first application we deduce that

$$(21) \qquad M(\varphi) \text{ of an automorphism } \varphi : F^n \to F^n \text{ is an integral matrix:}$$

$\operatorname{Aut} F^n$ is generated by the automorphisms

$$(22) \qquad O_i : a_i \mapsto a_i^{-1}; P_{ij} : \begin{array}{c} a_i \mapsto a_j \\ a_j \mapsto a_i \end{array}; U_{ij} : \begin{array}{c} a_i \mapsto a_i a_j \\ a_j \mapsto a_j \end{array}, s.[19].$$

(All listed indices are different; the remaining generators are fixed.)

Each of these generators stabilizes $Gr(a_i, a_j)$ and the $a_k, k \neq i, j$, so that the corresponding matrices can be determined by direct computations of the finitely many cases with $n \leq 5$. As these turn out to be integral, (21) holds.

As mentioned before in the introduction, all inner automorphisms of F^n are represented by the unit matrix in (14). To avoid this, we improve the representation obtained so far: *First* we apply a well-known trick (Magnus-Tretkoff [12]), namely to "add a blind variable" to the generators of F^n : The embedding

$$(23) \qquad \begin{array}{c} \iota : F^n \to F^{n+1} \\ a_i \mapsto a_i \end{array}$$

induces an embedding

$$\operatorname{Aut} F^n \hookrightarrow \operatorname{Stab}(Gr(a_1, \ldots, a_n), a_{n+1}) \subset \operatorname{Aut} F^{n+1}$$
$$\varphi \mapsto \tilde{\varphi}.$$

Thus we get a representation

$$(24) \qquad \begin{array}{c} \rho_2' : \operatorname{Aut} F^n \to GL\left(n + 1 + \binom{n+1}{2} + \binom{n+1}{3}, \mathbb{Z}\right) \\ \varphi \mapsto d_W \tilde{\varphi}^\xi \end{array}$$

the images of which operate on $\operatorname{Tan}_W \xi_{n+1}$.

The *second* modification uses the fact that every automorphism φ "commutes" with embedding ι,

$$\iota \circ \tilde{\varphi} = \varphi \circ \iota \, ,$$

whence

$$d_W \iota^\xi \bullet d_W \tilde{\varphi}^\xi = d_W \varphi^\xi \bullet d_W \iota^\xi \, .$$

Thus Ker $d_W \iota^\xi$ is invariant under the action of Aut F^n. Similarly for $\pi : F^{n+1} \to F^1$, $a_i \mapsto 1 (i = 1, \ldots, n)$, $a_{n+1} \to a_1$, we get $\tilde{\varphi} \circ \pi = \pi$, so that $Im d_W \pi^\xi$ is invariant too. Easy computations show that Ker $d_W \iota^\xi \cap Im d_W \pi^\xi$ is generated by the natural coordinate vectors with indices $(i, n+1)$ for $i = 1, \ldots, n$ and $(i, j, n+1)$ for $1 \le i < j \le n$. The goal of this paragraph is now given by the representation:

$$\rho_2 : \operatorname{Aut} F^n \to GL\left(n + \binom{n}{2}, \mathbb{Z}\right)$$

(25)

$$\varphi \mapsto d_W \tilde{\varphi}^\xi \bigg|_{\text{Ker } d_W \iota^\xi \cap Im \, d_W \pi^\xi} \, .$$

The following theorem determines its defect:

THEOREM 2. Ker $\rho_2 = $ Ker $\rho_2' = [K^n, K^n]$.

PROOF. The theorem is a consequence of the following three propositions. We state and prove them for ρ_2', but all arguments are likewise valid for the subrepresentation ρ_2.

1.) Ker $\rho_2' \subset K^n$;

2.) $\rho_2'(K^n)$ is abelian;

3.) ρ_2' maps the Magnus-generators of K^n onto free abelian generators of $\rho_2'(K^n)$.

The proof of these propositions results from the information on the shape of the matrices, given below. Using the method (18), (19), (20), every statement may be checked by computing a finite number of matrices.

For this purpose, we first have to conjugate the matrices:

$$M'(\tilde{\varphi}) = T_{n+1}^{-1} M(\tilde{\varphi}) T_{n+1}$$

where the conjugating matrix is of type

$$T_n =$$

with

$$a_{i,klm} = \begin{cases} -1\dots & i \in \{k,l,m\} \\ 0\dots & \text{otherwise}, \end{cases}$$

$$b_{ij,klm} = \begin{cases} 1\dots & \{i,j\} \subset \{k,l,m\} \\ 0\dots & \text{otherwise}. \end{cases}$$

It is immediate that the matrices $M(\iota_{i_1\dots i_k})$ of (18) don't change when being transformed to $T_{k+1}^{-1} M(\iota_{i_1\dots i_k}) T_{n+1}$.

Ad 1.) The columns of $M'(\tilde{\varphi})$ with indices $(1, n+1), \dots, (n, n+1)$ have entries zero except in the rows $(1, n+1), \dots, (n, n+1)$; the latter entries reproduce the matrix $\rho_1(\varphi)$ of (1).

Ad 2.) In [9] W. Magnus establishes the following set of generators of K^n:

$$
\begin{aligned}
& s_{i,k} : a_i \mapsto a_k a_i a_k^{-1} \quad (i, k = 1, \dots, n) \text{ and} \\
& \qquad\qquad a_j \mapsto a_j \quad (j \neq i) \\
(26) \\
& s_{i,k,l} : a_i \mapsto a_i a_k a_l a_k^{-1} a_\ell^{-1} \quad (i, k, l = 1, \dots, n; k \neq i \neq \ell) \\
& \qquad\qquad a_j \mapsto a_j \quad (j \neq i).
\end{aligned}
$$

All $s_{i,k}$ and $s_{i,k,l}$ turn out to have corresponding M'-matrices of type

(27)

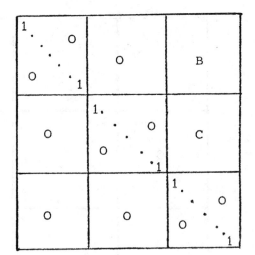

which clearly commute.

Ad 3.) For each element of (26), exactly one "characteristic" entry of the submatrix $(c_{r,n+1;s,t,n+1})_{\substack{1 \le r \le n \\ 1 \le s < t \le n}} \subset C$ (s.(27)) differs from zero, namely

$$c_{i,n+1;i,k,n+1} \text{ for } M'(s_{i,k})$$

and

$$c_{i,n+1;k,l,n+1} \text{ for } M'(s_{i,k,l}).$$

As an example, we list the matrices $M(\varphi)$ for the generating set $\{U_{12}, O_1, P_{12}\}$ of $\operatorname{Aut} F^2$:

$$M(U_{12}) = \begin{vmatrix} 0 & 0 & 1 \\ 0 & 1 & 0 \\ -1 & 0 & 2 \end{vmatrix}, M(O_1) = \begin{vmatrix} -1 & 0 & 0 \\ 0 & 1 & 0 \\ 0 & 2 & -1 \end{vmatrix}, M(P_{12}) = \begin{vmatrix} 0 & 1 & 0 \\ 1 & 0 & 0 \\ 2 & 2 & -1 \end{vmatrix}.$$

§ 3. The power series method. Many basic facts and notations of this paragraph are taken from chapter 5 of Magnus-Karrass-Solitar [11], see also Andreadakis [1].

Let $A = A(\mathbb{Z}, n)$ denote the set of infinite integral sums of "monomials", where each monomial is an element of the free semigroup with basis $x_1, \ldots, x_n$. By the usual power series operations (taking care of the noncommutativity of the x_i) A becomes a $\mathbb{Z}$-algebra. The *degree* of a monomial is its length as an x_i-word, and the elements of A with no terms of degree less than m constitute a two-sided ideal $A_m \subset A$, which is the m-th power of A_1. A may be topologized by taking the A_m as a fundamental system of neighbourhoods of O (s. [1]), which is induced by the metric given in [11], p.304. If each a_i is mapped onto $1 + x_i$ and a_i^{-1} onto the geometric series $1 - x_i + x_i^2 - x_i^3 + \ldots$, then the free group $F(a_1, \ldots, a_n)$ is multiplicatively embedded in A. We will identify F with its image henceforth.

Every mapping $x_i \mapsto Q_i \in A_1$ defines an (algebra-) endomorphism φ of $A(\mathbb{Z}, n)$ by extending to monomials and infinite sums, i.e. φ is the unique *continuous* endomorphism with $\varphi(x_i) = Q_i$. We denote by $E(A)$ the semigroup of these endomorphisms. If φ is an endomorphism of F, then $\varphi(1 + x_i) = 1 + Q_i, Q_i \in A_1$. Since F is a multiplicative subgroup of A, φ thus extends to an element of $E(A)$. This embedding of End F in $E(A)$ in particular yields that Aut F may be viewed as a subgroup of $E(A)$.

Now we use the $\mathbb{Z}$-module structure and the topology of $A(\mathbb{Z}, n)$: We assign to $\varphi \in E(A)$ an infinite $\mathbb{Z}$-matrix $M(\varphi)$, given by the expansions of the φ-images of every monomial as a power series. We first order the monomials by their degree and then lexicographically, and get

(28)   a faithful integral matrix representation ρ of $E(A)$, and thus of *End F* and *AutF*

by writing the coefficients of the φ-image of every monomial into a row of $M(\varphi)$. Once more our notational conventions agree with the traditional "from left to right" order of composition, s. (17).

$M(\varphi)$ has the shape
$M(\varphi) =$

(29)

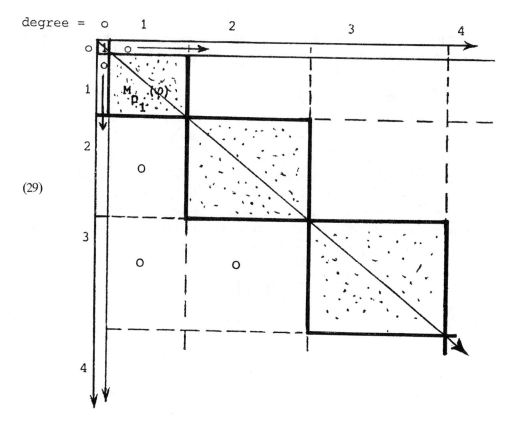

where $M_{\rho_1}(\varphi)$ belongs to the representation of $E(A)$, which restricts to the one of (1). Each dotted quadratic submatrix along the main diagonal is an m-fold tensor product of $M_{\rho_1}(\varphi)$ with itself, compare Magnus [10], page 271. In particular, the dotted matrices are invertible, resp. the unit matrix, if this is the case for $M_{\rho_1}(\varphi)$. This fact may be used to prove that

(30) $\varphi \in E(A)$ is bijective, if and only if $M_{\rho_1}(\varphi)$ is unimodular, in which case φ^{-1} also belongs to $E(A)$ (Theorem 10.3 of [1]).

Because of (29) we obtain finite dimensional integral matrix representations of $E(A)$, $\mathrm{End}\, F$ and $\mathrm{Aut}\, F$ by restricting $M(\varphi)$ to a left upper submatrix given by rows and columns up to a fixed degree m. This corresponds to passing from A to the truncated $\mathbb{Z}$-algebra A/A_{m+1}.

By an essential result of Magnus (compare [11], p.312 ff.), the kernel of $F \to A/A_{m+1}$ is precisely the term F_{m+1} in the lower central series (6). Thus the embedding of F into A yields an embedding of F/F_{m+1} into A/A_{m+1}. The number of monomials of degree r is n^r, and if we neglect the first row and column of (29), which don't contain any information, we may summarize some of these facts to

THEOREM 3. *The rows and columns of ρ of degree $1, \ldots, m$ constitute a representation $\rho_m : \mathrm{Aut}\, F^{(n)} \to \mathrm{Aut}\, F/F_{m+1} \overset{\sigma_m}{\hookrightarrow} GL(N, \mathbb{Z})$ with $N = n^1 + n^2 + \cdots + n^m$, where the first map is induced by the projection $F \to F/F_{m+1}$ and the second one is faithful.*

(Note that $\mathrm{Aut}\, F \to \mathrm{Aut}\, F/F_{m+1}$ is not an epimorphism in general, s. Andreadakis [1].)

As in § 2, we list the matrices of the standard generators $U : (a, b) \mapsto (ab, b)$, $O : (a, b) \mapsto (a^{-1}, b)$, $P : (a, b) \mapsto (b, a)$ of $\mathrm{Aut}\, F(a, b)$, which correspond to ρ_2. According to the convention proceeding (28) the monomials will be ordered as follows: $x_1, x_2, x_1^2, x_1 x_2, x_2 x_1, x_2^2$. U gives rise to the endomorphism $(x_1, x_2) \mapsto (x_1 + x_2 + x_1 x_2, x_2)$ of $A(\mathbb{Z}, 2)$ and thus to the matrix

$$M_{\rho_2}(U) = \left[\begin{array}{cc|cccc} 1 & 1 & 0 & 1 & 0 & 0 \\ 0 & 1 & 0 & 0 & 0 & 0 \\ \hline 0 & 0 & 1 & 1 & 1 & 1 \\ 0 & 0 & 0 & 1 & 0 & 1 \\ 0 & 0 & 0 & 0 & 1 & 1 \\ 0 & 0 & 0 & 0 & 0 & 1 \end{array}\right];$$

0 gives rise to $(x_1, x_2) \mapsto (-x_1 + x_1^2 - x_1^3 + \ldots, x_2)$ and

$$
M_{\rho_2}(0) = \left[
\begin{array}{cc|cc|cc}
-1 & 0 & 1 & 0 & 0 & 0 \\
0 & 1 & 0 & 0 & 0 & 0 \\
0 & 0 & 1 & 0 & 0 & 0 \\
0 & 0 & 0 & -1 & 0 & 0 \\
0 & 0 & 0 & 0 & -1 & 0 \\
0 & 0 & 0 & 0 & 0 & 1
\end{array}
\right];
$$

P with $(x_1, x_2) \mapsto (x_2, x_1)$ yields the matrix

$$
M_{\rho_2}(P) = \left[
\begin{array}{cc|cc|cc}
0 & 1 & 0 & 0 & 0 & 0 \\
1 & 0 & 0 & 0 & 0 & 0 \\
0 & 0 & 0 & 0 & 0 & 1 \\
0 & 0 & 0 & 0 & 1 & 0 \\
0 & 0 & 0 & 1 & 0 & 0 \\
0 & 0 & 1 & 0 & 0 & 0
\end{array}
\right].
$$

Remark: ρ_2 defined in (25) by means of the trace method is a representation of dimension $\frac{n+n^2}{2}$, whereas the power series method ends up with a ρ_2 of dimension $n + n^2$. Both have the same deficiency. They may be even closer related to each other.

(30) implies that many elements of $\text{End}\, F$, which are not in $\text{Aut}\, F$, are invertible in $E(A)$. Moreover, the following holds:

THEOREM 4. *If $\varphi \in \text{End}\, F$ induces an automorphism of F/F_2, the abelianiized free group, then φ induces an automorphism of F/F_m for all m.*

PROOF. We prove the cases $m \geq 2$ inductively. Let $\psi \in \text{End}\, F$ be an inverse of $\varphi \bmod F_m$, i.e., $\psi\varphi(a_i) = a_i \bullet w_i$, $w_i \in F_m, i = 1, \ldots, n$. Define $\lambda \in \text{End}\, F$ by $\lambda(a_i) = a_i \bullet w_i^{-1}$. Then by inspection of the expansions in $A(\mathbb{Z}, n)$, or by commutator calculus, we see that because of $m \geq 2$ for $u \in F_m : \lambda(u) = u \bullet u'$, $u' \in F_{m+1}$ holds. In particular, we have $\lambda(w_i) = w_i \bullet w_i', w_i' \in F_{m+1}$. Therefore, $\lambda\psi\varphi(a_i) = \lambda(a_i \bullet w_i) = a_i \bullet w_i^{-1} \bullet w_i \bullet w_i' = a_i \bullet w_i'$; hence $\lambda\psi$ is shown to be an inverse of $\varphi \bmod F_{m+1}$. $\square$

Theorem 4 is implicitly contained in Andreadakis [1] and intuitively supports his result that the natural map $\text{Aut}\, F \to \text{Aut}\, F/F_{m+1}$ is not surjective in general: as "many" endormorphisms of F induce automorphisms of F/F_{m+1}, it becomes more plausible that those of $\text{Aut}\, F$ don't suffice to generate $\text{Aut}\, F/F_{m+1}$.

We have included Theorem 4, because it may be helpful in the program described below and because its proof is analogous to that of a weakened version of Browning's theorem [3]:

THEOREM 5. *Let* $\mathbf{P} = \{a_1, \ldots, a_n; R_1, \ldots, R_n\}$ *be a perfect presentation, i.e.,* $R_i = a_i \bullet w_i, w_i \in F_m, m \geq 2$. *Then* $\mathbf{P}$ *is Q-equivalent to* $\mathbf{P}' = \{a_1, \ldots, a_n; R_1', \ldots, R_n'\}$, $R_i' = a_i \bullet w_i', w_i' \in F_{m+1}$.

PROOF. Replacing the a_j in $R_{i_o}(i_o \neq j)$ by w_j^{-1} –these are Q-transformations! – , we obtain, as in the proof of Theorem 4: $R_{i_o}' = a_{i_o} \bullet w_{i_o}'$, $w_{i_o}' \in F_{m+1}$. This process, performed successively to all R_i in an arbitrary order, yields the desired $\mathbf{P}'$. □

This version of Browning's theorem and its proof are part of our current algebraic efforts concerning the Andrews-Curtis-Conjecture, which we outline now.

I. *A first idea* is to use the $A(\mathbb{Z}, n)$-expansions of the relators R_i or the corresponding w_i of Theorem 5. Its proof reveals, how the first nonvanishing terms of degree > 0 of the w_i, the *deviations* ([11], p.313), can be pushed from m to $m + 1$. This process is not much more complicated than the tensor relations between the diagonal matrices in (29). We are presently studying, whether some (multilinear) invariant can be read off from it, which generalizes to a Q-invariant for perfect presentations. (We conjecture that Q-transformations between these are generated by elementary transformations of type $R_i \mapsto w \bullet R_i^{\pm 1} \bullet w^{-1}$ and $R_i \mapsto R_i \bullet [w, R_j], i \neq j$, compare (26)). An invariant which shows that not all w_i-deviations can be annihilated, probably will exhibit nontrivial $Q^{(*)}$-classes of balanced presentations of $\pi = 1$.

II. The *second idea* starts with Theorem 4 of [17], stating that a balanced presentation of $\pi = 1$ determines a set $w_1, \ldots, w_{2n}$ of generators of $F^n \oplus F^n$, which in the case of $Q^{(*)}$-triviality can be lifted to a basis $F^n * F^n = F(a_1, \ldots, a_n, b_1, \ldots, b_n)$, compare also [18], Theorem 3. Thus such a presentation determines a family $\{\varphi_\nu\}$ of endomorphisms of F^{2n}, which assign to $a_1, \ldots, a_n, b_1, \ldots, b_n$ some lift of $w_1, \ldots, w_{2n}$ as values, i.e., the $\varphi_\nu(a_i), \varphi_\nu(b_j)$ differ from a fixed choice $\varphi_o(a_i), \varphi_o(b_j)$ by elements of the normal closure $N([a_i, b_j]) = \ker(F^n * F^n \to F^n \oplus F^n)$.

Such a φ_ν induces an isomorphism in the abelianization of F^{2n} and thus by (30) is invertible in $E(A(\mathbb{Z}, 2n))$. *If for some balanced presentation* $\mathbf{P}$ *of* $\pi = 1$, *no member of the family* $\{\varphi_\nu\}$ *is invertible in* $\text{End } F^{2n}$, *then* $\mathbf{P}$ *is not* $Q^{(*)}$-*trivial*.

The Magnus expansion is likely to be a valuable tool to detect such "noninvertible families" unlike the Nielsen-reduction-method, which seems to be applicable only to a single φ_ν at a time.

The following two steps are relevant in this context. As shown by the subsequent discussion, step 1 is the harder part.

1.) Invert the φ_ν in $E(A)$;

2.) derive criteria to show that certain $E(A)$-elements are not contained in $\text{End } F$, and apply these criteria to the φ_ν^{-1}.

Ad 1.): Not only the matrices $M(\varphi_\nu)$, which are "quasi-triangular", should be considered, but also Theorem 4 and its proof, which allow to invert the φ_ν up to a certain degree by direct manipulations in a free group.

If $a_i \mapsto 1 + x_i, b_j \mapsto 1 + y_j$ in $A(\mathbb{Z}, 2n)$, then the $\varphi_\nu(x_i), \varphi_\nu(y_j)$ differ from the $\varphi_o(x_i), \varphi_o(y_j)$ by elements of the topological completion of the 2-sided ideal generated by the $(x_i y_j - y_j x_i)$. Thus, with "good inversion formulae", it may be possible to obtain at least enough information about infinitely many coefficients of the φ_ν^{-1} to proceed with step 2. But the necessary explicit computations are complicated.

Ad 2.): If $\varphi \in \operatorname{End} F \subset E(A)$, then each $\varphi(w), w \in F$ is a finite product of expressions $1 + x_i$ and geometric series $1 - x_i + x_i^2 - x_i^3 + \ldots$. Thus the expansion of $\varphi(w)$, although an infinite number of monomials may occur, will have

(31)   a finite upper bound for the number of index changes in the monomials with coefficient $\neq 0$.

(An *index change* is the transition of an x_i to $x_j, i \neq j$ in a monomial. Example: $x_1 x_1 x_2 x_2 x_2 x_3 x_1 = x_1^2 x_2^3 x_3 x_1$ has 3 index changes.)

(31) implies that, if we identify monomials with zero, which have an $x_i^r (r \geq 2)$ as a subword, then

(32)   the remaining terms of the $\varphi(w)$-expansion must be of finite number

compare [11], p.325, Problem 1.

Hence, we try to disprove the Andrews-Curtis-Conjecture by finding for each φ_ν^{-1} a word $w_\nu \in F^{2n}$, preferably a generator a_i, b_j, such that $\varphi_\nu^{-1}(w_\nu)$ violates (31) or (32). There exist various modifications of these tests, which may be helpful too. Moreover, one can derive *criteria for convergence* from the fact that for $\varphi \in \operatorname{End} F$, $w \in F$, $\varphi(w)$ is a finite product of terms $1 + x_i$ and geometric series $1 - x_1 + x_i^2 - x_i^3 + \ldots$.

III. Finally, let us turn to Theorem 1 of [18] and its terminology. We may weaken the notion of Q-equivalence to:

(33)   Two finite presentations $\mathbf{P} = \{a_i; R_j\}, \mathbf{Q} = \{a_i; S_j\}$ with the same generators and the same number of defining relators are called Q-equivalent up to degree m, if there exists a Q-transformation $\varphi \in W \subset \operatorname{Aut} F(a_i, b_j)$, which transforms the $F(a_i)/F_{m+1}(a_i)$-projection of the "vector" $\mathbf{P} = \{a_i; R_j\}$ into that of $\mathbf{Q} = \{a_i; S_j\}$.

This is clearly equivalent to saying that for each j the relators S_j of $\mathbf{Q}$ and S_j' of $\mathbf{Q}' = \varphi(\mathbf{P})$ agree up to terms of degree m in the Magnus expansion. By [18], Theorem 1 this is furthermore equivalent to the existence of another $\psi \in W$ that

(34) $$\sigma_{\mathbf{Q}'} = \varphi \circ \sigma_{\mathbf{P}} \circ \psi \text{ holds .}$$

For arbitrary $\mathbf{P} = \{a_i; R_j\}, \mathbf{Q} = \{a_i; S_j\}$ with the same generators and the same number of defining relators we may form $L(\mathbf{P}, \mathbf{Q}) \subset W$, the set of $\varphi \in W$ with $\varphi(\mathbf{P}) = \mathbf{Q}$, and $L_m(\mathbf{P}, \mathbf{Q})$, the set of $\varphi \in W$ transforming the $F(a_i)/F_{m+1}(a_i)$-projection of $\mathbf{P}$ into that of $\mathbf{Q}$.

Using (33) and (34) we can now state:

(35)

a) The $L_m(\mathbf{P}, \mathbf{Q})$ form a descending chain $L_1(\mathbf{P}, \mathbf{Q}) \supset L_2(\mathbf{P}, \mathbf{Q}) \supset \ldots$ with $\bigcap_{m=1}^{\infty} L_m(\mathbf{P}, \mathbf{Q}) = L(\mathbf{P}, \mathbf{Q})$;

b) $\mathbf{P}$ is Q-equivalent to $\mathbf{Q}$ up to degree m iff $L_m(\mathbf{P}, \mathbf{Q}) \neq \varnothing$;

c) $\mathbf{P}$ is Q-equivalent to $\mathbf{Q}$, iff $L(\mathbf{P}, \mathbf{Q}) \neq \varnothing$.

Likewise we may form

$L^*(\mathbf{P}, \mathbf{Q}) \subset W \times W$, the set of all pairs $(\varphi, \psi) \in W \times W$ with $\sigma_{\mathbf{Q}} = \varphi \circ \sigma_{\mathbf{P}} \circ \psi$, and $L_m^*(\mathbf{P}, \mathbf{Q}) \subset W \times W$, the set of all pairs $(\varphi, \psi) \in W \times W$ such that (34) holds for some $\mathbf{Q}'$, which agrees with $\mathbf{Q}$ up to terms of degree m.

(36) The $L_m^*(\mathbf{P}, \mathbf{Q})$ and $L^*(\mathbf{P}, \mathbf{Q})$ fulfill the same properties as stated in (35) for the $L_{(m)}$.

Moreover, an easy argument as in the proof of [18], Theorem 1 yields

(37) $L_{(m)}$ is the projection of $L_{(m)}^*$ to the first coordinate of $W \times W$.

Combining with Theorem 5, we get in particular:

THEOREM 6.   *Two perfect presentations* $\mathbf{P} = \{a_1, \ldots, a_n; R_1, \ldots, R_n\}$ *and* $\mathbf{Q} = \{a_1, \ldots, a_n; S_1, \ldots, S_n\}$ *fulfill* $L_m^{(*)}(\mathbf{P}, \mathbf{Q}) \neq \varnothing$ *for all* m. *They are Q-equivalent if and only if* $\bigcap_{m=1}^{\infty} L_m^{(*)}(\mathbf{P}, \mathbf{Q}) \neq \varnothing$.

Thus the analysis of the $L_m^{(*)}(\mathbf{P}, \mathbf{Q})$ in terms of the representations of Theorem 3 and their behaviour for $m \to \infty$ is our goal in this third approach. The use of the L_m^* enables us to work entirely within representations of *automorphisms of free groups*.

All three ideas "define an open end" of this paper. Readers, who are inspired, are invited to communicate with the authors.

REFERENCES

[1] Andreadakis, S., *On the automorphisms of free groups and free nilpotent groups*. Proc. London Math. Soc. (3) **15**, 239–268 (1965).

[2] Andrews, J. J. and Curtis, M.L., *Free groups and handlebodies*. Proc. Amer. Math. Soc. **16**, 192–195 (1965).

[3] Browning, W. J., *The effect of Curtis-Andrews-Moves on Jacobian Matrices of Perfect Groups*. Manuscript, Cornell University, Ithaca, N.Y.

[4] Burrow, M. D., *A complete survey of the representations of degree 3 of the automorphism group of the free group of rank 2*. Archiv Math. **38**, 208–216 (1982).

[5] Dunwoody, M., *The homotopy type of a two-dimensional complex*. Bull. London Math. Soc. 8, 282–285 (1976).

[6] Helling, H., *Diskrete Untergruppen von $SL_2(\mathbb{R})$. Inventiones math.* **17**, 217–229 (1972).

[7] Horowitz, R. D., *Characters of free groups represented in two-dimensional special linear group. Comm. Pure and Appl. Math.* **25**, 635–649 (1972).

[8] Horowitz, R. D., *Induced automorphisms on Fricke characters of free groups. Trans. Amer. Math. Soc.* **208**, 41-50 (1975).

[9] Magnus, W., *Über n-dimensionale Gittertransformationen. Acta Math.* **64**, 353–367 (1934).

[10] Magnus, W., *Beziehungen zwischen Gruppen und Idealen in einem speziellen Ring. Math. Annalen* **111**, 259–280 (1935).

[11] Magnus, W., Karrass, A. and Solitar, D., *Combinatorial group theory.* Interscience, New York (1966).

[12] Magnus, W. and Tretkoff, C., *Representations of automorphism groups of free groups.* In: Adian, S.I., Boone, W.W., Higman,G., eds.: Word Problems II, 255–259. The Oxford Book Studies in Logic and Foundations of Mathematics, Vol. 95, Amsterdam (1980).

[13] Magnus, W., *Rings of Fricke characters and automorphism groups of free groups. Math. Z.* **170**, 91–103 (1980).

[14] Magnus, W., *The uses of 2 by 2 matrices in combinatorial group theory. A survey.Resultate der Math.* **4**, 171–192 (1981).

[15] Merzlyakov, Y. I., *Integral representation of holomorphs of polycyclic groups. Algebra i Logika* **9**, 539–558 (1970) (in Russian; English Translation: *Algebra and logic* **9**, 326–337 (1970)).

[16] Metzler, W., *Über den Homotopietyp zweidimensionaler CW-Komplexe und Elementartransformationen bei Darstellungen von Gruppen durch Erzeugende und definerende Relationen. J. Reine Angew. Math.* **285**, 7–23 (1976).

[17] Metzler, W., *Äquivalenzklassen von Gruppenbeschribungen, Identitäten und einfacher Homotopietyp in niederen Dimensionen.* In: C.T.C. Wall ed.: Homological group theory, *Proc. Durham Symp., Lond. Math. Soc. Lecture Notes Series,* **36**, 291–326 (1979).

[18] Metzler, W., *On the Andrews-Curtis-Conjecture and related problems.* This volume.

[19] Nielsen, J., *Die Isomorphismengruppe der freien Gruppen.. Math. Ann.* **91**, 169-209 (1924).

[20] Whittemore, A., *On special linear characters of free groups of rank $n \geq 4$. Proc. Amer. Math. Soc.* **40**, 383-388 (1973).

JOHANN WOLFGANG GOETHE-UNIVERSITÄT
Current address: Frankfurt/Main, Fachbereich Mathematik

Contemporary Mathematics
Volume **44**, 1985

A NOTE ON COMMUTATORS AND SQUARES IN FREE PRODUCTS

RICHARD Z. GOLDSTEIN AND EDWARD C. TURNER

In [1], Griffiths showed that if $g = g_1 \ldots g_n \in G_1 * \ldots * G_n, 1 \neq g_i \in G_i$ a finitely presented group, then g can not be expressed as a product of fewer than n commutators. In this note, we show how the geometric ideas of [2] and [3] can be used to improve this as follows.

If for each i, g_i can be expressed minimally as a product of K_i commutators, then $g = g_1 \ldots g_n$ can be written minimally as a product of $k = k_1 + \ldots + k_n$ commutators.

We also obtain an analogous but somewhat more complicated statement for products of squares and combine these in Theorem C which describes when certain quadratic equations in free products can be solved. We assume familiarity with [3] and that $g_i \in G_i = <X_i | R_i>, g = g_1 \ldots g_n \in G_1 * \ldots * G_n$.

PRODUCTS OF COMMUTATORS

DEFINITION. *If $g \in [G, G]$, $C_G(g)$ denotes the least number of commutators of which g can be written as a product. If the subscript is omitted, g should be interpreted in the appropriate free group.*

From [3], we know that $C_G(g)$ can be obtained as follows: for each R-graph $\mathcal{P}$ for g, form the manifold $M_{\mathcal{P}}\{$ whose genus is by definition the genus of $\mathcal{P}\}$ — $C_G(g)$ is the least oriented genus of the manifolds $M_{\mathcal{P}}$. If $\mathcal{P}$ is any such R-graph, then $\mathcal{P}$ is the union of subgraphs $\mathcal{P}_i$ which are R_i graphs for the g_i. (See the figure.)

On the manifold level,

$$M_{\mathcal{P}} \cong M_{\mathcal{P}_1} \# M_{\mathcal{P}_2} \# \ldots \# M_{\mathcal{P}_n}$$

so that

$$\gamma_0(M_{\mathcal{P}}) = \gamma_0(M_{\mathcal{P}_1}) + \ldots + \gamma_0(M_{\mathcal{P}_n}).$$

Since minimizing the left side is equivalent to simultaneously minimizing all terms on the right side, we have proven

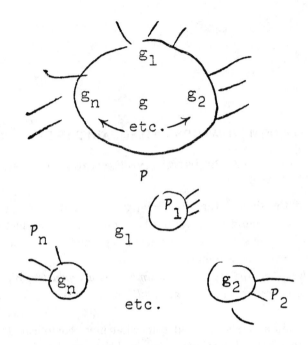

THEOREM A. *For* $g \in [G, G]$

$$C_{(g)} = C_{G_1}(g_1) + \ldots + C_{G_n}(g_n)$$

PRODUCTS OF SQUARES

DEFINITION. *If* $g \in G^2$, $S_g(g)$ *denotes the least number of commutators of which* g *can be written as a product. If the subscript is omitted, g should be interpreted in the appropriate free group.*

This case is slightly complicated by the fact that commutators are products of squares; viz.

$$[a, b] = (ab)^2 (b^{-1}a^{-1}b)^2 (b^{-1})^2.$$

This formula, together with

$$[a, b]c^2 = (ac^{-1}b)^2 (b^{-1}ca^{-1}b^{-1}cbc^{-1}b)^2 (b^{-1}c)^2$$

shows that for $g \in [G, G] \subset G^2$,

$$S_G(g) \leq 2C_G(g) + 1.$$

$[a^2, b] = a^2(b^{-1}ab)^2$ shows that the inequality may be strict. To deal with this, we extend the definitions of [3] slightly as follows: if P is an oriented R graph with genus $\gamma_0(P)$, then its *unoriented* genus is

$$\gamma_u(P) = 2\gamma_0(P) + 1.$$

Then in all cases, $\gamma_u(\mathcal{P})$ is the minimum genus of unoriented manifolds in which $\mathcal{P}$ embeds. (This was the point of view of [2].) The condition that $S_G(g) = 2C_G(g)+1$ means that some pairing graph of least *unoriented* genus is in fact an oriented graph.

As before, if $\mathcal{P}$ is a pairing graph for g, then

$$M_{\mathcal{P}} \cong M_{\mathcal{P}_1} \# \ldots \# M_{\mathcal{P}_n}$$

and

Case 1: If all of the $M_{\mathcal{P}_i}$ are oriented, then

$$\gamma_u(M_{\mathcal{P}}) = \gamma_u(M_{\mathcal{P}_1}) + \ldots + \gamma_u(M_{\mathcal{P}_n}) - n + 1$$

Case 2: If the number ℓ of the $M_{\mathcal{P}_i}$ which are unoriented is $< n$, then

$$\gamma_u(M_{\mathcal{P}}) = \gamma_u(M_{\mathcal{P}_1} + \ldots + \gamma_u(M_{\mathcal{P}_n}) - \ell.$$

(Observe that to minimize $\gamma_u(M_{\mathcal{P}})$ we want to minimize each $\gamma_u(M_{\mathcal{P}_i})$ and maximize ℓ.)

Thus we have proven

THEOREM B. *For $g \in G^2$ and ℓ = the number of $g_i \in [G_i, G_i]$ for which $S_{G_i}(g_i) = 2C_G(g_i)+1$,*
Case 1: *if $\ell = n$, then*

$$S_G(g) = S_{G_1}(g_1) + \ldots + S_{G_n}(g_n) - n + 1.$$

Case 2: *if $\ell < n$ then*

$$S_G(g) = S_{G_1}(g_1) + \ldots + S_{G_n}(g_n) - \ell.$$

QUADRATIC EQUATIONS

Now suppose we have a quadratic equation

$$\mathcal{W} = W_1 w_1 W_2 w_2 \ldots W_n w_n = 1$$

over $G = G_1 * \ldots * G_n$ in which the w_i represent the elements of distinct factors. Then any R-graph $\mathcal{P}$ for $\mathcal{W}$ is a union of subgraphs $\mathcal{P}_i$ as before, except that $\mathcal{P}$ may have several components – the components of $\mathcal{P}$ correspond to the components of the coinitial graph for $W = W_1 W_2 \ldots W_n$. The genus formula still holds, however, so that the proofs of Theorem A and B still work, providing the following generalization.

THEOREM C. *The quadratic equation*

$$\mathcal{W} = W_1 w_1 W_2 w_2 \ldots W_n w_n = 1$$

over $G_1 * \ldots * G_n$ with the w_i representing elements g_i of distinct factors has a solution iff one of the following holds:
   a) $\mathcal{W}$ is strictly quadratic, $g_i \in [G_i, G_i]$

and

$$C(W) \geq C_{G_1}(g_1) + \ldots + C_{G_n}(g_n).$$

b) W is non-strictly quadratic, there are exactly ℓ of the $g_i \in [G_i, G_i]$ for which
$S_{G_i}(g_i) = 2S_{G_i}(g_i) + 1$ *and*

Case 1: $\ell = n$ *and*

$$S(W) \geq S_{G_1}(g_1) + \ldots + S_{G_n}(g_n) - n + 1$$

Case 2: $\ell < n$ *and*

$$S(W) \geq S_{G_1}(g_1) + \ldots + S_{G_n}(g_n) - \ell.$$

BIBLIOGRAPHY

[1] H. B. Griffiths, *A note on commutators in free products I & II*, Proc. Camb. Philos. Soc. **50**, 1954, p. 178–188; **51** (1955), p. 245–251.

[2] R. Z. Goldstein and E. C. Turner, *Applications of topological graph theory to group theory*, Math. Zeit. **165** (1979), p. 1–10.

[3] R. Z. Goldstein and E. C. Turner, *Solving quadratic equations in groups*, preprint.

STATE UNIVERSITY OF NEW YORK AT ALBANY

Contemporary Mathematics
Volume **44**, 1985

RIGIDITY OF ALMOST CRYSTALLOGRAPHIC GROUPS

KYUNG BAI LEE* AND FRANK RAYMOND*

Let $A(n) = \mathbb{R}^n \circ GL(n, \mathbb{R})$ denote the group of affine motions of $\mathbb{R}^n$, and let $E(n) = \mathbb{R}^n \circ O(n)$ be the subgroup of rigid motions. A subgroup of $E(n)$ which acts on $\mathbb{R}^n$ without accumulation points and with compact fundamental domain is called a crystallographic group. A torsion-free crystallographic group acts on $\mathbb{R}^n$ freely and properly discontinuously. Thus these groups are exactly the fundamental groups of compact flat Riemannian manifolds. Such groups are completely understood by the celebrated work of Bieberbach (see [W], chapter 3):

THEOREM 1'. *Let* Γ *be an n-dimensional crystallographic group. Then* $\Gamma \cap \mathbb{R}^n$ *is a uniform lattice of* $\mathbb{R}^n$ *and* $\Gamma/\Gamma \cap \mathbb{R}^n$ *is finite.*

THEOREM 2'. *Let* $\psi : \Gamma \to \Gamma'$ *be an isomorphism between two crystallographic groups of dimension* n. *Then* ψ *is conjugation by an element of* $A(n)$.

Now let L be a connected and simply connected nilpotent Lie group. $A(L) = L \circ$ Aut (L) is called the affine group of L, acting on L by $(z, \alpha)x = z \bullet \alpha(x)$. Consider the linear connection on L defined by the left invariant vector fields. It is known [K-T] that $A(L)$ is the group of connection-preserving self-diffeomorphisms of L. Let C be any maximal compact subgroup of Aut(L). Any other maximal compact subgroup is conjugate to C. We shall call a discrete uniform subgroup of $L \circ C$ an *almost crystallographic group*. The torsion-free almost crystallographic groups are exactly the fundamental groups of compact infranilmanifolds. These manifolds are the almost flat manifolds in the diffeomorphism category. See [G], [F-H] and [R] for more details. Now the reason for using the term "almost" is almost obvious. The following generalization of Bieberbach's first theorem has been obtained by L. Auslander [A].

THEOREM 1. *Let* Γ *be an almost crystallographic group of* L. *Then* $\Gamma \cap L$ *is a uniform lattice of* L *and* $\Gamma/\Gamma \cap L$ *is finite.*

---

*Partially supported by NSF grant.

He also stated: Let Γ and Γ' be almost crystallographic groups for L (both $\Gamma, \Gamma' \subset L \circ C$ for some maximal compact subgroup C of $\mathrm{Aut}(L)$). Let $\psi : \Gamma \to \Gamma'$ be an isomorphism. Then ψ can be uniquely extended to a continuous automorphism ψ^* of $L \circ C$ onto itself.

Unfortunately, the last claim is not true. For crystallographic groups, Γ, Γ' (with $C = O(n)$), Theorem 2' in [A] is misstated, and is not true. In this paper, we shall prove the strongest possible generalization of Theorem 2':

THEOREM. *Let $\psi : \Gamma \to \Gamma'$ be an isomorphism between two almost crystallographic groups of L. Then ψ is conjugation by an element of $A(L)$.*

Let Γ be a torsion-free almost crystallographic group of L. Then $M = \Gamma \backslash L$ is a (compact) *infranilmanifold* with $\pi_1 M = \Gamma$. Let $Aff(M)$ (resp. $\mathcal{E}(M)$) denote the group of affine self-diffeomorphisms (resp. the H-space of self-homotopy equivalences) of M onto itself. $Aff_0(M)$ denotes the connected component of the identity. As M is aspherical, $\pi_0(\mathcal{E}(M))$ is naturally isomorphic to $\mathrm{Out}\,\Gamma$. So there is a natural homomorphism $\Psi : Aff(M) \to \mathcal{E}(M) \to \pi_0(\mathcal{E}(M)) \cong \mathrm{Out}\,\Gamma$. The theorem yields the following:

COROLLARY 1. *Homotopy equivalent infranilmanifolds are affinely diffeomorphic.*

COROLLARY 2. *For an infranilmanifold M, the sequence $1 \to \mathrm{Aff}_0(M) \to \mathrm{Aff}(M) \to \mathrm{Out}\,\pi_1 M \to 1$ is exact. In particular, any self-homotopy equivalence of M is homotopic to an affine diffeomorphism, and homotopic affine diffeomorphisms are isotopic.*

COROLLARY 3. *Let M be an infranilmanifold. A finite abstract kernel $G \to \mathrm{Out}\,(\pi_1 M)$ can be (effectively) geometrically realized as a group of affine self-diffeomorphisms of M if and only if it admits an extension E (such that the centralizer of $\pi_1 M$ in E is torsion-free).*

We follow the method of the original Bieberbach's proof. In the course we shall need the Theorem 1 of Auslander above. As expected, in order to obtain a rigidity, we need the vanishing of the first group cohomology.

We shall verify that the first cohomology of a finite group with coefficient in a connected, simply connected nilpotent Lie group (non-abelian coefficient) vanishes, generalizing the fact that $H^i(\mathrm{Finite}; \mathbb{R}^n) = 0$ for all $i > 0$.

Notation: Let G be a group, $a \in G$ and $A \subset G$. Then

$$\mu(a)(x) = a^{-1}xa \quad \text{for } x \in G,$$
$$Z(G) = \text{the center of } G,$$
$$C_G(A) = \text{the centralizer of } A \text{ in } G.$$

Let L be a connected and simply connected nilpotent Lie group, and Γ, Γ' be subgroups of $A(L)$. Let $N = \Gamma \cap L$, $N' = \Gamma' \cap L$. Then N and N' are the unique maximal normal nilpotent subgroups of Γ and Γ' by Theorem 1. In fact, they are maximal nilpotent and characteristic subgroups. Therefore, ψ induces an isomorphism of N onto N'. By [M], this isomorphism extends uniquely to an automorphism A of L.

Let $\psi(w, k) = (d, D)$ for $(w, k) \in \Gamma \subset A(L)$. For any $(z, 1) \in N$, $\psi(z, 1) = (Az, 1)$. Evaluating at $(w, k) \cdot (z, 1) \cdot (w, k)^{-1}$, we have $DAz = d^{-1} \cdot Aw \cdot Akz \cdot Aw^{-1} \cdot d$ for all $z \in N$. Therefore, $Dy = (d^{-1} \cdot Aw) \cdot AkA^{-1}y \cdot (Aw^{-1} \cdot d)$ for all $y \in N'$. Put $f(w, k) = Aw^{-1} \cdot d \in L$. Then $\psi(w, k) = (Aw \cdot f(w, k), \mu(f(w, k)^{-1}) \cdot AkA^{-1})$, where μ denotes the conjugation map. It is readily verified that f is independent of w. So

$$\psi(wk) = (Aw \cdot f(k), \mu(f(k)^{-1}) \cdot AkA^{-1})$$

for some map $f : Q = \Gamma/N \to L$. One can check that

$$f(hk) = AhA^{-1}(f(k)) \cdot f(h)$$

for all $h, k \in Q$. Suppose now that there exists $b \in L$ so that

$$f(k) = AkA^{-1}(b^{-1}) \cdot b$$

holds for all $k \in Q$. Then $\psi(w, k) = (Aw \cdot f(k), \mu(f(k)^{-1}) \cdot AkA^{-1}) = (b, \mu(b^{-1}) \cdot A) \cdot (w, k) \cdot (b, \mu(b^{-1}) \cdot A)^{-1}$ for all $(w, k) \in \Gamma$. Therefore, $\psi = \mu(b, \mu(b^{-1}) \cdot A)$.

Let us state our problem more formally. Let Q be a finite group, and let $\phi : Q \to \text{Aut}(L)$ be a homomorphism. A map $f : Q \to L$ is called a cocycle (with respect to ϕ) if

$$f(xy) = \phi(x)(f(y)) \cdot f(x)$$

for all $x, y \in Q$. On the set of cocycles we introduce an equivalence relation. We say $f_0 \sim f_1$ if and only if there is a $b \in L$ such that

$$f_1(x) = \phi(x)(b^{-1}) \cdot f_0(x) \cdot b$$

for all $x \in Q$. We denote by $H^1(Q; L)$ the set of equivalence classes of cocycles of Q into L. If L is abelian this is the usual cohomology group. See [CR-2, Appendix] for more details.

We go back to our problem now. Let $\phi : Q \to \text{Aut}(L)$ be the homomorphism given by $\phi(k) = AkA^{-1}$. If we show $H^1(Q; L) = 0$ then we shall have finished our proof. We show this by induction on the nilpotency of L. Let $f : Q \to L$ be a cocycle with respect to $\phi : Q \to \text{Aut}(L)$. Then ϕ naturally induces homomorphisms $\phi : Q \to \text{Aut}(Z(L))$ and $\overline{\phi} : Q \to \text{Aut}(\overline{L})$, where $Z(L) =$ center of L, and $\overline{L} = L/Z(L)$, because $Z(L)$ is characteristic in L. "Natural" means that the diagram

$$
\begin{array}{ccccccccc}
1 & \to & Z(L) & \to & L & \overset{p}{\to} & \overline{L} & \to & 1 \\
& & \downarrow \phi(x) & & \downarrow \phi(x) & & \downarrow \overline{\phi}(x) & & \\
1 & \to & Z(L) & \to & L & \overset{p}{\to} & \overline{L} & \to & 1
\end{array}
$$

is commutative for all $x \in Q$, where p is the quotient homomorphism.

If we put $\overline{f} = p \circ f : Q \to \overline{L}$, then $\overline{f}$ becomes a cocycle with respect to $\overline{\phi}$ because $f(xy) = \phi(x)(f(y)) \cdot f(x)$ for all $x, y \in Q$. By the induction hypothesis

$H^1(Q; \overline{L}) = 0$ as nilpotency $(\overline{L}) <$ nilpotency (L). Therefore there exists $\overline{a} \in \overline{L}$ such that

$$\overline{f}(x) = \overline{\phi}(x)(\overline{a}^{-1}) \cdot \overline{a}$$

holds true for all $x \in Q$.

Now choose any $a \in L$ such that $p(a) = \overline{a}$. Clearly $p(\phi(x)(a^{-1}) \cdot a) = p(f(x))$. Therefore, $\phi(x)(a^{-1}) \cdot a \cdot f(x)^{-1} \in Z(L)$ for all $x \in Q$. Let us define a new map $g : Q \to Z(L)$ by

$$g(x) = \phi(x)(a^{-1}) \cdot a \cdot f(x)^{-1}.$$

One can check that g is in fact a cocycle with respect to $\phi : Q \to \mathrm{Aut}(Z(L))$. As Q is finite and $Z(L)$ is a vector group, $H^1(Q; Z(L)) = 0$. Therefore there exists $c \in Z(L)$ for which $g(x) = \phi(x)(c^{-1}) \cdot c$ for all $x \in Q$. In other words, $\phi(x)(c^{-1}) \cdot c = \phi(x)(a^{-1}) \cdot a \cdot f(x)^{-1}$. Put $b = c^{-1}a \in L$. Then the above equality implies that $f(x) = \phi(x)(b^{-1}) \cdot b$. Thus we have proved $H^1(Q; L) = 0$ and this completes the proof of the theorem.

Now we explain the claim that we made after Theorem 1. Let $\Gamma = \Gamma' = \mathbb{Z}^2 \subset \mathbb{R}^2 \subset E(2)$ and let $\psi : \Gamma \to \Gamma'$ be the isomorphism given by the matrix $A = \left[\begin{smallmatrix} 1 & 2 \\ 1 & 3 \end{smallmatrix} \right]$. Suppose that there exists an extension ψ^* of ψ to an automorphism of $E(2) = \mathbb{R}^2 \circ O(2)$. Let $\psi^*(w, k) = (\psi_1(w, k), \psi_2(w, k))$. Then computation easily shows that $\psi_2(w, k) = AkA^{-1}$ and $\psi_1(w, k) = Aw \cdot f(k)$ for a cocycle $f : O(2) \to \mathbb{R}^2$. However, for $k \in O(2), AkA^{-1}$ is not an element of $O(2)$. Therefore ψ cannot be extended to an automorphism of $E(2)$.

In some nice cases, an isomorphism $\psi : \Gamma \to \Gamma'$ does extend to an automorphism ψ^* of $E(n)$. However, the extension is *not unique* in general. As a trivial example, take a non-zero element b of $\mathbb{R}^2$. Then $\mu(b, 1)$, the conjugation by $(b, 1) \in \mathbb{R}^2 \subset E(2)$, is an automorphism of $E(2)$ onto itself which restricts to the identity of a uniform lattice $\mathbb{Z}^2$ of $E(2)$. As $\mu(b, 1)$ is not the identity on $E(2)$, the identity map of $\mathbb{Z}^2$ has many distinct extensions which are continuous automorphisms of $E(2)$.

PROOF OF COROLLARY 1. Let $M = \Gamma \backslash L, M' = \Gamma' \backslash L'$ be infranilmanifolds, and let $h : M \to M'$ be a homotopy equivalence. We use h to denote the isomorphism of Γ onto Γ' induced by h. Let $N = L \cap \Gamma, N' = L' \cap \Gamma'$. As N and N' are the unique maximal normal nilpotent subgroups of Γ and Γ', h induces an isomorphism of N onto N'. Recall that N, N' are uniform lattices of L, L'. By Malcev, the isomorphism $h|_N$ extends uniquely to an isomorphism of L onto L'. This isomorphism preserves the connections defined by left invariant vector fields. Therefore, we may assume $L = L'$. Now we apply our theorem to the isomorphism $h : \Gamma \to \Gamma'$ between almost crystallographic groups of L. Let $h = \mu(\alpha)$ for $\alpha \in A(L)$. Then $(h, \alpha) : (\Gamma, L) \to (\Gamma', L)$ is a weak equivalence. That is, $\alpha(\sigma \cdot x) = h(\sigma) \cdot \alpha(x)$ for all $x \in L$ and $\sigma \in \Gamma$. Therefore, α induces a diffeomorphism α of M onto M' which is affine. Furthermore, α is homotopic to h.

PROOF OF COROLLARY 2. The natural homomorphism $\Psi : \mathrm{Aff}(M) \to \mathrm{Out}\,\Gamma$ $(\Gamma = \pi_1 M)$ is surjective by the Corollary 1. It remains to show that kernel (Ψ) is connected. Note that kernel (Ψ) is naturally isomorphic to $C_{A(L)}(\Gamma)/Z(\Gamma)$. We know that $Q = \Gamma/N$ (where $N = \Gamma \cap L$) acts on L. Let L^Q be the fixed point sets. Then $C_A(\Gamma) = \{(x, \mu(x^{-1}) \in A(L) : x \in L^Q\} \cong L^Q$, which is again a connected and simply connected nilpotent Lie group. Therefore, kernel $(\Psi) = L^Q/Z(\Gamma)$ is connected, and is isomorphic to $\mathrm{Aff}_0(M)$.

PROOF OF COROLLARY 3. Recall that we say an abstract kernel $\phi : G \to$ Out $\Gamma (\Gamma = \pi_1 M)$ is geometrically realized as a group of affine self-diffeomorphisms if there exists a homomorphism $\hat\phi : G \to \mathrm{Aff}\,(M)$ such that $\Psi \circ \hat\phi = \phi$. We use exactly the same method as in [LR1, Theorem 6]. Suppose ψ has an extension $1 \to \Gamma \to E \to G \to 1$. As $N = \Gamma \cap L$ is characteristic in Γ, it is normal in E. By [M] the pair (N, L) has the unique automorphism extension property. Then there exists a commutative diagram with exact rows

$$1 \to N \to E \to E/N \to 1$$
$$\downarrow \quad \downarrow \quad \downarrow =$$
$$1 \to L \to \mathcal{L} \to E/N \to 1$$

where all the vertical maps are injective. As L is divisible, the bottom sequence splits so that $\mathcal{L} = L \circ (E/N)$. The splitting gives rise to a homomorphism of E/N into Aut (L), which in turn yields a homomorphism τ of $\mathcal{L}$ into $A(L)$. Certainly τ has a finite kernel, so $\tau|\Gamma$ is injective. By the Theorem, we may assume that $\tau|\Gamma$ is the identity (by composing with the inverse of the conjugation by an element of $A(L)$ which is an extension of $\tau|\Gamma$). Then $\tau(E)$ acts on L as affine diffeomorphisms. As this action restricts to the original deck transformation of Γ on L, $\tau(E)/\Gamma$ acts on $M = \Gamma \backslash L$ as a group of affine diffeomorphisms. If $C_E(\Gamma)$ is torsion-free, one can show that τ itself is injective. In this case, $\tau(E)/\Gamma = G$ and G acts effectively on M as a group of affine self-diffeomorphisms. The converse is proved in [LR1].

Remark. A generalized version of the main theorem of this paper will appear in [KLR] where the Seifert construction with nilpotent group as a kernel is discussed.

REFERENCES

[A] L. Auslander, *Bieberbach's theorems on space groups and discrete uniform subgroups of Lie groups*, Ann. Math. **71**, (1960), 579–590.

[CR] P. E. Conner and F. Raymond, *Manifolds with few periodic homeomorphisms*, Proc. Second Conference on Compact Transformation Groups, Part II, *Springer LN in Math.* **299**, (1972), 227–264.

[FH] F. T. Farrell and W. C. Hsiang, *Topological characterization of flat and almost flat Riemannian manifolds $M^n (n \neq 3, 4)$*, American J. Math. **105**, (1983), 641–672.

[G] M. Gromov, *Almost flat manifolds*, J. Differential Geometry **13**, (1978), 231–241.

[H]   S. Helgason, *Differential Geometry and Symmetric Spaces*, Academic Press, New York, 1962.

[K-T]   F. W. Kamber and Ph. Tondeur, *Flat manifolds with parallel torsion*, J. Differential Geometry, **2**, (1968), 385–389.

[KLR]   Y. Kamishima, K. B. Lee and F. Raymond, *The Seifert construction and its application to infranilmanifolds*, to appear in Quarterly J. Math., Oxford.

[LR-1] K. B. Lee and F. Raymond, *Topological, affine and isometric actions on flat Riemannian manifolds*, J. Differential Geometry **16**, (1981), 255–269.

[LR-2] K. B. Lee and F. Raymond, *Ibid II*, Topology and Appl. **13**, (1982), 295–310.

[LR-3] K. B. Lee and F. Raymond, *Geometric realization of group extensions by the Seifert Construction*, Contemporary Mathematics, Vol. 33, Amer. Math. Soc., Providence, R.I., 1984, pp. 353–411.

[M]   A. Malcev, *On a class of homogeneous spaces*, Translations AMS **39**, (1941).

[R]   E. Ruh, *Almost flat manifolds*, J. Differential Geometry **17** (1982), 1–14.

[W]   J. A. Wolf, *Spaces of constant curvature*, Publish or Perish, 1977.

PURDUE UNIVERSITY (CURRENTLY, UNIVERSITY OF OKLAHOMA) AND UNIVERSITY OF MICHIGAN

Contemporary Mathematics
Volume **44**, 1985

FINITE GRAPHS AND FREE GROUPS*

JOHN R. STALLINGS

Introduction

Free groups can be represented as fundamental groups of graphs. It is useful to examine certain categorical constructions in graph-theory and to apply them to group-theory. I have also found the notion of immersion of graphs to be particularly useful.

Several results come out very easily. Howson's theorem says that the intersection of two finitely generated subgroups of a free group is finitely generated; S. M. Gersten has shown that H. Neumann's bound on the rank of the intersection can be proved using an Euler characteristic computation in the graph-theoretic picture. Marshall Hall's theorem, roughly that finitely generated subgroups of free groups are closed sets in the profinite topology, also has an easy proof. The most difficult theorem that I have been able to prove using graph-theory is a consequence of a paper of Greenberg's; it states that if A and B are finitely generated subgroups of a free group, and (denoting the join of A and B by $A \vee B$) if $A \cap B$ is of finite index in both A and B, then $A \cap B$ is of finite index in $A \vee B$.

The results of this note are explained more fully in [8].

1. Graphs and paths

1.1. The category of graphs which I use is described by Serre [7]. Briefly, a *graph* Γ consists of two sets: E, the set of the *directed edges*, and V, the set of *vertices*. The set E is endowed with a fixed-point-free involution, $e \to \bar{e}$, *reversal* of orientation; and there are two functions $\iota, \tau : E \to V$, *initial* and *terminal vertex*, such that $\tau(e) = \iota(\bar{e})$. A *map of graphs* $\Gamma \to \Delta$ is a pair of structure-preserving functions, edges to edges, and vertices to vertices. (In Gersten's formulation [2], an edge can be collapsed to a vertex, and this has both advantages and disadvantages. However, here I use Serre's rules, and edges never degenerate under a map.)

There are two functors, "edges" and "vertices", from graphs to sets. Every category-theoretic construction is preserved by these functors, and so I claim that all such constructions are "obvious". In particular, the *pullback* of two

*Partly supported by NSF grant MCS 80-2858..

maps $f_1 : \Gamma_1 \to \Delta$ and $f_2 : \Gamma_2 \to \Delta$ always exists and can be constructed by constructing the pullbacks of the edges and vertices. The *pushout* of two maps $g_1 : \Gamma \to \Delta_1$, $g_2 : \Gamma \to \Delta_2$ exists if and only if there are orientations (selection of an edge out of each pair $\{e, \bar{e}\}$) of $\Gamma, \Delta_1, \Delta_2$ that are preserved by g_1 and g_2, in which case the pushout can be constructed by constructing in set-theory the pushouts of the edges and vertices.

1.2. A *path* in Γ is an n-tuple of edges $e_1 e_2 \ldots e_n$ such that $\tau(e_i) = \iota(e_{i+1})$; it has *length* n and connects $\iota(e_1)$ to $\tau(e_n)$; there is also, for each vertex v, a path Λ_v of length 0 from v to v. Certain paths p and q can be concatenated, giving pq. This produces a small category, $P(\Gamma)$, the *category of paths* in Γ; the set of "objects" in $P(\Gamma)$ is V.

A *round-trip* ("aller-retour" in Serre) is a path of the form $e\bar{e}$. A path containing no round-trip is said to be *reduced*. The equivalence relation on $P(\Gamma)$ generated by omitting round-trips is *homotopy*, denoted by $\sim$. The set of homotopy-classes of paths is a small category $\pi(\Gamma)$, the *fundamental groupoid* of Γ; in $\pi(\Gamma)$ every class of paths has an inverse, and so $\pi(\Gamma)$ is a groupoid. The subset of $\pi(\Gamma)$ consisting of elements beginning and ending at a fixed vertex v is a group, $\pi_1(\Gamma, v)$, the *fundamental group* of Γ based at v.

1.3. It is fairly easy to show that every homotopy-class of paths contains a unique reduced path.

Another classical result is that $\pi_1(\Gamma, v)$ is a free group. A basis of this free group is constructed by choosing a maximal tree T in the component Γ_0 of Γ containing v, and an orientation of Γ_0; for each edge e of $\Gamma_0 - T$ selected by the orientation, we consider the homotopy class of the loop that begins at v, goes through T to $\iota(e)$, across e to $\tau(e)$, and back through T to v; the free basis of $\pi_1(\Gamma, v)$ consists of all such homotopy-classes.

2. Coverings and immersions

2.1. The *star* of a vertex v in a graph Γ consists of all edges e such that $\iota(e) = v$. Call this $St(v, \Gamma)$. A map of graphs $f : \Gamma \to \Delta$ produces, for each vertex of Γ, a function

$$f_v : St(v, \Gamma) \to St(f(v), \Delta).$$

A map f is called an *immersion* if, for all vertices v, f_v is injective. We call f a *covering*, resp. a *locally surjective map*, if, for all vertices v, f_v is bijective, resp. surjective. (The term "immersion" was suggested by R. Kirby to be better than "locally injective"; the parallel terms for "covering" and "locally surjective" map would, I suppose, be "bimersion" and "surmersion.")

2.2. Coverings behave, in graph theory, exactly as covering projections behave in topology (a covering projection is a local product with discrete fibre). There are lifting theorems, existence of coverings corresponding to subgroups of the fundamental group, universal coverings, covering translations, etc.

The following theorem can be proved by constructing the covering corresponding to the join in the theorem (recall that $A \vee B$ is the subgroup generated by $A \cup B$) and applying lifting theorems.

2.3. *Let*

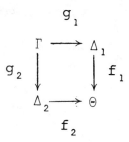

be a pushout diagram, where Γ *is connected. Let* v *be a vertex of* Γ, *the images of which in* $\Delta_1, \Delta_2, \Theta$ *are, respectively,* v_1, v_2, w. *Then*

$$\pi_1(\Theta, w) = f_1 \pi_1(\Delta_1, v_1) \vee f_2 \pi_1(\Delta_2, v_2).$$

2.4. In a graph Γ, an *admissible* pair of edges (e_1, e_2) satisfies $\iota(e_1) = \iota(e_2)$ and $e_1 \neq \bar{e}_2$. Given such, we can identify $\tau(e_1)$ to $\tau(e_2)$, e_1 to e_2, $\bar{e}_1$ to $\bar{e}_2$, and obtain a graph $\Gamma/[e_1 = e_2]$, the result of *folding* e_1 and e_2. This is a particularly simple, and therefore easy to construct, type of pushout. An application of **2.3**, or an explicit computation of fundamental groups, shows that the folding map $\epsilon : \Gamma \to \Gamma/[e_1 = e_2]$ is surjective on fundamental groups.

If $f : \Gamma \to \Delta$ is any map, and Γ is a finite graph, then f can be factored through a finite series of foldings and an immersion:

$$\Gamma = \underbrace{\Gamma_0 \xrightarrow{\varepsilon_1} \Gamma_1 \xrightarrow{\varepsilon_2} \cdots \to \Gamma_k}_{\text{foldings}} \underbrace{\xrightarrow{f_k} \Delta}_{\text{immersion}}$$

This enables us to represent any finitely generated subgroup of $x\pi_1(\Delta, w)$ as the π_1-image of an immersion.

2.5. Immersions behave to a fair degree like coverings. In particular, *if* $f : \Gamma \to \Delta$ *is an immersion, then*

(a) *For any reduced path* p *in* Γ, fp *is a reduced path in* Δ.

(b) *If* p *and* q *are paths in* Γ *starting at the same vertex, and if* $fp = fq$, *then* $p = q$.

(c) $f : \pi_1(\Gamma, v) \to \pi_1(\Delta, f(v))$ *is injective.*

(d) *If* Θ *is a connected graph with vertex* v, *and* $g_1, g_2 : \Theta \to \Gamma$ *are maps of graphs such that* $g_1(v) = g_2(v)$ *and* $fg_1 = fg_2$, *then* $g_1 = g_2$.

2.6. THEOREM. *Let*

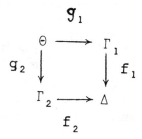

be a pullback diagram of graphs, where f_1 and f_2 are immersions. Then $f_3 = f_1 g_1 = f_2 g_2 : \Theta \to \Delta$ is an immersion.

Let v_1, v_2 be vertices of Γ_1, Γ_2, respectively, such that $f_1(v_1) = f_2(v_2) = w$. Let v_3 be the corresponding vertex of Θ. Then

$$f_3 \pi_1(\Theta, v_3) = f_1 \pi_1(\Gamma_1, v_1) \cap f_2 \pi_1(\Gamma_2, v_2).$$

PROOF. The crucial point is this:

$$\text{Let } \alpha \in f_1 \pi_1(\Gamma_1, v_1) \cap f_2 \pi_1(\Gamma_2, v_2).$$

Then there are reduced loops, based at v_1 and v_2, say p_1 in Γ_1, p_2 in Γ_2, such that $f_1 p_1$ and $f_2 p_2$ belong to α. Now, by **2.5.(a)** these are both reduced, and because there is a unique reduced path in the homotopy class α, $f_1 p_1 = f_2 p_2$. Because of the pullback property, there is a path p_3 in Θ mapping to p_1 and p_2; this path p_3 is a loop, and then $f_3 p_3$ belongs to α, so that $\alpha \in f_3 \pi_1(\Theta, v_3)$.

2.7. COROLLARY. *(Howson's theorem [5]) If A_1 and A_2 are finitely generated subgroups of a free group B, then $A_1 \cap A_2$ is finitely generated.*

PROOF. Using foldings as in **2.4**, A_1 and A_2 can be represented by immersions of finite graphs Γ_1, Γ_2 into a graph Δ with $B = \pi_1(\Delta, w)$. Their pullback Θ is a finite graph, and so the component Θ_0 of Θ containing the base-point is finite. By **2.6**, $A_1 \cap A_2$ is represented by the immersion $\Theta_0 \to \Delta$, and so is finitely generated.

2.8. I refer to Gersten [2] for the proof of H. Neumann's inequality [6]. This inequality states, where $r(G)$ is the rank of G, if $A \cap B$ is non-trivial, then

$$r(A \cap B) - 1 \leq 2(r(A) - 1)(r(B) - 1).$$

Gersten's proof involves a clever computation of the Euler characteristic of the component Θ_0 of the pullback representing $A \cap B$.

Neumann expressed the hope that "2" can be replaced by "1"; this seems to be a difficult combinatorial problem, and there is some hope that graph-theory will clarify it.

2.9. Every immersion can be extended to a covering. A particular case of this is:

THEOREM. *Let $f : \Gamma \to \Delta$ be an immersion, where Γ has only finitely many vertices and Δ has only one vertex. Then there is a graph Γ' containing Γ, with $\Gamma' - \Gamma$ consisting only of edges, and an extension $f' : \Gamma' \to \Delta$ of f, such that f' is a covering*

PROOF. Choose an orientation of Δ, so that $\pi_1(\Delta)$ is free on the set of chosen edges $\{e\}$. For each such edge, if e_1 is an edge of Γ with $f(e_1) = e$, define $R_e(\iota(e_1)) = \tau(e_1)$. Then R_e is a well-defined injective function of a subset of V, the vertices of Γ, into V; this happens because f is an immersion. Since V is finite, R_e extends to a permutation $S_e : V \to V$. These permutations S_e represent

$\pi_1(\Delta)$ into the symmetric group on V; corresponding to this representation we construct a covering $f' : \Gamma' \to \Delta$ extending f.

2.10. Corollary (Marshall Hall's theorem [4],[1]):

If A is a finitely generated subgroup of a free group F, and $\{\beta_1, \ldots, \beta_k\} \subset F - A$ is a finite set, then there exists a subgroup S of finite index in F such that $A \subset S, \{\beta_1, \ldots, \beta_k\} \subseteq F - S$, and such that A is a free factor of S.

PROOF. We find an immersion $f : \Gamma \to \Delta$, where Γ is a finite graph, Δ is a graph with only one vertex, $\pi_1(\Delta) = F$, and $f\pi_1(\Gamma, v) = A$, and where there are non-closed paths $p_1, \ldots, p_n$ in Γ st. rting at v, representing $\beta_1, \ldots, \beta_k$. Using **2.9.**, extend f to a covering $f' : \Gamma' \to \Delta$. Define $S = f'\pi_1(\Gamma', v)$, and read off the conclusions of the theorem.

3. Consequences of Greenberg's paper

3.1. Greenberg [3] proved some theorems about Fuchsian groups, which yield a remarkable consequence that I prove using graph theory.

3.2. A *cyclically reduced circuit* in Γ is a reduced loop $e_1 \ldots e_n$ such that $e_1 \neq \bar{e}_n$. A *core-graph* Γ is a connected graph such that every edge belongs to some cyclically reduced circuit. An *essential edge* of Γ is an edge belonging to a cyclically reduced circuit; the *core* of Γ is the smallest subgraph containing all essential edges. The core of Γ has, modulo a change of base-point, the same fundamental group as Γ; it can be thought of as Γ with all extraneous trees shaved off.

If $f : \Gamma \to \Delta$ is an immersion, Γ connected, and Δ a connected core-graph, and if the fundamental group situation is of finite index, then f is a covering.

3.3. THEOREM. *Suppose that A and B are finitely generated sub-groups of a free group F such that $A \cap B$ is of finite index both in A and in B. Then $A \cap B$ is of finite index in $A \vee B$.*

PROOF. Represent F by $\pi_1(\Delta)$, and A and B by immersions of finite connected graphs $\Gamma_1, \Gamma_2, f_1 : \Gamma_1 \to \Delta, f_2 : \Gamma_2 \to \Delta$. Then by **2.6**, $A \cap B$ is represented by a component Γ_3 of the pullback of f_1 and f_2. If $A \cap B$ is trivial, the theorem is obvious; hence, suppose $A \cap B \neq \{1\}$. By conjugating the situation within F, arrange the basepoint of Γ_3 to lie within the core of Γ_3. Shave off trees from Γ_3, Γ_1, and Γ_2 to obtain core-graphs. Hence suppose $\Gamma_1, \Gamma_2, \Gamma_3$ are finite core-graphs, and

$$
\begin{array}{ccc}
 & g_1 & \\
\Gamma_3 & \longrightarrow & \Gamma_1 \\
g_2 \downarrow & \searrow f_3 & \downarrow f_1 \\
\Gamma_2 & \longrightarrow & \Delta \\
 & f_2 &
\end{array}
$$

is a commutative diagram, f_1, f_2 are immersions, Γ_3 a subgraph of the pull-back. Since $A \cap B$ is of finite index in A and in B, both g_1 and g_2 are coverings. Now we shall show that the pushout of g_1 and g_2 is also a covering, by a strange device. Let $h : \tilde{\Gamma}_3 \to \Gamma_3$ be the universal covering. Then $g_1 h$ and $g_2 h$ are universal coverings.

Let $G(g_1 h)$ be the set of covering translations of $g_1 h$; and similarly for $G(g_2 h)$. These consist of certain graph-automorphisms $\sigma : \tilde{\Gamma}_3 \to \tilde{\Gamma}_3$, for which it is easily seen that $f_3 h \sigma = f_3 h$, so that σ is a "translation" of the immersion $f_3 h$.

Let $G = G(g_1 h) \vee G(g_2 h)$, a group of automorphisms of $\tilde{\Gamma}_3$. Because of **2.5(d)** applied to the immersion $f_3 h : \tilde{\Gamma}_3 \to \Delta$, G acts freely on $\tilde{\Gamma}_3$.

Let $\Theta = \tilde{\Gamma}_3/G$, $\tilde{g}_1 = g_1 h$, $\tilde{g}_2 = g_2 h$; then we get a commutative diagram.

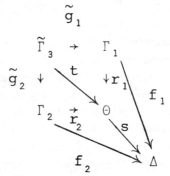

Because G acts freely on $\tilde{\Gamma}_3$, t is a covering. Hence, since $\tilde{g}_1$ is a covering, r_1 is a covering.

Now Θ represents $A \vee B$. Because Γ_1 is a finite graph, r_1 is a finite-sheeted covering. And so $r_1 g_1 : \Gamma_3 \to \Theta$ is a finite-sheeted covering. This implies that $A \cap B$ is of finite index in $A \vee B$.

REFERENCES

[1] Burns, R. G., *A note on free groups*, Proc. Amer. Math. Soc. **23** (1969), 14–17.

[2] Gersten, S. M., *Intersections of finitely generated subgroups of free groups and resolutions of graphs*, Inv. math. **71** (1983), 567–591.

[3] Greenberg, L., *Discrete groups of motions*, Canad. J. Math. **12** (1960), 414–425.

[4] Hall, M., Jr., *Coset representations in free groups*, Trans. Amer. Math. Soc. **67** (1949), 421–432.

[5] Howson, A. G., *On the intersection of finitely generated groups*, J. London Math. Soc. **29** (1954), 428–434.

[6] Neumann, H., *On the intersection of finitely generated free groups*, Publ. Math. Debrecen. **4** (1956), 36–39; addendum, 5 (1957), 128.

[7] Serre, J.-P., *Arbres, Amalgames SL_2*, Astérisque No. **46**, Société Math. de France (1977).

[8] Stallings, J. R., *Topology of finite graphs*, Inv. Math. **71** (1983), 551–565.

MATHEMATICS DEPARTMENT

UNIVERSITY OF CALIFORNIA, BERKELEY

Contemporary Mathematics
Volume 44, 1985

A TOPOLOGICAL PROOF OF A THEOREM
OF BRUNNER AND BURNS ABOUT M. HALL GROUPS

C. L. TRETKOFF* AND M. D. TRETKOFF**

§1. Introduction

The purpose of the present paper is to give a simple topological proof of the following, somewhat more precise, version of a theorem of A. M. Brunner and R. G. Burns [1]:

THEOREM. *Suppose A is a Frobenius group with kernel N and complement J. Moreover, suppose τ is an isomorphism of J onto a proper subgroup, K, of a finite group B. Let H be a finitely generated subgroup of infinite index in G, the amalgamated free product of A and B with J and K identified by τ. Then H is a proper free factor of a subgroup H^* of finite index in G. Moreover, $H^* = H * M$, where $M = M_1 * \ldots * M_s$ and each M_ν is conjugate in G to N.*

We recall that a finite group A is called a *Frobenius group* with *complement* J provided J is a proper subgroup and $J \cap xJx^{-1} = 1$ for all $x \notin J$. It is a theorem of Frobenius that such groups possess a normal subgroup $N \lhd A$, called the *Frobenius kernel*, such that $A = JN$ and $J \cap N = 1$. A famous theorem of J. Thompson asserts that N is nilpotent and yields the corollary that N is unique and that J is unique up to conjugation (see, for example, Huppert, [4], chapter V, Satz 8.17).

Groups satisfying the conclusion of the Brunner-Burns theorem are called *M. Hall groups* in honor of M. Hall, who first proved that free groups have this property [3]. R. G. Burns [2] subsequently demonstrated that the collection of M. Hall groups is closed under the formation of free products. His proof utilized the details of the proof of the Kurosh subgroup theorem. Finally, Brunner and Burns discovered their theorem and gave a long and technical proof of it. In particular, they employed the details from the proofs of the Karras-Solitar subgroup theorems [5], [6] for amalgamated free products and the theorem of

*This research was supported in part by a grant from The City University of New York PSC-CUNY Researach Award Program.
**Supported in part by National Science Foundation Grant MCS-8103453.

Burns [2] cited above. They reasoned by contradiction, beginning with a minimal counter-example. It is unclear whether their proof yields the structure of M as we have presented it; they do not mention it explicitly.

Our proof utilizes the notions of fundamental group, covering spaces and (the simplest form of) van Kampen's theorem, which may be found in introductory algebraic topology texts (see, for example, Massey [7]). The subgroup theorems of Kurosh and Karrass-Solitar are completely avoided. From the theory of finite groups, we only use the existence of the Frobenius kernel; in particular, we do not employ the uniqueness of the Frobenius complement. Finally, we note that additional results can be seen upon recasting our proof in the context of the Bass-Serre theory of groups acting on trees.

§2. A topological description of the hypothesis and its consequences

It will be convenient to introduce a third group C and monomorphisms $\varphi : C \to A$ and $\psi : C \to B$ with J and K as their respective images and $\tau = \psi\varphi^{-1}$. Since any group may be represented as the fundamental group of a CW-complex with a single O-cell, which we employ as the base point, we shall write

$$A = \pi_1(X, P), \ B = \pi_1(Y, Q) \text{ and } C = \pi_1(U, z_0).$$

Moreover, U may be selected so there are cellular maps $\Phi : U \to X$ and $\Psi : U \to Y$ which induce the given monomorphisms, that is, $\Phi_* = \varphi$ and $\Psi_* = \psi$.

Next, let Z denote the quotient space obtained from $X \cup Y \cup \{U \times [0,1]\}$ by identifying $(u, 0)$ with $\Phi(u)$ and $(u, 1)$ with $\Psi(u)$ for all $u \in U$. Here, $[0,1]$ is the interval $0 \leq t \leq 1$. Denoting the point of Z representing (u, t) by $[u, t]$, we may embed U in Z by sending u to $[u, 1/2]$. Now, by van Kampen's theorem, $G = \pi_1(Z, z_0)$.

The subspaces $L = \{[u, t] | u \in U, 0 \leq t \leq 1/2\}$ and $R = \{[u, t] | u \in U, 1/2 \leq t \leq 1\}$ play an important role in the sequel. We also set $E = L \cup R, Z(P) = X \cup L$ and $Z(Q) = Y \cup R$.

PROPOSITION I. *Let $r : Z(N) \to Z(P)$ be the covering defined by $N \lhd A$. Then $r^{-1}(U)$ consists of a single simply connected component.*

PROOF. The components of $r^{-1}(U)$ are simply connected because their fundamental groups are isomorphic to $J \cap N = 1$. Thus each of these components is equivalent to the universal covering of U and therefore contains $|J|$ points above z_0. But $r^{-1}(z_0)$ consists of $|A : N| = |J|$ points, so there is only one such component. This completes our proof.

We shall call $r^{-1}(U)$ the *face* of $Z(N)$.

Next, let $p : \tilde{Z} \to Z$ be a covering with $p_*\pi_1(\tilde{Z}, \tilde{z}_0) = H$ and $p(\tilde{z}_0) = z_0$.

PROPOSITION II. *Both $p^{-1}(X) = \{X_i\}, i \in I$, and $p^{-1}(Y) = \{Y_j\}, j \in J$, have infinitely many path components. Each of these components meets only finitely many components of $p^{-1}(E)$.*

PROOF. Since $|G : H| = \infty$, $\tilde{Z}$ is an infinite sheeted covering of Z, and $p^{-1}(P)$ is an infinite set. Now, $X_i \cap p^{-1}(P)$ is finite because X_i is a covering of a space,

X, with finite fundamental group. It follows that I is infinite. Since $P \in E, X_i$ has nonvoid intersection with finitely many components of $p^{-1}(E)$. The same argument applies to $\{Y_j\}, j \in J$. This completes our proof.

We shall denote the component of $p^{-1}(Z(P))$ containing X_i by $Z(P_i)$; the components $Z(Q_j)$, $j \in J$, are defined similarly. Next, we denote the components of $p^{-1}(E)$ having nonvoid intersection with X_i and Y_j by $E_{ijk}, k \in K(i,j)$, and we refer to the subspaces $U_{ijk} = p^{-1}(U) \cap E_{ijk}$ as *faces* of $Z(P_i)$ and $Z(Q_j)$.

A key ingredient in our proof of the Brunner-Burns Theorem is a combinatorial description of $\tilde{Z}$ as a union of the $Z(P_i)$ and $Z(Q_j)$. For this purpose, it is convenient to introduce a bipartite graph, Γ, with vertices $P_i, i \in I$, and $Q_j, j \in J$. The edges, $e_{ijk}, k \in K(i,j)$, of Γ joining P_i to Q_j are in one to one correspondence with the components, E_{ijk}, of $p^{-1}(E)$ having nonvoid intersection with X_i and Y_j. We view these edges as simple oriented paths $e_{ijk}(t)$, $0 \leq t \leq 1$, beginning at P_i and ending at Q_j. Finally, we introduce the midpoints, $m_{ijk} = e_{ijk}(1/2)$, as a third set of vertices and consider Γ to be a directed tripartite graph with edges $e_{ijk}(t)$, $0 \leq t \leq 1/2$, and $e_{ijk}(t)$, $1/2 \leq t \leq 1$. Proposition II implies that Γ has infinitely vertices of the form P_i and Q_j and that each vertex is incident with a finite number of edges.

There is a continuous mapping $f : \tilde{Z} \to \Gamma$ given by $f(X_i) = P_i$, $f(Y_j) = Q_j$ and $f(\tilde{z}) = e_{ijk}(t)$, where $p(\tilde{z}) = [u, t]$. Obviously, the 1-skeleton $\tilde{Z}^{(1)}$, of $\tilde{Z}$ is mapped onto Γ by f, and every loop in Γ based at $m_0 = f(\tilde{z}_0)$ is the image of a loop in $\tilde{Z}^{(1)}$ based at $\tilde{z}_0$. Since $\pi_1(\tilde{Z}, \tilde{z}_0)$ is finitely generated, there are loops λ_ν, $1 \leq \nu \leq N$, in $\tilde{Z}^{(1)}$ representing generators of $\pi_1(\tilde{Z}, \tilde{z}_0)$ whose images, δ_ν, $1 \leq \nu \leq N$, represent generators of $\pi_1(\Gamma, m_0)$. Let Δ denote the subgraph of Γ spanned by the vertices occurring in these δ_ν; obviously, Δ is a finite connected graph. Finally, we select a maximal tree $T \subset \Gamma$ and a maximal tree, $\tilde{T}$, in the 1-skeleton of $f^{-1}(T)$. Since all the vertices of $\tilde{Z}^{(1)}$ belong to $\tilde{T}$, the latter is a maximal tree in $\tilde{Z}^{(1)}$. Of course, $f(\tilde{T}) = T$.

We now introduce important decompositions of Γ and T. Namely, let $\Gamma = \Delta \cup T_1^* \cup T_2^* \cup \ldots$ and let $T = \cup T_n^*$, $n = 0, 1, 2, \ldots$ where $T_0^* = T \cap \Delta$ and T_n^* is the union of T_{n-1}^* and the edges of T with a vertex in T_{n-1}^*. Of course, T_n^* is a finite connected tree. Moreover, we have

$$T = T_1^* \cup T_1 \cup \ldots \cup T_m \cup T_{m+1} \cup \ldots \cup T_{m+n}$$

where the T_ν are the closures in T of the connected components of the complement of T_1^*. Each T_ν is an infinite tree. We simplify notation by setting $m_\nu = m_{i(\nu), j(\nu), k(\nu)} = T_1^* \cap T_\nu$ and suppose that $P_\nu = P_{i(\nu)}$ belongs to T_ν if $1 \leq \nu \leq m$ and that $Q_\nu = Q_{j(\nu)}$ belongs to T_ν if $m + 1 \leq \nu \leq m + n$.

If $S \subset T$ is a connected subtree, there is a unique irreducible path, $\lambda(S)$, in $\tilde{T}$ joining $\tilde{z}_0$ to $f^{-1}(S)$. We shall denote the union of $\lambda(S)$ and the $Z(P_i)$ and $Z(Q_j)$ with P_i and Q_j in S by $Z^*(S)$. In case S consists of the single vertex $m_{ijk}, Z^*(m_{ijk})$ shall denote the union of $\lambda(S)$ and the face $U_{ijk} = f^{-1}(m_{ijk})$. In the sequel we shall write $\pi_1(S)$ in place of $\pi_1(Z^*(S), \tilde{z}_0)$; in particular $\pi_1(\Gamma) = \pi_1(\tilde{Z}, \tilde{z}_0)$. Finally, we note that $\pi_1(m_{ijk})$ is embedded in $\pi_1(P_i)$ and $\pi_1(Q_j)$ because Φ_* and Ψ_* are monomorphisms.

PROPOSITION III. $Z(P_i), P_i \notin T_5^*$ and $Z(Q_j), Q_j \notin T_3^*$ are simply connected.

PROOF. Setting $\Delta^* = \Delta \cup T_1^*$ and applying van Kampen's theorem, we see that $\pi_1(\Gamma)$ is the amalgamated free product of $\pi_1(\Delta^*)$ and the $\pi_1(T_\nu)$, $1 \leq \nu \leq m+n$, with the subgroups corresponding to $\pi_1(m_\nu)$ identified. Thus $\pi_1(\Delta^*)$ is a subgroup of $\pi_1(\Gamma)$. But $\pi_1(\Gamma)$ is generated by the homotopy classes of loops, λ_ν, belonging to $Z^*(\Delta^*)$, so $\pi_1(\Gamma) = \pi_1(\Delta^*)$. It follows that $\pi_1(T_\nu) = \pi_1(m_\nu)$, $1 \leq \nu \leq m+n$.

Suppose that $1 \leq \nu \leq m$. We have

(i) $T_\nu = st(P_\nu) \cup T_\nu^{(1)} \cup \ldots \cup T_\nu^{(n(\nu))}$

where $st(P_\nu)$ is the union of the edges of T_ν incident with P_ν and the $T_\nu^{(\mu)}$ are the closures of the connected components of the complement of $st(P_\nu)$ in T_ν. Observing that $Z^*(P_\nu) = Z^*(st(P_\nu))$ and applying van Kampen's theorem, we see that $\pi_1(T_\nu)$ is the amalgamated free product of $\pi_1(P_\nu)$ and $\pi_1(T_\nu^{(\mu)})$, $1 \leq \mu \leq n(\nu)$, with the subgroups corresponding to $\pi_1(m_\nu^{(\mu)})$ identified. Here, $m_\nu^{(\mu)} = st(P_\nu) \cap T_\nu^{(\mu)}$. Thus, $\pi_1(P_\nu)$ and the $\pi_1(T_\nu^{(\mu)})$ are subgroups of $\pi_1(T_\nu) = \pi_1(m_\nu) \leq \pi_1(P_\nu)$. It follows that $\pi_1(T_\nu) = \pi_1(m_\nu) = \pi_1(P_\nu)$ and $\pi_1(T_\nu^{(\mu)}) = \pi_1(m_\nu^{(\mu)})$.

Since $\pi_1(P_\nu) = \pi_1(m_\nu)$, any loop in $Z(P)$ which lifts to a closed loop in $Z(P_\nu)$ is homotopic to a loop in U. Thus, $Z(P_\nu)$ is a covering of $Z(P)$ defined by a subgroup J_ν of J. If $J_\nu \neq 1$, the face $U_\nu = f^{-1}(m_\nu)$ is not simply connected. Nevertheless, the faces $U_\nu^{(\mu)} = f^{-1}(m_\nu^{(\nu)})$ are simply connected because their fundamental groups project to $J \cap x J_\nu x^{-1} \subset J \cap x J x^{-1}$, $x \notin J$, and the latter group is 1. Thus, $\pi_1(m_\nu^{(\mu)}) = \pi_1(T_\nu^{(\mu)}) = 1$.

In analogy with (i), we write

(ii) $T_\nu^{(\mu)} = st(Q_\nu^{(\mu)}) \cup T_{\nu,\mu}^{(1)} \ldots \cup T_{\nu,\mu}^{n(\nu,\mu)}$,

where the $T_{\nu,\mu}^{(\omega)}$ are the closures of the connected components of the complement of $st(Q_\nu^{(\mu)})$ in $T_\nu^{(\mu)}$. Thus, $\pi_1(T_\nu^{(\mu)})$ is the amalgamated free product of the subgroups $\pi_1(Q_\nu^{(\mu)})$ and $\pi_1(T_{\nu,\mu}^{(\omega)})$. But $\pi_1(T_\nu^{(\mu)}) = 1$, so $\pi_1(Q_\nu^{(\mu)}) = 1 = \pi_1(T_{\nu,\mu}^{(\omega)})$. Iterating this procedure, we see that $\pi_1(P_i) = 1 = \pi_1(Q_j)$ for all P_i and Q_j in $T_\nu^{(\mu)}$. Clearly, no P_i or Q_j belongs to $T_3^* \cap T_\nu^{(\mu)}$.

Next, we suppose that $m+1 \leq \nu \leq m+n$ and we now repeat our construction with Q_ν in place of P_ν and $P_\nu^{(\mu)}$ in place of $Q_\nu^{(\mu)}$. We find that $\pi_1(T_\nu) = \pi_1(m_\nu) = \pi_1(Q_\nu)$ and that $\pi_1(T_\nu^{(\mu)}) = \pi_1(m_\nu^{(\mu)})$. If $\pi_1(m_\nu^{(\mu)}) = 1$, the previous argument applies and, once again, $\pi_1(P_i) = 1 = \pi_1(Q_j)$ for all P_i and Q_j in $T_\nu^{(\mu)}$. However, since B need not be a Frobenius group, we may have $\pi_1(m_\nu^{(\mu)}) \neq 1$. Nevertheless, we still find that $\pi_1(T_\nu^{(\mu)})$ is the amalgamated free product of $\pi_1(P_\nu^{(\mu)})$ and the $\pi_1(T_{\nu,\mu}^{(\omega)})$, identifying the subgroups corresponding to $\pi_1(m_{\nu,\mu}^{(\omega)})$, where $m_{\nu,\mu}^{(\omega)} = st(P_\nu^{(\mu)}) \cap T_{\nu,\mu}^{(\omega)}$. Reasoning as before, we conclude successively that $\pi_1(T_\nu^{(\mu)}) = \pi_1(P_\nu^{(\mu)})$, $\pi_1(m_{\nu,\mu}^{(\omega)}) = 1$ and $\pi_1(P_i) = 1 = \pi_1(Q_j)$ for all P_i and Q_j in $T_{\nu,\mu}^{(\omega)}$. It is clear from our construction that no P_j belongs to $T_5^* \cap T_{\nu,\mu}^{(\omega)}$ and that $Q_j \in T_\nu$, $Q_j \notin T_3^*$ imply $Q_j \in T_{\nu,\mu}^{(\omega)}$.

§3. A Geometric Construction and Proof of the Theorem

Let $S \subset \Gamma$ denote the union of Δ, T_4^* and all edges of T incident with $m_{\nu,\mu}^{(\omega)}$, $m + 1 \le \nu \le m + n$. Thus, S is a finite connected graph and $Z^*(S)$ is the union of finitely many $Z(P_i)$ and $Z(Q_j)$. If $Q_j \in S$, $Q_j \notin T_3^*$ we apply Proposition III to conclude that $Z(Q_j)$ and, therefore, each of its faces is simply connected. Moreover, $Z(Q_j)$ has exactly one face which is also a face of $Z(P_i)$ for some $P_i \in S$. The remaining faces of $Z(Q_j)$ will be called *free faces*.

Now, let g be a homeomorphism of a CW-complex V onto $Z^*(S)$ and let $p' = pg$. If U_{ijk} is a free face of $Z^*(S)$, we call $V_{ijk} = g^{-1}(U_{ijk})$ a *free face* of V. The restriction, p'_{ijk}, of p' to V_{ijk} is the universal covering projection onto U. Corresponding to each V_{ijk}, we now construct a covering $r_{ijk} : Z_{ijk} \to Z(P)$ equivalent to $r : Z(N) \to Z(P)$. According to Proposition I, $r_{ijk}^{-1}(U)$ consists of a single connected component which is homeomorphic to the universal covering of U and, therefore, to V_{ijk}. Using this homeomorphism, Z_{ijk} may be attached to V along the face V_{ijk} and the mapping p' may be extended to all points of Z_{ijk}.

Let W be the space obtained by attaching all the Z_{ijk} to V along the finite collection of free faces V_{ijk} and let q denote the extension of p' to W. It is clear that q presents W as a finite sheeted covering of Z. Thus, W is defined by a subgroup, H^*, of finite index in $G = \pi_1(Z, z_0)$. Moreover, the loops λ_ν which represent generators of $\pi_1(\tilde{Z}, \tilde{z}_0)$ are contained in $V \subset W$, so H is a subgroup of H^*. Finally, we may apply van Kampen's theorem to conclude that H^* is the free product of H and a finite set of groups in one to one correspondence with the free faces V_{ijk}. Each of these groups is conjugate in G to N. This completes our proof of the Brunner-Burns Theorem.

ACKNOWLEDGMENT

The authors would like to thank the Institute for Advanced Study, Princeton, New Jersey and its staff for their kind hospitality during the time this paper was written.

CAROL TRETKOFF
  Department of Computer Science
  Brooklyn College
  Bedford Avenue and Avenue H
  Brooklyn, NY 11210
MARVIN TRETKOFF
  Mathematics Department
  Stevens Institute of Technology
  Castle Point Station
  Hoboken, NJ 07030

REFERENCES

[1] A. M. Brunner and R. G. Burns, *Groups in which every finitely generated subgroup is almost a free factor*, Can. J. Math., *XXXI*, **6** (1979), 1329–1338, *Corrigenda*, Can. J. Math. *XXXII*, **3**, (1980), 766.

[2] R. G. Burns, *On finitely generated subgroups of free products*, J. Austral. Math. Soc.,**12** (1971), 358–364.

[3] M. Hall, Jr., *Coset representations in free groups*, Trans. Amer. Math. Soc., **67** (1949), 421–432.

[4] B. Huppert, *Endliche Gruppen I*, Springer-Verlag, 1967.

[5] A. Karrass and D. Solitar, *The subgroups of a free product of two groups with an amalgamated subgroup*, Trans. Amer. Math. Soc., **150** (1970), 227–255.

[6] A. Karrass and D. Solitar, The free product of two groups with a malnormal amalgamated subgroup, Can. J. Math. **23** (1971), 933–959.

[7] W. S. Massey, *Algebraic Topology, An Introduction*, Springer-Verlag, New York, 1977.

[8] M. Tretkoff, *Covering space proofs in combinatorial group theory*, Communications in Algebra,**3(5)**, (1975), 429–457.

[9] M. Tretkoff, *A Topological approach to the theory of groups acting on trees*, Journal of Pure and Applied Algebra, **16**, (1980), 323–333.

KNOT THEORY

Contemporary Mathematics
Volume **44**, 1985

THE MURASUGI SUM IS A NATURAL GEOMETRIC OPERATION II

DAVID GABAI

The Murasugi sum or generalized plumbing is an operation which under the appropriate circumstances associates to two oriented surfaces R_1 and R_2 in S^3 a new oriented surface R in S^3. In [G-1] we showed that if R_1 and R_2 possess certain geometric properties, then R also possesses these properties. We gave a counterexample to the converse of one of those properties. In this paper, we give positive proofs to the converse of the other results of [G-1].

THEOREM. *If R is a Murasagi sum of R_1 and R_2 and R is a surface of minimal genus for the oriented link $L = \partial R$, then for $i = 1, 2$ R_i is a surface of minimal genus for the oriented link $L_i = \partial R_i$.*

Seifert observed this result for the case of connected sums.

THEOREM. *If R is a Murasagi sum of R_1 and R_2 and $L = \partial R$ is a fibred link with fibre R, then for $i = 1, 2$ $L_i = \partial R_i$ is a fibred link with fibre R_i.*

This result was proven algebraically in [G-1]. We give a geometric proof here. We then observe how to compute the monodromy of L from that of L_1 and L_2, an observation made independently by Morton and Melvin [M]. All these results follow from the main result which can be roughly stated as follows.

THEOREM. *Nice codimension one foliations on $S^3 - \overset{\circ}{N}(L)$ restrict to nice codimension one foliations on $S^3 - \overset{\circ}{N}(L_i)$ $i = 1, 2$.*

Nice means transversely oriented, a prescribed surface (e.g., R, R_1, or R_2) is a leaf, no Reeb components and perhaps every leaf is compact, or the space of leaves is especially simple, or the foliation is smooth.

§0

Definition. The oriented surface $R \subset S^3$ is a *Murasugi sum* of oriented surfaces R_1 and R_2 in S^3 if

1) $R = R_1 \cup_D R_2$ $D = 2n$ gon
2) $R_1 \subset B_1, R_2 \subset B_2$ where $B_1 \cap B_2 = S$ a 2-sphere, $B_1 \cup B_2 = S^3$ and $R_1 \cap S = R_2 \cap S = D$.

We refer the reader to [G-1] for a picture of a Murasugi sum.

Notation. If S is a subspace of M then $N(S)$ denotes a product neighborhood of S in M. $\overset{\circ}{E}$ denotes the interior of E.

§1

The main result of this paper is Theorem 1.1. In this section, we use 1.1 to help prove Theorems 1.2 and 1.3. The proof of 1.1 is given in §2.

THEOREM 1.1. *Let R_i $i = 1, 2$ be oriented surfaces in S^3 and $L_i = \partial R_i$. Let R be a Murasugi sum of R_1 and R_2 with $\partial R = L$. Suppose that there exists a transversely oriented codimension one foliation $\mathcal{F}$ defined on $S^3 - \overset{\circ}{N}(L)$ such that R is a leaf of $\mathcal{F}$, $\mathcal{F}$ is transverse to $\partial(N(L))$ and $\mathcal{F}$ and $\mathcal{F}|\partial(N(L))$ have no Reeb components, then for $i = 1, 2$ there exists a codimension one transversely oriented foliation $\mathcal{F}_i$ defined on $S^3 - \overset{\circ}{N}(L_i)$ such that $\mathcal{F}_i$ is transverse to $\partial(N(L_i))$ and $\mathcal{F}$ and $\mathcal{F}|\partial(N(L_i))$ have no Reeb components and R_i is a leaf of F_i. Furthermore*

A) If the map

$$p: S^3 - (\overset{\circ}{N}(L) \cup R) \to \text{Space of leaves of} \mathcal{F}|S^3 - (\overset{\circ}{N}(L) \cup R)$$

is a fibration over S^1, then $\mathcal{F}_i$ $i = 1, 2$ can be constructed (if $R_i \neq D^2$) so that the map

$$p_i: S^3 - (\overset{\circ}{N}(L_i) \cup R_i) \to \text{Space of leaves of} \mathcal{F}_i|S^3 - (\overset{\circ}{N}(L_i) \cup R_i)$$

is a fibration over S^1. p, p_1, p_2 are the quotient maps which contract leaves to points.

B) If each leaf of $\mathcal{F}$ is compact, then $\mathcal{F}_i$ $i = 1, 2$ can be constructed so that each leaf is compact.

C) If $\mathcal{F}$ is C^∞, then each $\mathcal{F}_i$ is C^∞.

THEOREM 1.2. *Let R_1 and R_2 be oriented surfaces in S^3. Let R be a Murasugi sum of R_1 and R_2, $L = \partial R$, $L_i = \partial R_i$. If R is a minimal genus surface for L, then for $i = 1, 2$ R_i is a minimal genus surface for L_i.*

PROOF. By [G-2] there exists a C^∞ transversely oriented foliation $\mathcal{F}$ of $S^3 - \overset{\circ}{N}(L)$ such that $\mathcal{F}$ is transverse to $\partial N(L)$, $\mathcal{F}$ and $\mathcal{F}|\partial N(L)$ have no Reeb components and R is a leaf of $\mathcal{F}$. By Theorem 1.1 there exists C^∞ transversely oriented foliations $\mathcal{F}_i$ $i = 1, 2$ of $S^3 - \overset{\circ}{N}(L_i)$ such that $\mathcal{F}_i$ is transverse to $\partial N(L_i)$, $\mathcal{F}_i$ and $\mathcal{F}_i|\partial N(L_i)$ have no Reeb components and R_i is a leaf. By [T-2], R_i is a minimal genus surface for L_i. □

Remark. The converse of 1.2 was proven in [G-1]. These results generalize Seifert's 1934 result that if S is a connected sum of S_1 and S_2, then each S_i is a minimal genus surface if and only if S is a minimal genus surface.

THEOREM 1.3. *Let R_1 and R_2 be oriented surfaces in S^3. Let R be a Murasugi sum of R_1 and R_2, $L = \partial R$ and $L_i = \partial R_i$. If L is a fibred link with fibre R, then for $i = 1, 2$ L_i is a fibred link with fibre R_i.*

PROOF. This is exactly part B of Theorem 1.1. □

Remark. In [G-1] we proved this result algebraically and gave a geometric proof of the converse. The converse was proven algebraically by Stallings [S] in 1975.

If L is a fibred link whose fibre R is a Murasugi sum of R_1 and R_2, then it follows from the proof of Theorem 1.1 or equivalently from the construction of Theorem 4 of [G-1] that if we know what the fibration of L_1 and L_2 look like, then we know exactly what the fibration of L looks like. In particular, we can read the monodromy of L from that of L_1 and L_2.

COROLLARY 1.4. *Suppose that R is a Murasugi sum of R_1 and R_2, $\partial R_i = L_i$, $\partial R = L$, and L_i is a fibred link with monodromy $f_i : R_i \to R_i$ such that $f_i|\partial R_i = $ id. Then L is a fibred link with fibre R and monodromy $f = f'_2 \circ f'_1 : R \to R$ where $f'_i|R_i = f_i$ $f'_i|R - R_i = $ id.*

Remark. The monodromy is calculated by "pushing" out from the $+$ side to the $-$ side. We assume that the $+$ side of R faces B_1. If it faced B_2 then $f = f'_1 \circ f'_2$.

PROOF. If J is a fibred link with fibre T and $\vec{X}$ is a non-singular vector field transverse to the fibres and pointing out from the $+$ side of T then $\vec{X}$ determines a homeomorphism $g : T \to T$ homotopic to any representative of the monodromy. Now choose non-singular vector fields $\vec{X_i}$ $i = 1, 2$ on $S^3 - \overset{\circ}{N}(L_i)$ transverse to the fibres whose return map gives f_i. Using the construction of Theorem 4 of [G-1], we get a vector field $\vec{X}$ on $S^3 - \overset{\circ}{N}(K)$ transverse to the fibres.

$\vec{X}$ satisfies the following properties. For $i, j \in \{1, 2\}$, $i \neq j$, $S^3 - (\overset{\circ}{N}(K) \cup W_i)$ is homeomorphic to $S^3 - (\overset{\circ}{N}(L_j) \cup V_j)$ where V_j is a very small neighborhood of $R_j \cap B_1 \cap B_2$ and W_i is a very small neighborhood of B_i. Under this identification one constructs $\vec{X}$ so that $\vec{X}|S^3 - (\overset{\circ}{N}(L) \cup W_i) = \vec{X_j}|S^3 - (\overset{\circ}{N}(L_j) \cup V_j)$. The singular foliation on $N(B_1 \cap B_2) - \overset{\circ}{N}(L) = W$ induced from the fibration on $S^3 - \overset{\circ}{N}(L)$ is pictured in Figure 4 of [G-1]. This foliation has two tangencies of index $-n$ with $B_1 \cap B_2$, if R was obtained by summing along a $2n+2$ - gon. We construct $\vec{X}$ so that $\vec{X}|W$ is close to the "canonical" transverse vector field. $\vec{X}$ points into B_1 at the tangency of R and $B_1 \cap B_2$ and points into B_2 at the other tangency. Since $W \cup W_1 \cup W_2 = S^3 - \overset{\circ}{N}(L)$ we have completely described $\vec{X}$. By following the trajectories of $\vec{X}$ the result follows. □

Remark. If $x \notin f_1^{-1}(N(D) \cap R)$, then the positive trajectory through x does not leave B_1 until it crosses some point of $f_1^{-1}(N(D) \cap R)$.

If $x \notin N(D) \cap R$, then the positive trajectory through x does not leave B_2 until it crosses some point of $N(D) \cap R$.

This corollary has been independently obtained by Morton and Melvin [M].

§2

Proof of Theorem 1.1. Let $N = (S^3 - \overset{\circ}{N}(L)) - \overset{\circ}{N}(R)$. N is a manifold with corners, i.e., $\partial N = R^+ \cup R^- \cup A$ where R^+ corresponds to the $+$ side of R and

A is the set of annuli $\partial N(L) - \overset{\circ}{N}(R)$. Similarly let $N_i = (S^3 - \overset{\circ}{N}(L_i)) - \overset{\circ}{N}(R_i)$ with $\partial N_i = R_i^+ \cup R_i^- \cup A_i$ $i = 1, 2$. Assume that R is oriented so that the $+$ side of D points into the ball containing R_1 where D is the disc along which R_1 and R_2 were summed. Let $E = (S^2 - D) \cap N$ where S^2 is the sphere separating R_1 and R_2. Let B_i be the ball bounded by S^2 containing R_i.

Key observation (Figure 1). If $P_i = (N - \overset{\circ}{N}(E)) \cap B_i$ then after a small isotopy

$$N_1 = P_1$$
$$R_1^- = R^- \cap P_1$$
$$R_1^+ = R^+ \cap P_1 \cup (N(E) \cap P_1 - \overset{\circ}{N}(E \cap R^-))$$

and

$$N_2 = P_2$$
$$R_2^+ = R^+ \cap P_2$$
$$R_2^- = R^- \cap P_2 \cup (N(E) \cap P_2 - \overset{\circ}{N}(E \cap R^+)).$$

Now $\mathcal{F}$ induces a codimension 1 transversely oriented foliation $\mathcal{G}$ on N so that $\mathcal{G}$ is transverse to A, tangent to $R^+ \cup R^-$, $\mathcal{G}$ and $\mathcal{G}|A$ have no Reeb components and $\mathcal{F}$ is obtained from $\mathcal{G}$ by gluing R^+ to R^-. By construction E is transverse to $\mathcal{G}$ in a neighborhood of ∂E. By Thurston [T-1] or Rousserie [R] and the Poincaré Hopf index formula one can isotope E so that E is transverse to $\mathcal{G}$ except along a finite number of saddle tangencies whose indices add up to $n - 1$, if E is a $2n$ gon . Here we use the fact that $\mathcal{F}$ has no Reeb components. We will now construct the desired foliation on $S^3 - \overset{\circ}{N}(L_1)$. The other case is similar. Define $\mathcal{G}_1 = \mathcal{G}|N_1$. $\mathcal{G}_1$ is a singular foliation. Our goal is to heal the scar.

LEMMA. *If x is a point of E tangent to $\mathcal{G}_1$, then the normal to $\mathcal{G}_1$ at x points out of N_1.*

PROOF. It follows from the Poincaré Hopf formula that if M is a 3-manifold and $\vec{X}$ is a non-singular vector field pointing normal to ∂M, then $\chi(\partial_+) = \chi(\partial_-)$

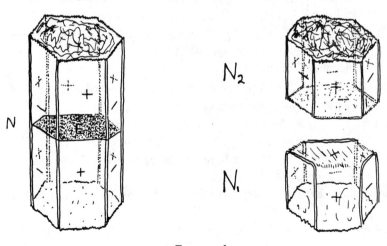

FIGURE 1

where $\partial_+(\partial_-)$ denotes the union of those ∂ components where $\vec{X}$ points out (in). More generally, if $\vec{X}$ is non-singular and is tangent to ∂M along a union of annuli, pointing out along T^+ and in along T^-, then $\chi(T^+) = \chi(T^-)$. By perturbing the normal vector field to $\mathcal{G}_1$ we can find a vector field $\vec{X}$ as above so that $(R^+ \cap N_1) \subset T^+$ and $(R^- \cap N_1) \subset T^-$. The result now follows by observing that

$$\chi(R^- \cap N) = \chi(R_1^-) = 1/2\chi(\partial N_1)$$

and

$$\chi(T^-) < \chi(R^- \cap N_1)$$

if some normal pointed in at a tangency of E with $\mathcal{G}_1$.   □

We now prove the theorem by induction on the number of tangencies of E with $\mathcal{G}_1$.

Case 1. The number of tangencies equals zero.

PROOF. Obtain $\mathcal{F}_1$ by gluing R_1^+ to R_1^-. $\mathcal{F}_1$ is the desired foliation except in the case when $\mathcal{F}$ satisfies hypothesis A and every leaf of $\mathcal{F}_1$ is compact. In that case if $R_1 \neq D^2$, then we can perturb the foliation slightly (see [G-1]) to get the correct foliation.   □

Case 2. The number of tangencies equals one.

PROOF. Figure 2 shows how to obtain a foliation $\mathcal{G}_2$ from $\mathcal{G}_1$ tangent to ∂N_1 along $R_1^+ \cup R_1^-$ and transverse to ∂N_1 along A_1. The desired topological foliation on $S^3 - \overset{\circ}{N}(L_1)$ is obtained by gluing R_1^+ to R_1^-.   □

If hypothesis A of 1.1 holds, then there is a map

$$p : N_1 - (R^+ \cup R^-) \to S^1$$

which is constant on leaves and an immersion on transverse arcs. If such an arc has an endpoint on R^+ or R^-, then this immersion is of infinite degree. There is thus a unique way to glue to obtain $\mathcal{G}_1$ and a map p_1 which extends p; hence the quotient

$$p_1' : N_1 - (R_1^+ \cup R_1^-) \to \text{ Space of leaves of } \mathcal{G}_2 | N_1 - (R_1^+ \cup R_1^-)$$

is a fibration over S^1 or $(0,1)$. If the former holds, then the desired foliation is obtained by gluing R_1^+ to R_1^-. If the latter holds, then the foliation obtained by gluing R_1^+ to R_1^- is a foliation by compact leaves. As before, we perturb this fibration slightly to get the correct foliation.

If hypothesis B holds, then there is a map

$$p : N_1 \to I$$

which is constant on leaves of $\mathcal{G}_1$ and an immersion on transverse arcs. There is a unique way to obtain $\mathcal{G}_2$ and a map which extends p which is constant on leaves. The desired foliation is obtained by gluing R_1^+ to R_1^-.

If $\mathcal{G}$ is C^∞, then there are many ways to perform the gluing to obtain a smooth foliation $\mathcal{G}_2$ on N_1. The foliation obtained by gluing R_1^+ to R_1^- is not necessarily smooth; however, a smooth one can be obtained as follows.

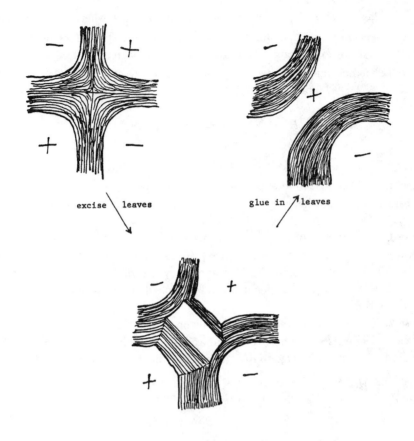

FIGURE 2

Let $g_1, \ldots, g_n$ be a set of pairwise disjoint (except for the basepoint) simple closed curve generators of $\pi_1(R_1^+)$.   For each i there exists a map $f_i : [-1, 0] \to [-1, 0]$ such that $f_i = id$ in a neighborhood of -1 and $h_i : [-1, a) \to [-1, b)$ is smooth where $h_i|[-1, 0] = f_i$ and $h_1|[0, a)$ is a germ of the holonomy map corresponding to g_i on R_1^+. Since $\partial S \neq \varnothing$ we can easily construct a foliation $\mathcal{H}$ on $R_1^+ \times [-1, 0]$ so that the leaves are transverse to $t \times [-1, 0]$, $t \in R_1^+$ and for each i the holonomy map corresponding to g_i is given by f_i. Extend $\mathcal{H}$ and $\mathcal{G}_2$ to the manifold $N_1' = N_1 \cup R_1^+ \times [-1, 0]$ with $R_1^+ \times 0$ identified with R_1^+. By repeating this procedure we obtain a smooth foliation on a manifold with corners homeomorphic to N_1 which has the product foliation on neighborhoods of R_1^+ and R_1^-. The desired foliation is obtained by gluing R_1^+ to R_1^-.

These foliations have no 3-dimensional Reeb components, since the above procedures do not create leaves with empty boundary. There are no 2-dimensional Reeb components, since every boundary leaf intersects a transverse closed curve. $\square$

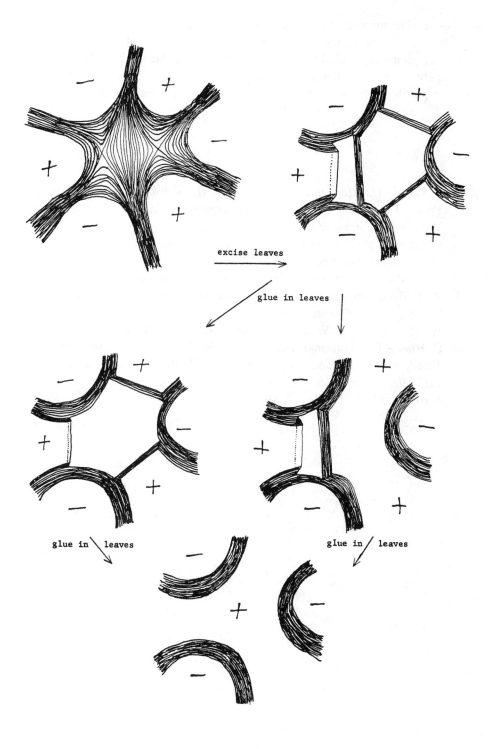

FIGURE 3

Case 3. The number of tangencies is $k > 1$.

PROOF. The proof follows exactly as in case two except that we need to perform k gluings instead of one to obtain the non-singular foliation $\mathcal{G}_2$. This can be done by successively healing outermost singularities. In figure 3 we illustrate the two possible pathways for the case of two tangencies each of index -1. □

BIBLIOGRAPHY

[G-1] D. Gabai, *The Murasugi Sum is a Natural Geometric Operation*, to appear, *Proc. of Conf. on Low Dimensional Topology*, AMS New Contemporary Mathematics Series.

[G-2] D. Gabai, *Foliations and the Topology of 3-Manifolds*, to appear, *Journal of Differential Geometry*, 1983.

[M] H. R. Morton, *Fibred Knots with a given Alexander polynomial*, preprint.

[T-1] W. P. Thurston, *Foliations of Manifolds which are Circle Bundles*, Thesis, University of California, Berkeley, 1972.

[T-2] W. P. Thurston, *A Norm on the Homology of 3-Manifolds*, preprint.

[R] R. Rousserie, *Plongements dan les Variétés Feiulletees et Classification de Feuilletages sans Holonomie*, IHES **43**, (1973), 101–142.

[S] J. Stallings, *Constructions of Fibred Knots and Links*, Proc. Symp. Pure Math., AMS **27**, (1975), 315–319.

INSTITUTE FOR ADVANCED STUDY, PRINCETON

Contemporary Mathematics
Volume 44, 1985

THE ARF INVARIANT OF CLASSICAL KNOTS

Louis H. Kauffman

1. Introduction

This paper discusses a geometrical interpretation of the Arf invariant of a classical knot. By a classical knot, I mean an embedding of a circle in the (oriented) three dimensional sphere S^3. Such an embedding is assumed to be tame—hence this paper can be read in either the differentiable or piecewise linear categories. Two knots are said to be *equivalent* if they represent ambient isotopic embeddings. The primary difference between classical and higher dimensional knots lies in the classical susceptibility to diagrammatic representation. Thus classical knot theory can be regarded as a purely combinatorial subject where knots are represented by knot diagrams such as that shown for the trefoil knot in Figure 1.

It is assumed that the reader is familiar with this representation, and with the theorem that ambient isotopic knots have diagrams related by a sequence of Reidemeister moves (see [9]).

The Arf invariant of a knot is usually defined as the Arf invariant of an associated $\mod -2$ quadratic form (See [10]). As such, it assigns 0 or 1 to each knot, separating them into two classes. It is known that the Arf invariant is an invariant of concordance. This means that it must vanish on slice knots and hence on ribbon knots. (It is unknown whether every slice knot is ribbon. Here slice means differentiably slice—that the knot bounds a differentiably locally flat disk properly embedded in the four ball D^4.)

A ribbon knot is a knot that bounds an immersed disk Δ in S^3 having only ribbon singularities. A typical ribbon singularity is illustrated in Figure 2. This geometry will be discussed further in section 2.

We invert the historical road to the Arf invariant by giving a combinatorial move Γ with the following properties:

1. Ribbon diagrams can be reduced to unknotted circles by finitely many applications of Γ and the Reidemeister moves.
2. Any knot diagram can be transformed via Γ and ambient isotopy (Reidemeister moves) to either the trefoil diagram or to the unknot. The trefoil and the unknot are not transforms of each other.

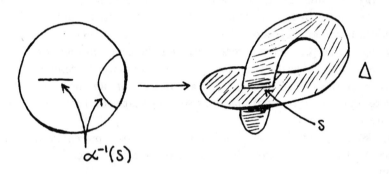

Trefoil Diagram

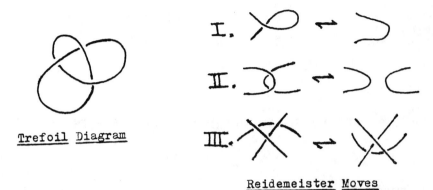

Reidemeister Moves

FIGURE 1

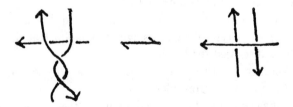

$\alpha^{-1}(s)$

FIGURE 2

We show that the Γ move generates an equivalence relation identical to that produced by the Arf invariant. The move itself is a strand—switch of the form

The details are in section 2.

It might seem that this begins a fruitful line of approach to new invariants vanishing on ribbon knots: Look for combinatorial moves that make ribbons fall apart! I have, however, been unable to produce any moves of this sort that generate relations different from either the trivial relation or the Arf invariant.

Nevertheless, we do now have a simple geometric form for the Arf invariant, and it may pave the way toward "higher-order" combinatorial invariants of concordance.

The paper is organized as follows: Section 2 explains the details of the Γ-move and an equivalent formulation denoted pass-equivalence. Section 3 uses the Conway polynomial (See [2], [6]) to prove that the trefoil and the unknot are not pass-equivalent. This proof is strictly elementary and it leads to an algorithm for computing the pass-class from the mod -2 reduction of a sum of linking numbers associated with the knot diagram (Proposition 3.4). Section 4 recalls the definition of the Arf invariant and proves the agreement between pass-equivalence and the Arf invariant. Section 5 states a conjecture about Arf invariants of curves embedded in the spaning surface of a slice knot. We use the techniques of this paper to illustrate how this conjecture could be used, if it is true, to prove that certain algebraically slice knots are not slice.

Some of the material in this paper may be found in the author's monograph [7]. I would like to thank William Browder for introducing me to the Arf invariant (albeit in different guise). The notion of Γ equivalence is due to my students Steve Winker, Bob Brandt and Thaddeus Olzyk.

2. Pass Equivalence

Recall that a *ribbon knot* $K \subset S^3$ is a knot that bounds an immersed disk $\Delta \subset S^3$ having only ribbon singularities. A ribbon singularity is a transverse

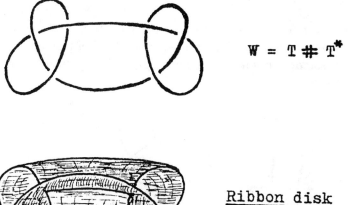

$$W = T \mathbin{\#} T^*$$

Ribbon disk
bounding W.

FIGURE 3

self intersection (of the disc Δ) consisting of an arc s. When the immersion is represented by a map $\alpha : D^2 \to S^3$ ($\alpha(D^2) = \Delta$) then $\alpha^{-1}(s)$ is a disjoint union of two arcs on D^2, one contained in the interior, the other touching the boundary transversely at its endpoints. Each arc is embedded in the pre-image. Figure 2 illustrates a typical ribbon singularity. In Figure 3, we depict a ribbon disc for the connected sum of a trefoil T with its mirror image T^*.

The connected sum, $K \# K^*$, of a knot with its mirror image is always ribbon. The proof can be divined by contemplating Figure 3. Note that diagrammatically the mirror image is obtained by reversing all crossings.

A knot is said to be *slice* if it bounds a smooth disk embedded in the four-ball D^4. Ribbon knots are slice since the singularities of the immersed spanning disk can be removed through the fourth dimension. It is not known whether every slice knot is ribbon.

Here is a scheme for detecting non-ribbon knots:

1. Devise a procedure that unknots ribbons.
2. Show that this procedure does not unknot every knot.

The simplest candidate for such a procedure is a move of type

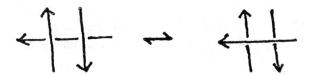

Such a move can be used to remove ribbon singularities, and hence it can unknot ribbon knots. Unfortunately, repeated applications of this move combined with ambient isotopy can unknot any knot (exercise!). We modify accordingly to form the move:

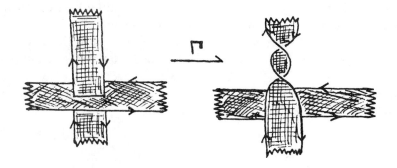

Removing a ribbon singularity via Γ.

FIGURE 4

Γ switches a single strand with respect to oppositely oriented parallel strands, and it places a 360° twist in the parallel strands. For precision, we let the direction of the single strand determine the sense of the 360° twist. Thus

We now have that Γ is well-defined and that two applications of the move returns the original configuration:

Here $\sim$ denotes ambient isotopy.

Definition 2.1. Call two knots $\underline{\Gamma}\text{-equivalent}$ (denoted $\sim$) if a diagram for one knot can be obtained from a diagram for the other knot by a finite combination of Δ-moves and ambient isotopies.

PROPOSITION 2.2.. *If K is a ribbon knot, then K is Γ-equivalent to the unknot* $(K \underset{\Gamma}{\sim} 0)$.

PROOF. Apply the procedure indicated in Figure 4 until a knot is obtained
that bounds a disk without singularities. This is the unknot.

In order to show that there are knots outside the Γ-class of the unknot, it
is convenient to compare Γ-equivalence with a second relation that I call pass-
equivalence.

Definition 2.3. A *pass-move* on a knot diagram is a move of one of the following
two forms:

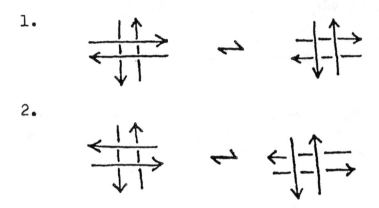

In any pass-move there are two pairs of parallel strands and each pair consists
in two oppositely oriented lines.

Two knots are *pass-equivalent* if there is a finite sequence of pass-moves and
ambient isotopies leading from K to K'. Pass-equivalence of K and K' is denoted
by $K \underset{p}{\sim} K'$.

PROPOSITION 2.4.. *Two knots K and K' are pass-equivalent if and only if they
are Γ-equivalent.*

PROOF. We first show that a Γ-move can be accomplished by pass-equivalence:

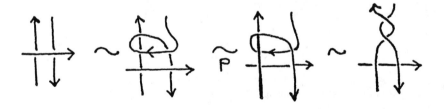

Then we show that a pass-move can be accomplished by Γ-equivalence:

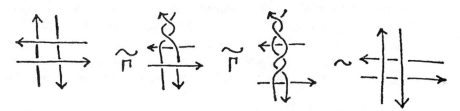

This completes the proof.

PROPOSITION 2.5. *Any knot K is pass-equivalent either to the trefoil knot $T = \bigcirc\!\!\!\!\!\!\bigcirc$ or to the unknot $U = \bigcirc$.*

PROOF. Any knot $K \subset S^3$ spans an orientable surface $F(\subset S^3)$. This surface can be represented as a standardly embedded disk with attached bands that are knotted twisted and linked. The edges of any band are oppositely oriented, and hence we may *pass bands* according to the schema

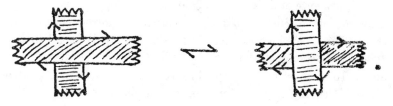

while preserving the pass-class of the surface boundary. By passing bands in this fashion, we can reduce the surface to a connected sum of surfaces of the forms

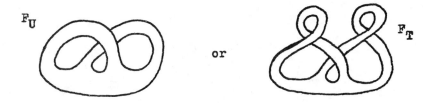

or

The bands of these surfaces are not twisted or linked with another, these complexities having been removed through the following processes:

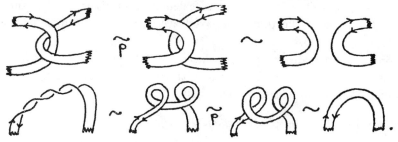

The surface F_T has boundary the trefoil, and F_U has boundary the unknot. This proves that K is pass-equivalent to a possibly empty sum of trefoils.

We also see that any K is pass-equivalent to its mirror image K^*. (Represent K as the boundary of a disk with attached bands and pass every band.) Since $K \# K^*$ is ribbon (compare Figure 3), we have

$K \# K \underset{p}{\sim} K \# K^* \underset{p}{\sim} 0$. Hence $K \# K^* \underset{p}{\sim} 0$ (using Proposition 2.4). Therefore, any connected sum of trefoils reduces to either one trefoil or to the unknot. This completes the proof.

Proof that T is not pass-equivalent to U is deferred to the next section. Given this fact, we define the A-invariant, $A(K)$, of a knot K. This invariant will be identified with the Arf invariant in section 4.

Definition 2.6.
$$A(K) = 0 \text{ if } K \underset{p}{\sim} U$$
$$A(K) = 1 \text{ if } K \underset{p}{\sim} T$$

where U and T are the unknot and the trefoil, respectively. We regard the values 0 and 1 as members of the field of two elements Z_2. Thus $1 + 1 = 0$.

3. Pass Equivalence and Conway Polynomial.

We continue our discussion of the A-invariant by introducing the *Conway polynomial*. This is a refined version of the Alexander polynomial that may be understood purely combinatorially. See [1], [2], [6], [7].

Axiom 1. To each oriented knot or link K, there is associated a polynomial $\nabla_K(z) \in Z[z]$, the ring of polynomials in z with integer coefficients. If K and K' are ambient isotopic then $\nabla_K = \nabla_{K'}$.

Axiom 2. If K is the unknot, then $\nabla_K = 1$.

Axiom 3. If three links $K, \overline{K}$ and L are related so that they differ only at the site of one crossing, as indicated below, then $\nabla_K - \nabla_{\overline{K}} = z \nabla_L$.

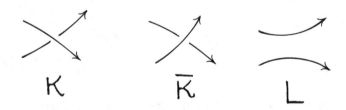

$$K \qquad\qquad \overline{K} \qquad\qquad L$$

These axioms suffice for a recursive calculation of $\nabla_K(z)$. That they are consistent is a more delicate matter. We leave discussion of consistency to the references quoted above.

In order to calculate anything, we need the following lemma. Recall that a link is *split* if it possesses a diagram with two non-empty pieces, each piece contained in one of two disjoint planar disks.

LEMMA 3.1. *If L is a split link, then $\nabla_L = 0$.*

PROOF. See [6].

Figure 5 illustrates the calculation of $\nabla_T(z)$ where T is the trefoil knot.

$$\nabla_K - \nabla_{\overline{K}} = Z\nabla_L , \ \nabla_T = \nabla_K , \ \nabla_{\overline{K}} = \nabla_U = 0 .$$
$$\nabla_L - \nabla_{\overline{L}} = Z\nabla_U , \ \nabla_{\overline{L}} = 0 \ \text{(split)}$$
therefore $\nabla_L = Z$
therefore $\nabla_T = 1 + Z^2 .$

In general we may write $\nabla_K(z) = a_0(K) + a_1(K)z + a_2(K)z^2 + \cdots$ where $a_n(K) \in Z$ is a topological invariant of K for each $n = 0, 1, 2, 3, \ldots$.

Let the crossings of K be indexed by the variable i, and S_iK denote the result of switching the ith crossing, E_iK the result of eliminating the ith crossing. Let $e_i(K)$ denote the sign of the ith crossing, as shown in Figure 6. We then have the more explicit version of axiom 3:

$$\nabla_K - \nabla_{S_iK} = e_i(K)z\nabla_{E_iK}.$$

and the coefficient equations:

$$a_{n+1}(K) - a_{n+1}(S_iK) = e_i(K)a_n(E_iK)$$
$$a_0(K) - a_0(S_iK) = 0.$$

LEMMA 3.2.
a) $a_0(K) = \begin{cases} 1 & \text{if } K \text{ has one component.} \\ 0 & \text{otherwise} . \end{cases}$

b) $a_1(K) = \begin{cases} Lk(K) & \text{if } K \text{ has two components.} \\ 0 & \text{otherwise.} \end{cases}$

Here $Lk(K)$ denotes the linking number of a two-component link K.

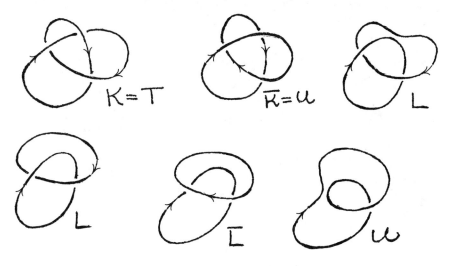

FIGURE 5

PROOF. Since $a_0(K) = a_0(S_iK)$ we see that, for a knot K, $a_0(K)$ has the same value as $a_0(U)$ (The unknot can be obtained by switching a subset of crossings of K.). Hence $a_0(K) = a_0(U) = 1$ for a knot K. If K has more than one component, then K can be changed to a split link by switching crossings. Therefore, $a_0(K) = a_0$ (split-link) $= 0$. This completes the proof of a).

To prove b) recall the definition of linking number

$$Lk(K) = (1/2)\sum_{i \in I} e_i(K)$$

where I is the set of crossings of strands of different components of a two-component link K. It follows immediately that $Lk(K) - Lk(S_iK) = e_i(K)a_0(K)$. Thus Lk and a_1 satisfy identical formulas, and these formulas suffice for calculation (recursively) in terms of values of Lk and a_1 on split links, and values of a_0 on unknots and split links. Thus $a_1(K) = Lk(K)$ when K is a two-component link.

If K does not have two components, then E_iK will never have one component. Hence $a_1(K) - a_1(S_iK) = e_i(K)a_0(E_iK) = 0$. This implies that $a_1(K) = 0$, and completes the proof of Lemma 3.2.

THEOREM 3.3. *If the knots K and K' are pass-equivalent then $a_2(K) \equiv a_2(K')$* *(modulo 2).*

Hence the trefoil and the unknot are not pass-equivalent. The A-invariant is the mod-2 reduction of the second coefficient of the Conway polynomial.

PROOF. It suffices to show that $a_2(K)$ and $a_2(K')$ have the same parity when K and K' differ by a single pass-move. Suppose the pass has the form below with crossings labelled $1, 2, 3, 4$.

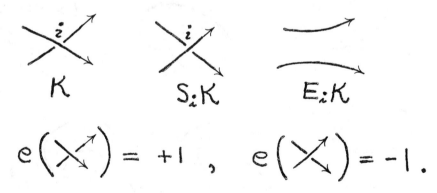

FIGURE 6

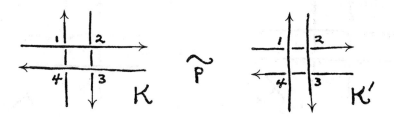

Then $K' = S_4 S_3 S_2 S_1 K$. Let $K_1 = S_1 K$, $K_2 = S_2 K_1$, $K_3 = S_3 K_2$, $K_4 = S_4 K_3$ and let $X_1 = E_1 K$, $X_2 = E_2 K_1$, $X_3 = E_3 K_2$, $X_4 = E_4 K_3$. Note that the signs of crossings $1, 2, 3, 4$ are $+1, -1, +1, -1$ respectively. Hence

$$a_2(K) - a_2(K_1) = a_1(X_1)$$
$$a_2(K) - a_2(K_2) = -a_1(X_2)$$
$$a_2(K_2) - a_2(K_3) = a_1(X_3)$$
$$a_2(K_3) - a_2(K_4) = -a_1(X_4).$$

Therefore $a_2(K) - a_2(K') = a_1(X_1) - a_1(X_2) + a_1(X_3) - a_1(X_4)$.

Hence we must show that the sum of linking numbers on the right-hand side of this equation is even.

In order to determine linking number contributions, we must determine the forms of connection of ⚎ that produce a single component (hence knot) diagram. These are listed in Figure 8. Each drawing in Figure 8 is schematic for a class of more complex diagrams with this form of connection. Figure 9 illustrates a specific diagram with connection form a). Figure 10 shows shema for $X_1, \ldots, X_4$ that correspond to form A of Figure 8. In Figure 10, crossings between different components are circled. Note that each crossing appears twice among four diagrams. From this it is easy to see that $a_2(K)$ and $a_2(K')$ have the same parity for forms of type A (of Figure 8). Similar analysis applies to the remaining forms in Figure 8, and to the other type of pass-move. This completes the proof of Theorem 3.3.

COROLLARY 3.4. *Let K be a knot, and let $S_1, \ldots, S_n$ be a set of crossing exchanges so that $S_n \ldots S_1 K = U$ is unknotted. Let $K_1 = S_1 K$ and $K_r = S_r K_{r-1}$ for $r = 2, \ldots, n$. Let $X_1 = E_1 K$ and $X_r = E_r K_{r-1}$ for r in this same range. Then the A-invariant of K has formula $A(K) \equiv \sum_{r=1}^{n} Lk(X_r)$ (modulo 2).*

PROOF. This follows from the fact that $a_2(K)$ and $A(K)$ are congruent modulo 2.

LOUIS H. KAUFFMAN

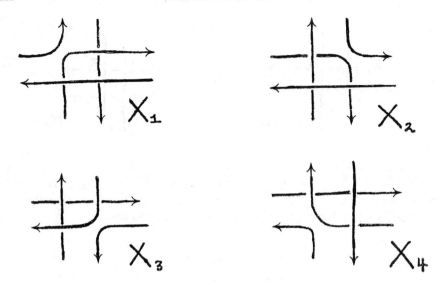

FIGURE 7

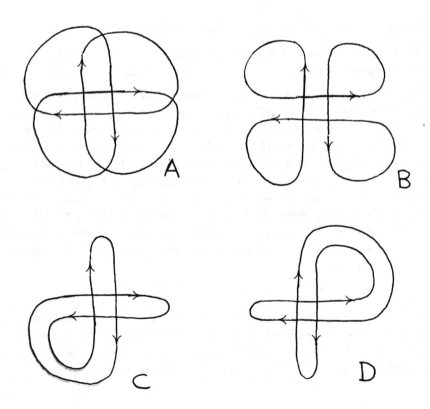

FIGURE 8

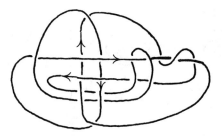

FIGURE 9

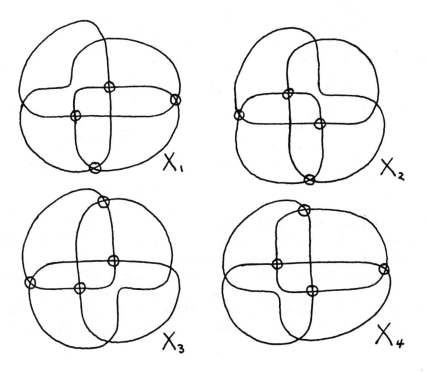

FIGURE 10

Example 1. $K = T$.

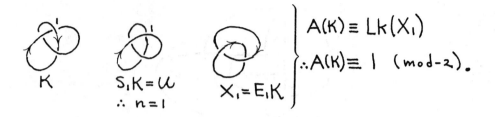

$$A(K) \equiv Lk(X_1)$$
$$\therefore A(K) \equiv 1 \pmod{2}.$$

$S_1 K = U$
$\therefore n = 1$

$X_1 = E_1 K$

Example 2.

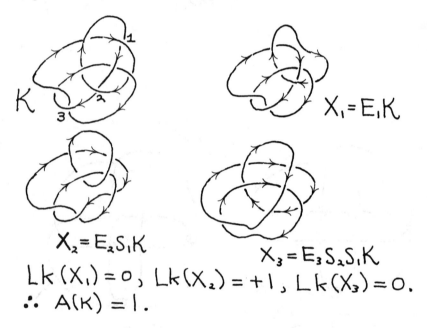

K

$X_1 = E_1 K$

$X_2 = E_2 S_1 K$

$X_3 = E_3 S_2 S_1 K$

$$Lk(X_1) = 0, \; Lk(X_2) = +1, \; Lk(X_3) = 0.$$
$$\therefore A(K) = 1.$$

4. The A-Invariant is the Arf Invariant

We now recall the definition of the Arf invariant, $ARF(K)$, of a knot K. It is the Arf invariant of a mod-2 quadratic form associated with the knot.

A mod-2 quadratic form $q : V \to Z_2$ is a function from a finite dimensional vector space V over the field Z_2 to the field Z_2. It must satisfy that $q(x+y) - q(x) - q(y) = <x, y>$ is bilinear in x and y. The Arf invariant is defined for non-degenerate forms (this means that the associated bilinear form is non-degenerate) by the formula:

$$Arf(q) = \begin{cases} 1 & \text{if} q \text{ takes the value 1 on a majority of elements of } V. \\ 0 & \text{if } q \text{ takes the value 0 on the majority of elements of } V. \end{cases}$$

A mod-2 form is associated with a knot K by way of the *Seifert Pairing* $\ominus$: $H_1(F) \times H_1(F) \to Z$. Here $F \subset S^3$ is an oriented spanning surface for the knot K. The pairing is defined by the formula $\ominus(x, y) = Lk(x^+, y)$ where Lk denotes linking number in S^3 and x^+ is the result of pushing the cycle x into the complement of F along the positive normal direction to F.

By defining $q(x) \equiv \ominus(x, x)$ (modulo-2), we obtain a mapping $q : H_1(F) \otimes Z_2 \to Z_2$. This is the quadratic form of the knot, and one defines $ARF(K) = Arf(q)$. (See [8],[10].)

A key geometric property of the Arf invariant that follows from these definitions is the

THEOREM 4.1. *Suppose that K is a knot, that $\overline{K}$ is obtained from K by switching a single crossing, and that the two-component link L is obtained by elmininating this crossing.*

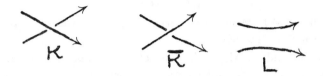

Then $ARF(K) - ARF(\overline{K}) \equiv Lk(L)$ (modulo-2).

PROOF. Omitted. See [7].

It follows at once from this result that $ARF(K)$ is the mod-2 reduction of the second degree coefficient of the Conway polynomial (denoted $a_2(K)$ in the last section). The proof is by recursive comparison as in Lemma 3.2. Consequently, the A-Invariant and the ARF-invariant are identical.

5. A Conjecture

We close with a conjecture about slice knots and Arf invariants. Let F be an oriented spanning surface for a slice knot $K \subset S^3$. Call a curve α embedded in $F(\alpha \subset F)$ a *null curve* if $\ominus([\alpha], [\alpha]) = 0$ and $[\alpha] \neq 0$ where $[\alpha]$ is the homology class of α and $\ominus$ is the Seifert pairing for this surface, as described in Section 4. Each null curve $\alpha \subset F \subset S^3$ gives an embedded curve $\alpha \subset S^3$, hence a knot in S^3. As such, each null curve has an ARF invariant.

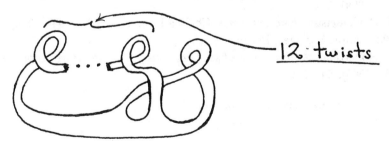

FIGURE 11

CONJECTURE.   *There exists a null curve α on any spanning surface F for a slice knot K such that $ARF(\alpha) = \underline{0}$.*

If this conjecture is true, it can be used to give direct verifications that certain algebraically slice knots are not slice. For example, let K be as shown in Figure 11. K is the boundary of the indicated surface. This is the simplest of the examples of non-slice, algebrically slice knots given by Casson and Gordon in [1.5]. We leave it as an exercise for the reader to determine the null curves on this surface (there are two) and to compute that both curves have ARF invariant equal to one.

REFERENCES

[1]   R. Ball and M.L. Mehta, *Sequence of inariants of knots and links*, (to appear in Journal de Physique).

[1.5] A Casson and C. McA. Gordon, *On slice knots in dimension three*, In *Geometric Topology*, R.J.Milgram (editor), *Proceedings of Symposia Pure Mathematics XXXII*, Amer. Math. Soc., Providence 1978, pp. 39–53.

[2]   J. H. Conway, *An enumeration of knots and links and some of their algebraic properties*, Computational Problems in Abstract Algebra, Pergamon Press, New York (1970), 329–358.

[3]   R. H. Fox, *A quick trip through knot theory*, Topology of Three-Manifolds, Prentice Hall (1962), 120–167.

[4]   R. H. Fox and J. W. Milnor, *Singularities of 2-spheres and 4-space and cobordism of knots*, Osaka J. Math., **3**, (1966), 257–267.

[5]   L. H. Kauffman and T. F. Banchoff, *Immersions and mod-2 quadratic forms*, Amer. Math. Monthly **84**, (1977), 168–185.

[6]   L. H. Kauffman, *The Conway polynomial*, Topology **20**, (1980), 101–108.

[7]   L. H. Kauffman, *Formal Knot Theory*, (to appear in Princeton Lecture Notes).

[8]   J. Levine, *Knot cobordism groups in codimension two*, Comm. Math. Helv. **44**, (1969), 119–244.

[9]   K. Reidemeister, *Knotentheorie*, Chelsea Pub. Co., N.Y. (1948), copyright 1932, Julius Springer, Berlin.

[10] R. Robertello, *An invariant of knot cobordism*, Comm. Pure Appl. Math. **18**, (1965), 543–555.

THE UNIVERSITY OF ILLINOIS AT CHICAGO

Contemporary Mathematics
Volume 44, 1985

THE UNKNOTTING NUMBER OF A CLASSICAL KNOT

W. B. Raymond Lickorish

From any presentation of a classical knot k, new knots can be created by changing some of the cross-overs from overpasses to underpasses. The unknotting number of k, $u(k)$, is the minimum number of such changes needed to create the unknot, the minimum being taken over all possible sets of changes in all possible presentations of k. This is a very intuitive invariant: Thinking of k as a simple closed loop moving about (smoothly) in 3-space, $u(k)$ is the minimum number of times k must pass through itself to become unknotted. The unknotting number is not easy to calculate; no algorithm seems to exist for such calculation. Recently, [N], Y. Nakanishi produced a survey of methods of calculation. He gave a table of unknotting numbers for the eighty-four prime knots with at most nine crossings, but this table contained twenty question marks. The purpose of this paper is to add one more method to those used by Nakanishi. He used the inequality $|\sigma(k)| \leq 2u(k)$ of K. Murasugi [M], where $\sigma(k)$ is the signature of k, and the inequality (related to [W]) $m(k) \leq u(k)$, where $m(k)$ is the minimal size of a presentation matrix for the Alexander module of k. An upper bound for $u(k)$ is, in any simple case, best determined by intelligent experiment. The technique explained here has been developed jointly with J. R. Rickard who has extended and deepened it (see [R]). The method consists of an examination of the linking form on the first homology of the double cover of S^3 branched over the knot.

Standard notations will be used. In particular, M_k will denote the double cover of S^3 branched over a knot k. The surgery notation of Rolfsen will be used, so that the 3-manifold obtained by p/q-surgery on a knot k is created by removing a solid torus neighbourhood of k and re-inserting it so that the curve representing p meridians and q longitudes of k becomes nulhomotopic in the new manifold. Thus the lens space $L_{p,q}$ is obtained by p/q-surgery on the unknot. The next result is essentially well known ([Mo],[B] give a full discussion) but as it is crucial to what follows a proof is included for completeness.

LEMMA 1. *If k has unknotting number equal to one, then M_k is obtained by $n/2$-surgery on some knot in S^3, n being an odd integer.*

PROOF. Let A be the annulus $\{re^{i\theta} : 1 \le r \le 2\} \subset \mathbb{C}$, and let $\tau : A \to A$ be the twisting homeomorphism defined by

$$\tau(re^{i\theta}) = re^{i(\theta + 2\pi r)}.$$

Let ρ be the rotation about the real axis of the solid torus $T = A \times [-1, 1] \subset \mathbb{C} \times \mathbb{R}$ given by $\rho(re^{i\theta}, t) = (re^{-i\theta}, -t)$. Define a homeomorphism $\overline{h}$ from the boundary of T to itself by

$$\overline{h}(re^{i\theta}, 1) = (\tau re^{i\theta}, 1)$$
$$\overline{h}(re^{i\theta}, -1) = (\tau^{-1} re^{i\theta}, -1)$$

$\overline{h}$ being the identity on the remainder of ∂T. This $\overline{h}$ commutes with $\rho|\partial T$ and so induces a homeomorphism on the quotient space $h : \partial T/\rho \to \partial T/\rho$.

Now, the quotient T/ρ is a 3-ball and the projection $p : T \to T/\rho$ is a double covering branched over two standard (straight) arcs spanning T/ρ. The pair $(T/\rho,$ two arcs) may be identified with the pair $(D^2,$ two points$) \times I$, with the annulus $(A \times \{-1, 1\})/\rho$ corresponding to $\partial D^2 \times I$. (This can be imagined by considering the top half of T, namely $A \times [0, 1]$, and folding $A \times \{0\}$ along the real axis, glueing $re^{i\theta}$ to $re^{-i\theta}$; the real axis becomes the two arcs.) Under the identification $h : \partial T/\rho \to \partial T/\rho$ is the identity on $D^2 \times \partial I$ and a copy of τ on $\partial D^2 \times I$. Although $\overline{h}$ does not extend over the whole of T, h extends over T/ρ to give a homeomorphism which, when regarded as operating on $D^2 \times I$ maps the (two points) $\times I$ to a pair of arcs containing a complete twist as shown in Figure 2.

Suppose now that k is a knot in S^3, and that a copy of $(D^2,$ two points$) \times I$ is removed from the pair (S^3, k) and replaced using the homeomorphism h. The knot k changes by the insertion of a full twist in a pair of parallel strands in the manner of Figure 2. The double cover of S^3 branched over the new knot is $(M_k - \operatorname{Int} T) \cup_{\overline{h}} T$. But if m is a meridian curve of T, $\overline{h}(m)$ intersects m algebraically twice. Thus $(M_k - \operatorname{Int} T) \cup_{\overline{h}} T$ is the manifold obtained from M_k

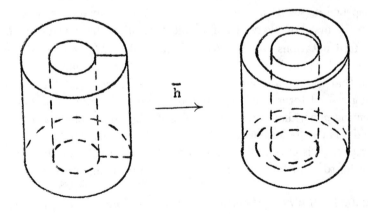

FIGURE 1

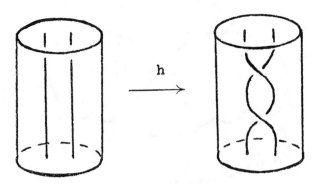

FIGURE 2

by $n/2$-surgery, for some odd n, on a core curve of T. In any presentation any crossing may be changed by a suitable use of the move of Figure 2, and, as the double cover of S^3 branched over the unknot is another copy of S^3, the result now follows at once.

Recall that the *linking form* of an oriented 3-manifold M with finite first homology is a symmetric bilinear map

$$\lambda : H_1(M) \times H_1(M) \to \mathbb{Q}/\mathbb{Z}$$

that can be defined as follows: If x and y are 1-cycles in M, nx bounds a 2-chain c for some integer n. Then $\lambda([x],[y]) = \frac{1}{n}(c \cdot y)$ where $c \cdot y$ is the intersection number of c and y.

LEMMA 2. *Let M be the 3-manifold obtained by p/q-surgery on a knot k in S^3. Then $H_1(M)$ is cyclic of order $|p|$ with a generator g represented by a meridian of k, and $\lambda(g,g) = q/p \in \mathbb{Q}/\mathbb{Z}$.*

PROOF. Let X be the complement of the interior of a tubular neighbourhood of k, and regard g as a meridian in X. Then, the sum of p copies of g is homologous in X to the sum of p meridians and q longitudes. The latter class is represented by a curve that bounds a disc in the solid torus $M - Int\ X$. Thus pg bounds a chain meeting g algebraically in q points, so $\lambda(g,g) = q/p$.

Note that the lens space $L_{p,q}$ has precisely the linking form mentioned in Lemma 2, as is seen by taking k to be the unknot.

THEOREM. *The knot 7_4 has unknotting number 2.*

PROOF. The notation 7_4 is that of Alexander and Briggs [A,B]; the knot is shown in Figure 3 and in the terminology of Conway [C] it is the rational knot $3\ 1\ 3$. Let k denote this knot. As $3 + (1 + 3^{-1})^{-1}$ is $15/4$ M_k is the lens space $L_{15,4}$, so $H_1(M_k)$ is cyclic of order 15 with a generator a such that the linking

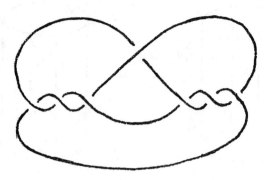

FIGURE 3

form is defined by $\lambda(a,a) = 4/15$. If, however, $u(k) = 1$, then, by Lemmas 1 and 2, $H_1(M_k)$ has another generator b such that $\lambda(b,b) = \pm 2/15$. Thus $a = tb$ for some integer t, so $\pm 2t^2/15 = 4/15$ in $\mathbb{Q}/\mathbb{Z}$. Hence $t^2 \equiv \pm 2$ modulo 15. But this is not true; neither $+2$ nor -2 is a quadratic residue modulo 15. Hence $u(k) \geq 2$. However, a moment's contemplation of Figure 3 confirms that two crossing changes suffice to create the unknot, so $u(k) = 2$.

The method used to prove the above theorem is to compare the linking form on $H_1(M_k)$ for a given knot k with the linking form that would arise if the knot had a certain unknotting number. This has been considered in much greater depth by J. R. Rickard in [R] where he succeeds in evaluating the unknotting numbers of six knots, each with nine or fewer crossings, which could not be determined by the methods used in compiling the table of [N]. In the discussion here all questions of orientation have been avoided, but these are discussed in some detail in [R] where emphasis is laid upon the fact that a crossing change may be from a positive crossing to a negative one, or vice versa. Thus a sign can be allocated to the procedure of changing a crossing; some knots can be unknotted with a single crossing change of one sign but not the other.

REFERENCES

[A,B] J. W. Alexander and G. B. Briggs, *On types of knotted curves*, Ann. of Math. **28** (1927), 562–586.

[B] S. A. Bleiler, *Knots prime on many strings.* (to appear).

[C]   J. H. Conway, *An enumeration of knots and links, and some of their algebraic properties*. Computational Problems in Abstract Algebra, Pergamon Press (1969) 329–358.

[Mo] J. M. Montesinos, *Surgery on links and double branched coverings of S^3*. Ann. of Math. Studies **84**, (1975), 227–259.

[Mu] K. Murasugi, *On a certain numerical invariant of link types*, Trans. Amer. Math. Soc. **117**, (1965), 387–422.

[N]   Y. Nakanishi, *A note on unknotting number*, Math. Sem. Notes, Kobe Univ. **9**, (1981), 99–108.

[R]   J. R. Rickard, to appear.

[W]   H. Wendt, *Die gordische Auflösung von Knoten*. Math. Zeit. **42**, (1937), 680–696.

DEPARTMENT OF PURE MATHEMATICS

CAMBRIDGE, ENGLAND

Contemporary Mathematics
Volume 44, 1985

A GENERAL POSITION THEOREM
FOR SURFACES IN EUCLIDEAN 4-SPACE*

Bruce Trace

0. INTRODUCTION

The primary purpose of this paper is to study the geometry involved in the problem of smoothly slicing classical knots. The approach taken here is non-traditional in the sense that the problem of slicing a knot is shown to be equivalent to a problem involving proper immersions of the 2-disk in the 3-ball. The proof that these problems are equivalent boils down to showing that, under certain framing conditions on the knots projection, the Morse (or Kearton-Lickorish) positioning of the slicing disk can be strengthened so as to eliminate twists which might appear in the embedded 1-handles. Since when working with maps of bounded manifolds the assumption that the maps be proper often is useful in proving general theorems—the approach taken in this paper to slicing knots appears to be at least philosophically more appealing then the traditional approach of constructing nonproperly immersed 2-disks in the 3-sphere which can be desingularized by being pushed into the 4-ball. Upon proving this general position theorem, the remainder of the paper serves roughly as a language dictionary, translating aspects of the traditional approach to slicing knots into the context of this paper. Also included is a short digression into immersion theory where the techniques used in proving the above general position theorem give rather quick proofs of certain existence of immersions results predicted by Smale [S] and Hirsch [H2] but not requiring the heavy machinery.

1. SOME DEFINITIONS, NOTATION AND BASIC FACTS

Unless otherwise stated, we shall be working in the smooth ($\equiv C^{\infty}$) or locally flat piecewise linear category. For the dimensions considered here, these categories are equivalent and will be used interchangeably. For the most part, the definitions and notation used here are those used in Rourke and Sanderson, "An Introduction to Piecewise Linear Topology", and Hirsch, "Differential Topology". The one exception is that a map, $f : M \to N$, between manifolds is *proper* provided $f^{-1}(\partial N) = \partial M$ and f is transverse to ∂N.

*Research supported in part by NSF Grant #MCS-810-3387..

Let $\mathbb{R}^4_-$ denote those points in $\mathbb{R}^4$ whose last coordinate is nonpositive. $\mathbb{R}^3$ is the natural boundary of $\mathbb{R}^4_-$, and it is occasionally useful to view $\mathbb{R}^4_-$ as $\mathbb{R}^3 \times (-\infty, 0]$. We will view $\mathbb{R}^2$ as the (x, y)-plane in $\mathbb{R}^3$. We use π to denote the projection $\mathbb{R}^3 \to \mathbb{R}^2$.

We will think of a knot as being an isotopy class of an embedded S^1 in $\mathbb{R}^3$. Traditionally, a knot is viewed by a representative which is in general position with respect to π, i.e. the singular set of $\pi|$ (this representative) consists of transverse double points. For the purposes of this paper, it is useful to view such a representative via a pair of maps (b_f, f) where f is a general position map of $S^1 \to \mathbb{R}^2$ and b_f is a map $S^1 \to \mathbb{R}^1$ such that $(b_f \times f)(S^1)$ is a representative of the knot. Essentially all b_f does is determine over and under crossings. In particular, b_f desingularizes f—so we shall call b_f a *blow-up* of f. More generally, given a map $f : M \to N$ with non-empty singular set and a map $b_f : M \to \mathbb{R}^1$ such that $b_f \times f$ is an embedding of M into $\mathbb{R}^1 \times N$ then we will call b_f a *blow-up* of f.

We recall that the 1st-homology of the complement of an embedded S^1 in $\mathbb{R}^3$ is infinite cyclic. These homology classes can all be realized by pushing an oriented copy of the embedded S^1 off the original along the various normal vector fields to the initial embedding. These normal vector fields are called *framings*. Evidently the framings of an embedded S^1 in $\mathbb{R}^3$ are in one-to-one correspondence with $\mathbb{Z}$ up to isotopy. The framing with the property that the resulting pushoff is null-homologous in the complement of the embedded S^1 is called the 0-*framing*.

Given a representative, (b_f, f), of a knot we will call the restriction of $\frac{\partial}{\partial z}$ to $b_f \times f(S^1)$ the *natural framing of* (b_f, f). If the natural framing of (b_f, f) is the 0-framing of $b_f \times f(S^1)$ we will call the representative, (b_f, f), *normally framed*. We remark that if we give $b_f \times f(S^1)$ the orientation induced by f where S^1 is given the standard orientation one can compute the natural framing associated to (b_f, f) as follows: We look down on $b_f \times f(S^1)$ from high on the z-axis. From this vantage point, up to rigid motions of $\mathbb{R}^2$, double points fall into two types, ⤬ and ⤬. We assign a -1 to ⤬ and a $+1$ to ⤬ then sum up. The resulting integer is the natural framing associated to (b_f, f) under the correspondence with $\mathbb{Z}$.

A knot is *slice* provided any of its representatives bounds a smooth, properly embedded 2-disk in $\mathbb{R}^4$. It is useful to recall some of the basic facts from Morse theory for understanding slicing disks. The approach to Morse theory used here is the version due to Kearton-Lickorish [K-L].

Suppose that F is a compact surface properly embedded in $\mathbb{R}^4_-$. According to Kearton-Lickorish, there is an ambient isotopy, $h : \mathbb{R}^4_- \to \mathbb{R}^4_-$ such that $h(F)$ has the following properties:

A) there exists a finite number of values $t_1, \ldots t_n$ in $\mathbb{R}^1$ such that $0 = t_0 > t_1 > \cdots > t_n$, called the *critical values* of $h(F)$, such that $h(F) \cap (\mathbb{R}^3 \times (t_{i+1}, t_i)) = [h(F) \cap (\mathbb{R} \times \frac{t_{i+1} + t_i}{2})] \times (t_i + 1, t_i)$ for $i = 0, \ldots, n - 1$.

B) a 2-cell appears in $h(F) \cap (\mathbb{R}^3 \times t_i)$, for $i = 1 \cdots, n$, these 2-cells are handles in a handle decomposition of $h(F)$. Figure 1 depicts some basic examples of these embedded handles.

We note that 1-handles are of two types. Type I 1-handles join two components into a single component as we increase in the 4th coordinate. Type II 1-handles split a component into two components as we increase in the 4th coordinate. It is a well-known result (due originally to Milnor and Fox I believe) that the above positioning can be strengthened in the sense that, as we increase in the 4th coordinate 0-handles appear first, then Type I 1-handles appear, then Type II 1-handles and finally 2-handles appear. If a surface is positioned as in the previous statement, we'll call the surface *well-positioned*.

2. THE MAIN THEOREM

In this section we prove

THEOREM 1. *Suppose K is a slice knot and (b_f, f) is a normally framed representative of K. Then there exists a proper immersion $F : D^2 \to \mathbb{R}^3_-$ and a blow-up map $B_F : D^2 \to \mathbb{R}^1$ such that $(B_F \times F)|\partial D^2 = b_f \times f$.*

PROOF. Initially, we let $\Delta \subset \mathbb{R}^4_-$ denote a slicing disk of some representative of K. The goal is to replace Δ by a new slicing disk, $|\Delta_*$, such that $b_F \times f(S^1) = \partial \Delta_*$ and $\Delta_* \cap (\mathbb{R}^3 \times t)$ projects to $\mathbb{R}^2 \times t$ via an immersion for each $t \in (-\infty, 0]$. Note that such Δ_* gives rise to the desired pair B_F and F.

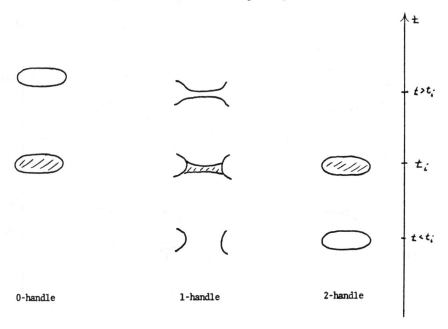

0-handle   1-handle   2-handle

FIGURE 1

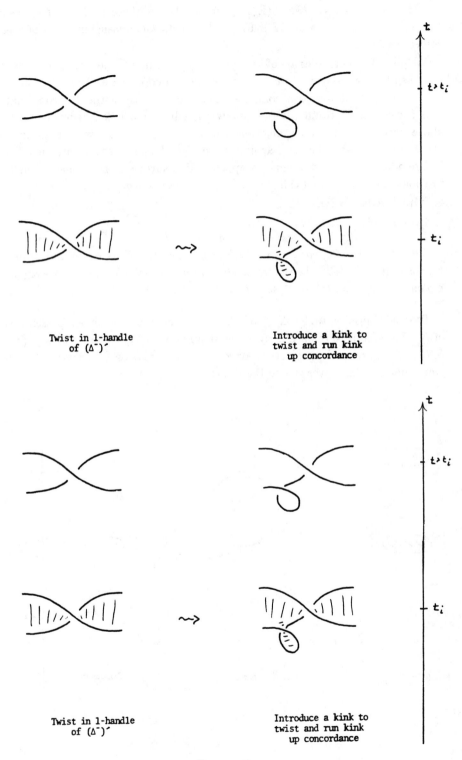

FIGURE 2

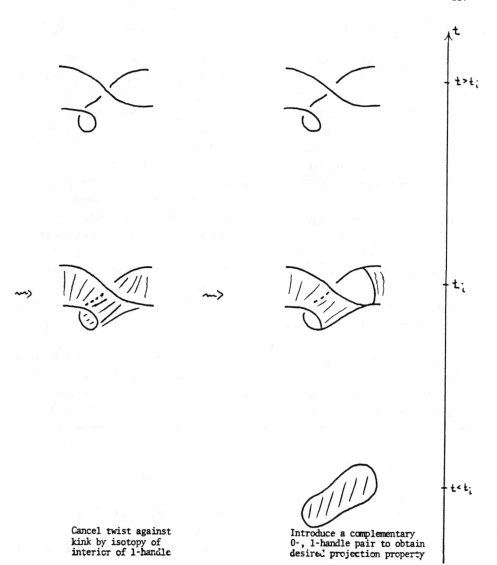

FIGURE 2 (continued)

We assume initially that Δ is well-positioned. The proof is broken into 3 steps.

STEP 1: Let t_- and t_+ denote values of $\mathbb{R}$ such that $t_- < t_+$ and all the $0-$ and type I 1 $-$handles of Δ appear below t_-, all the type II 1-handles and 2-handles appear above t_+. Let $\Delta^- = \Delta \cap \mathbb{R}^3 \times [-\infty, t_-]$ and $\Delta^+ = \Delta \cap \mathbb{R}^3 \times [t_+, 0]$. In this step we change Δ^- to Δ^-_* and Δ^+ to Δ^+_* where $\Delta^{\pm}_* \cap (\mathbb{R}^3 \times t)$ projects to $\mathbb{R}^2 \times t$ via an immersion for all t, $\Delta^{\pm}_* \cap (\mathbb{R}^3 \times t_+)$ is isotopic to $\Delta^{\pm} \cap (\mathbb{R}^3 \times t_+)$ in $\mathbb{R}^3 \times t_+$ and $\Delta^{\pm}_* \cap (\mathbb{R}^3 \times 0)$ is isotopic to $\Delta^+ \cap (\mathbb{R}^3 \times 0)$ in $\mathbb{R}^3 \times 0$.

We first show how to change Δ^- into Δ^-_*. this is accomplished in two stages. The first stage is to alter Δ^- to $(\Delta^-)'$ has the property that 0-handles will project into the appropriate $\mathbb{R}^2$'s via an immersion except where twists might appear in 1-handles. The second stage is to eliminate these twists—thus obtaining Δ^-_*.

To obtain (Δ^-_*) we begin low on the 4th dimensional axis and work our way up. If we come to a 0-handle at the level t, we find an isotopy of $\mathbb{R}^3 \times t$ which flattens this 0-handle out and fixes any circles which might be rising from previous levels. If g is the end diffeomorphism of this isotopy we reposition the disk by leaving it fixed below the $(t-\epsilon)$-level for ϵ small but positive, use the isotopy to reposition the disk in $\mathbb{R}^3 \times [t - \epsilon, t]$ and then use $g \times$(identity) to reposition the disk in $\mathbb{R}^3 \times [t, t_-]$.

The same isotopy argument works when we come to a 1-handle. Although in this case it is a little bit more difficult to construct the desired isotopy which flattens out the 1-handle except for some twists. To construct this isotopy one can first find an isotopy which puts the core arc of the 1-handle in general position with respect to the projection into the plane and fixes end points of the arc. We can then take a tubular neighborhood of the projected arc in the plane and lift it to a horizontal I-bundle over the arc in $\mathbb{R}^3$—by this we mean that each fiber of the tubular neighborhood lifts as a horizontal line segment. Then we push the 1-handle into this horizontal I-bundle as much as possible. We can push the 1-handle into this horizontal I-bundle everywhere except where twists occur. This gives rise to the isotopy of the appropriate $\mathbb{R}^3$-level which is then extended over $\mathbb{R}^3 \times (-\infty, t_-]$ as in the 0-handle case. This is the construction of $(\Delta^-)'$.

The construction of $(\Delta^+)'$ is analogous to that of $(\Delta^-)''$. The only difference is that we work from $\mathbb{R}^3 \times 0$ down to $\mathbb{R}^3 \times t_+$. Also, one might have to put $\Delta^+ \cap (\mathbb{R}^3 \times 0)$ into general position with respect to projection into $\mathbb{R}^2 \times 0$. This completes the 1st stage of Step 1. Figure 2 depicts how to eliminate the twists which might appear in 1-handles in $(\Delta^-)'$.

Applying the technique of Figure 2 to $(\Delta^-)'$ some finite number of times we obtain Δ^-_*. By turning the technique depicted in Figure 2 upside down and applying it to (Δ^+) we obtain Δ^+_*. This completes Step 1.

STEP 2: In this step we show how to fill in the gap between $\Delta^-_* \cap (\mathbb{R}^3 \times t_-)$ and $\Delta^+_* \cap (\mathbb{R}^3 \times t_+)$ with an imbedded annulus having the required property that its projection into the appropriate plane at each level is an immersion. To accomplish this we will have to alter Δ_*^+ by introducing an arcs worth of kinks on occasion—but the altered Δ_*^+ will retain the required projection property. The main tool used is the theorem of Reidemeister that the various representatives of a knot can all be obtained from a given representative by a finite sequence of Reidemeister moves as depicted in Figure 3.

First note that $\Delta_*^- \cap (\mathbb{R}^3 \times t_-)$ and $\Delta_*^+ \cap (\mathbb{R}^3 \times t_+)$ when viewed in $\mathbb{R}^3$ represent the same knot. This is because no handles appear in $\Delta \cap (\mathbb{R}^3 x[t_-, t_+])$ and the fact that $\Delta_*^\pm \cap (\mathbb{R}^3 \times t_\pm)$ is isotopic to $\Delta^\pm \cap (\mathbb{R}^3 \times t_\pm)$. Thus we could use the Reidemeister moves to fill in this gap—unfortunately the first Reidemeister move violates the required projection property.

To side-step this problem, we begin at $\Delta^+ \cap (\mathbb{R}^3 t_+)$ and work our way down to $\Delta_*^- \cap (\mathbb{R}^3 \times t_-)$ using the Reidemeister moves. Whenever we must introduce a kink we introduce the kink at $\Delta_*^+ \cap (\mathbb{R}^3 \times 0)$ and run this kink down through Δ_*^+ and also through the annulus constructed up to the appropriate point in time—so it appears that we have introduced the desired kink. To eliminate a kink, we introduce an oppositely framed kink to $\Delta_*^+ \cap (\mathbb{R}^3 \times 0)$, we then run this kink down next to the kink we want to eliminate and cancel these kinks using the technique depicted in Figure 2. This completes Step 2.

STEP 3: To this point in time we have constructed a slicing disk for some representation of K with the required projection property. We now show how to make this representative of K become (b_f, f).

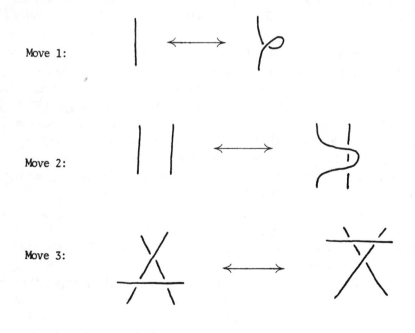

Move 1:

Move 2:

Move 3:

FIGURE 3

For this view $(b_f \times f)(S^1)$ as lying in $\mathbb{R}^3 x1$. Repeating the argument of Step 2, we can fill in the gap between $\mathbb{R}^3 x0$ and $\mathbb{R}^3 \times 1$ with an embedded annulus having the required projection property in $\mathbb{R}^3 \times [0,1]$. This is done at the expense of altering $(b_f \times f)(S^1)$ by adding on some kinks. We claim the framings assigned to these kinks sum to zero. This is because as we pass through the critical levels of the slicing disk constructed for $(b_f \times f)(S^1)$ +(kinks) we observe that the projection property implies that the natural framing below the critical point is the same as the natural framing above the critical point. (We note that in general we aren't working with knots but rather links in this setting. Take framing to mean the sum of the signed double points associated to the link.) Since below the slicing disk the framing is 0 we must have that $(b_f \times f)(S^1)$+ (kinks) is normally framed. By hypothesis $(b_f \times f)(S^1)$ is also normally framed, thus the sum of the framings associated to the kinks is 0. It is now a relatively straight forward exercise using the technique of Figure 2 and the second and third Reidemeister moves to remove these kinks. So by a reparameterization of the 4th coordinate, the proof of Theorem 1 is complete.

We conclude this section with some remarks. Initially, we note that by introducing complementary $0-$, $1-$ and 2-handles to the slicing disk constructed in Theorem 1, we can assume $F(D^2)$ not only satisfies the conclusion of the theorem, but also that it is in general position. A second observation, coming from Steps 1 and 2, is that every isotopy class of ribbon disks in $\mathbb{R}^4_-$ can be realized by some (B_F, F). Where the singular set of F consists of transverse double points. I suspect that, due in particular to Step 2, the resulting slicing disk built in the proof of Theorem 1 is not isotopic to the slicing disk we begin with. It would be interesting to show every isotopy class for a slicing disk in $\mathbb{R}^4_-$ can be realized by some (B_F, F).

Even though Theorem 1 is stated and proved for disks, the proof works just as well for orientable surfaces where each boundary component is normally framed. The only delicate point here is in Step 3 where we show $b_f x f(S') +$ (kinks) is normally framed. The problem is that, if a surface has multiple boundary components, the kinks which cancel may appear in distinct components. This problem can be eliminated by introducing kinks at the completion of Step 1 so each boundary component is normally framed and then running all the kinks obtained in Step 2 to a fixed component of the boundary.

We next observe that the proof of Theorem 1 immediately implies that a necessary and sufficient condition for a general positioned S' in $\mathbb{R}^2$ to bound an immersed 2-disk in $\mathbb{R}^3_-$ is that the number of transverse double points be even. The idea is to begin with such an immersion, say $f: S' \to \mathbb{R}^2$, then let $b_f: S' \to \mathbb{R}'$ be a blow up such that $b_f \times f(S')$ is the unknot. The pair (b_f, f) might not be normally framed so kinks might have to be introduced. Then we can isotope $b_f \times f(S') +$ (kinks) to the standard projection of the unknot as in Theorem 1. We can cap off the unknot with a disk. The only problem left is to get rid of the kinks. For this we will have to change the framings associated to the kinks—but since we are not working for an embedding but rather an immersion in $\mathbb{R}^3_-$ the singularities introduced by changing unders to overs do not matter. So, upon arranging the framings of the kinks to sum to zero, the kinks can be cancelled off. We note that with a little work this observation can also be deduced from the Whitney-Graustein Theorem, or Smale [S].

The above observation immediately implies that, if (b_f, f) is normally framed, then f extends to an immersion $F : D^2 \to \mathbb{R}^3_-$. In fact, if the natural framing associated to (b_f, f) is even, we have such an extention. It is now a rather straightforward exercise to show that a framing for a knot can be realized by a normal vector field in the tangent space of an immersed disk spanning the knot if and only if the framing is even. A second application of this observation is that it implies that any proper map of a compact surface into a 3-manifold can be ϵ-approximated by an immersion. This is done basically using p.1. approximation, then cutting out disjoint 3-cells containing the image of interior vertices, and then introducing arcs worth of kinks to get the number of double points in the boundaries of the removed 3-cells to be even. In the case where the surface has boundary the above construction is all that is necessary. If the surface is closed, a standard singular set argument allows one to conclude that upon fixing all but one intersection with boundaries of removed 3-cells the final boundary intersection must have an even number of double points. Thus, allowing for the final immersed disk. It is mildly interesting to note that the tools used in the second application were available to Papakyriakopolous at the time he proved the Loop and Sphere theorems. Knowing initially that the disk or sphere is immersed saves the troubles involved with branch points.

The last comment of this section deals with the traditional approach to slicing a knot via spanning the knot with an immersed disk then pushing the immersed disk into $\mathbb{R}^4$ so as to obtain a slicing disk. Note that the obstructions to finding a blow-up for F are the same as for performing the pushing operation. From this it follows that an upper bound for the number of times an immersed disk with the push-in property to transversely intersect the knot is the #double points of f for any normal framing (b_f, f) of the knot.

3. A SLICING THEOREM

Let (b_f, f) denote a representation for the knot K. We presently answer the question of when f extends to an immersion $F : D^2 \to \mathbb{R}^3_-$ with blow-up B_F extending b_f such that singular set of F consists of transverse double points. The pair (B_F, F) in this case gives rise to what might be interpreted as a ribbon disk. Although, as we shall see, not every projection of a ribbon knot, even the unknot, can always extend to (B_F, F) where the singular set of F contains no triple point. Clearly, for the pair (B_F, F) to exist we must be able to view the singular set of f as the singular set of F restricted to ∂D^2. Thus, the singular set of F would have to consist of pairs of arcs embedded in D^2 matching overcrossings and undercrossings of a pair of double point of f—with all such arcs being mutually disjoint. The existence of such a collection of arcs is almost the desired necessary and sufficient condition for slicing (b_f, f) in this manner. We must also weave in a notion related to winding number. To see how winding numbers come into play we initially study the question of when (b_f, f) extends to an immersed surface which can blown-up to an embedding where the surfaces's singular set consists of transverse double points.

Let $\underline{X} = \{(b_f, f) | (b_f, f)$ is a representative for some link having a finite number of components$\}$. We define an equivalence relation on $\underline{X}$, $\sim$, as follows: We set $(b_{f_0}, f_0) \sim (b_{f_1}, f_1)$ provided there exists an orientable surface M^2 and a proper immersion $F : M^2 \to \mathbb{R}^2 \times I$ such that $(B_F \times F)|F^{-1}(\mathbb{R}^2 \times i) = b_{f_i} \times f_i$ for $i = 0, 1$, and the singular set of F consists of transverse double points. Note that that under disjoint union $\underline{X}/\sim$ is an abelian group.

THEOREM 2. *The group $\underline{X}/\sim$ is isomorphic to $\bigoplus_{\mathbb{Z}} \mathbb{Z}$.*

PROOF. We begin by defining a map $\Phi : \underline{X} \to \bigoplus_{\mathbb{Z}} \mathbb{Z}$. Let $(b_f, F) \in \underline{X}$ and give S^1 its usual counterclockwise orientation. Then f induces an orientation on $f(\amalg S^1) \subset \mathbb{R}^2$. (Here $\amalg S^1$ is the disjoint union of S^1's, the number of which equals the number of components of the link represented by (b_f, f).) We cut $f(\amalg S^1)$ into a collection of oriented circles and figure eights in $\mathbb{R}^2$ in much the same way that one constructs a Seifert surface for the link i.e.

at each double point. We associate to each figure eight a pair of integers, (m, n), as follows: Fix a figure eight. Take a proper ray whose initial point is the double point of this figure eight and orient this ray be denoting the positive direction going towards ∞. Every time this ray crosses an oriented circle we assign a plus or minus one via

(up to motions of $\mathbb{R}^2$). The integer m is the sum of these plus and minus ones. The integer n corresponding to this figure eight is the framing of the corresponding double point in $(b_f \times f)(S^1)$.

Let $\{(m_i, n_i)\}$ be the collection of ordered pairs of integers associated to the figure eights obtained by cutting $f(S^1)$. We then define $\Phi((b_f, f))$ by setting the kth-coordinate of $\Phi((b_f, f))$ to be $\sum_{m_i = k} n_i$ for $k \in \mathbb{Z}$.

Note that Φ is surjective and assuming Φ respects $\sim$ it is a homomorphism. We now show that Φ respects $\sim$. For this suppose $(b_{f_0}, f_0) \sim (b_{f_l}, f_l)$ via (B_F, F) and M^2. If necessary, we can alter $F(M)^2$ slightly by Morse theory and by introducing maxima and minima so that the critical points of $F(M^2)$ are embedded handles or immersed 1-handles as depicted in Figure 4.

Now as we pass through the critical points of $F(M^2)$ we note that the image of Φ remains unchanged. Indeed, this is clear for critical points involving embedded handles—at immersed 1-handles the pair of double points introduced or deleted have the same m value and opposite n values since we can blow up the singular set of $F(M^2)$. It follows that $\Phi(b_{f0}, f_0)) = \Phi(b_{f_1}, f_1))$.

It follows that Φ induces the desired isomorphism provided $\Phi((b_{f0}, f_0)) = \Phi((b_{f1}, f_1))$ implies $(b_{f0}, f_0) \sim (b_{f1}, f_1)$. It suffices to show this in the case $\Phi((b_f, f)) = 0$. To accomplish this we first note that by introducing embedded 1-handles we can realize the cutting operation used in defining Φ—thus obtaining (b_{f_1}, f_1) where $(b_f, f) \sim (b_{f_1}, f_1)$. Now embed arcs joining figure eights which cancel under Φ. Since figure eights joined in this manner have the same m-value we can introduce embedded 1-handles so that these arcs miss the oriented circles, obtaining (b_{f_2}, f_2) where $(b_{f1}, f_1) \sim (b_{f2}, f_2)$. We then cancel geometrically these paired figure eights using the technique depicted in Figure 5, obtaining (b_{f_3}, f_3) where $(b_{f_2}, f_2) \sim (b_{f_3}, f_3)$.

To complete the construction we note that upon cancelling figure eights we are left only with oriented circles which are capped off with disks, implying $(b_{f_3}, f_3) \sim \varnothing$. This completes the proof of Theorem 2.

It is now clear how the obstruction for the immersed disk suggested in the opening paragraph of this section must be modified to obtain a necessary condition for (b_f, f) to extend to a slicing disk. The double points of f which are paired off by the paired arcs in the singular set of F must cancel under Φ. By an

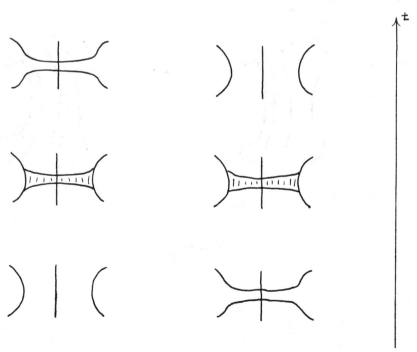

FIGURE 4

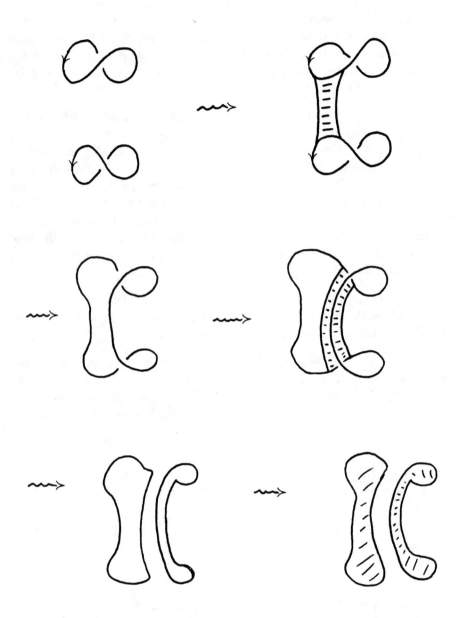

FIGURE 5

outermost arc argument, we obtain that these conditions are also sufficient for constructing a slicing disk for (b_f, f). To be more precise, we observe

THEOREM 3. *Let (b_f, f) represent a link of k components. View the singular set of f as lying in the boundary of $\overset{\amalg}{k} D^2$. Suppose there exist pairwise disjoint properly embedded arcs A_{ij} in $\overset{\amalg}{k} D^2$, for $1 \le i \le \dfrac{\text{# double points of } f}{2}$ and $j = \pm 1$, satisfying*

A.) each A_{ij} joins points in the singular set of f,

B.) $f|\partial A_{ij}$ is an embedding and $f(\partial A_{i,+1}) = f(\partial A_{i,-1})$

C.) $A_{i,-j}$ joins undercrossing points and $A_{i,+1}$ joins overcrossing points for all i, and

D.) the double points of f joined by the A_{ij} cancel under Φ for all i.

Remark: The reason for stating Theorem 3 for (b_f, f) representing a link is for purposes of induction in the proof.

PROOF OF THEOREM 3. Suppose A_{ij} is an outermost in arc appearing in some component $\overset{\amalg}{k} D^2$. Because A_{ij} is outermost, the picture of $f(\overset{\amalg}{k} S)$ near the double points paired by $A_{i,\pm 1}$ looks like $\dashv\vdash$. Because these double points have the same m-value, the vertical arcs are oriented in opposite directions. Because these double points are of opposite framing, both vertical arcs are overs or unders. Thus the process

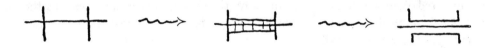

corresponds to cutting $\overset{\amalg}{k} D^2$ along the $A_{i,\pm j}$ since the $A_{i,\pm j}$ join overs and unders. The process depicted above shows how to get F started and because overs and unders were paired we may also start B_F. Induction yields the Theorem.

We remark that Theorem 3 is false if we replace 2-disks with punctured 2-disks. This can be seen by taking

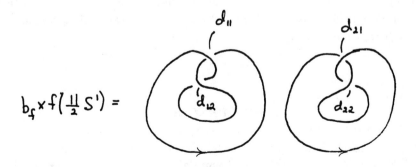

and pairing d_{11} with d_{22} and d_{12} with d_{21} on an annulus. (Of course the pairing of d_{11} with d_{21} and d_{12} with d_{22} can be realized by an immersed annulus with a blow-up map.)

4. FINAL REMARKS

Much work appears necessary to strengthen Theorem 3 to cases where extensions of f require triple points. The work of S. Kaplan's [K] involving the Arf invariant of a knot can be realized in the context of proper immersions. Basically this means given a representative for a knot (b_f, f) we can find a proper immersion $F : D^2 \to \mathbb{R}^3_-$ such that the arcs in the singular set of F join overcrossings or undercrossings of $b_f \times f(S^1)$. The obstruction then to finding a B_F extending b_f lines in the triple points of F. The number of triple points which can't be blown up (mod 2) is the Arf invariant for $b_f \times f(S^1)$. It is conceivable that in the context of proper immersions Kaplan's interpretation of Arf invariant could perhaps be strengthened to a $\mathbb{Z}$-invariant. The intuitive reason would be separation properties of a properly immersed disk in $\mathbb{R}^3_-$.

It is known by the work of Smale, [S], that any two proper immersions $D^2 \to \mathbb{R}^3_-$ are regular homotopic through proper immersions fixing the boundary. The difficulty then in strengthening Kaplan's result appears to lie in the fact that the regular homotopy might not preserve the property that the obstruction to blowing up the immersions lives solely in triple points. Perhaps there are ways of getting around this difficulty.

A second approach to strengthening Theorem 3 would be via constructing immersed Whitney disks in $\mathbb{R}^3_-$. Initially, this would appear as difficult as the original problem. However, using $F(D^2)$ is properly immersed, the boundary curve of the would-be Whitney disk can be slid up into $\mathbb{R}^2 \subset \mathbb{R}^3_-$. This curve is usually simpler to work with than $f(S^1)$. Of course, even if this curve can be sliced the Whitney disk might not cancel the double points–due to other intersections with $F(D^2)$. Perhaps the Smale regular homotopy theorem is useful for this second problem.

Finally, we remark that the context of proper immersions with blow up maps might be useful in the slice implies ribbon conjecture. It is true that given any normally framed representative (b_f, f) for a ribbon knot we can find (B_F, F) extending (b_f, f) such that $F(D^2)$ has no maxima. Since the index of critical points of $F(D^2)$ coincides with those of $B_F \times F(D^2)$ we can build a ribbon disk for $b_f \times f(S^1)$ from $F(D^2)$.

REFERENCES

[H1] M. W. Hirsch, *Differential Topology*, Springer-Verlag, New York, 1976.

[H2] M. W. Hirsch, *Immersions of Manifolds*, Trans. AMS **93**,(1959), 242–276.

[K] S. Kaplan, *Constructing framed 4-manifolds with given almost framed boundaries*, Trans. AMS **254**, (1979), 237–263.

[K-L] C. Kearton, W.B.R. Lickorish, *Piecewise linear critical levels and collaps-ing, Trans. AMS* **170**, (1972), 415–424.

 [R] D. Rolfsen, *Knots and Links*, Publish or Perish, Inc., Berkeley, CA, 1976.

[R-S] C. P. Rourke, B. J. Sanderson, *Introduction to Piecewise-Linear Topology*, Springer-Verlag, New York, 1972.

 [S] S. Smale, *A classification of immersions of the two-sphere, Trans. AMS* **90**, (1958), 281–290.

UNIVERSITY OF ALABAMA

TUSCALOOSA, ALABAMA

3-MANIFOLDS

Contemporary Mathematics
Volume 44, 1985

ON THE EQUIVARIANT DEHN LEMMA

Allan L. Edmonds

The purpose of this note is to sketch a new, purely topological proof of the following theorem of W. Meeks and S.-T. Yau [3]:

EQUIVARIANT DEHN LEMMA. *Let G be a finite group acting by orientation-preserving, piecewise linear homeomorphisms on an orientable 3-manifold M. Suppose that $C \subset \partial M$ is a simple loop such that $C \simeq 0$ in M and such that for each $g \epsilon G$ either $gC = C$ or $gC \cap C = \emptyset$ and G acts freely on $G(C) = \cup g(C)$. Then there is an embedded 2-disk $D \subset M$ such that $\partial D = C$ and for all $g \epsilon G$ either $gD = D$ or $gD \cap D = \emptyset$.*

Full details of the new proof will appear in [1]. One should note that one can drop the orientability hypothesis if it is assumed that the group action has singular set of dimension at most one. Also, the hypothesis that G acts freely on $G(C)$ is of a technical nature only: if the action on $G(C)$ is not free, then replace C by a boundary component of a small annular neighborhood of C in ∂M.

To simplify the exposition here *we assume henceforth that G is a cyclic group of prime order p.* The force of this assumption is that the proof separates into two distinct cases:

$$\text{I. } \quad gC \cap C = \emptyset, \text{ all } g \neq e; \qquad \text{II. } \quad gC = C, \text{ all } g.$$

Proof in Case I. This is the easier of the two cases, but illustrates most of the ideas.

First note that it suffices to prove that there is a G-immersion $\phi : G \times D^2 \to M$ such that $\phi(G \times D) \subset M - F$, where F denotes the fixed point set of the action, G acts on $G \times D$ by left translation in the first factor, and $\phi|G \times \partial D$ parametrizes $G(C)$. For one can apply the ordinary Dehn Lemma to the induced map of orbit spaces $D \to M^* - F^*$ to obtain an embedding which lifts to the desired equivariant disk.

To start the argument, one applies the ordinary Dehn Lemma to obtain a PL G-immersion $\phi : G \times D \to M$ with the following properties:

(a) $\phi|G \times S^1$ parametrizes $G(C)$ homeomorphically

---

*Supported in part by an NSF grant.

(b) ϕ is transverse to the fixed point set F

(c) $\phi^{-1}(F) \cap \Sigma(\phi | e \times D) = \varnothing$, where $\Sigma(\phi | e \times D) = \{x \, \epsilon \, e \times D : \phi^{-1}\phi(x) \neq x\}$ is the singular set of $\phi | e \times D$.

To start with, one can, of course, choose $\Sigma(\phi | e \times D) = \varnothing$; in later steps $\phi^{-1}(F)$ consists of isolated points, while $\Sigma(\phi | e \times D)$ is 1-dimensional, in general.

If $\#\phi^{-1}(F) > 0$ we show how to reduce it, maintaining properties (a)–(c).

Triangulate M and $G \times D$ very finely, so the actions and the map ϕ are simplicial, with $\phi(G \times D)$ a full subcomplex.

Let $N_1 = N(\phi(G \times D), M)$ be the equivariant simplicial neighborhood of $\phi(G \times D)$ in M. The fixed set of G on N_1 consists of arcs transverse to $\phi(G \times D)$.

Suppose first that $\pi_1(N_1) = 0$. Then $H_1(\partial N_1) = 0$ and hence $\partial N_1 = \cup S^2$. The action on $\cup S^2$ is perfectly standard, and one may choose a 2-disk $\Delta \subset \partial N_1$ such that $\partial \Delta = \phi(e \times S^1)$ and $\Delta \cap \phi(G \times S^1) = \partial \Delta$. Then $G \times D \cong G(\Delta)$ is the desired equivariant disk.

When $\pi_1(N_1) \neq 0$ we construct an equivariant tower of universal covering spaces. The basic construction is an *elementary* tower:

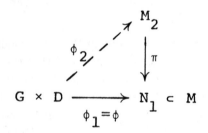

Here $\pi : M_2 \to N_1$ is the universal covering, with triangulation induced from N_1. One lifts $\phi_1 | e \times D$ to $\phi_2 : e \times D \to M_2$. Then, choosing $x_1 \epsilon F_1 = F \cap N_1$ and $x_2 \epsilon \pi^{-1}(x_1) \cap \phi_2(e \times D)$, there is a unique G action on M_2 such that x_2 is a fixed point and π is a G-map. Now extend ϕ_2 to $\phi_2 : G \times D \to M_2$ by equivariance. Let $N_2 = N(\phi_2(G \times D), M_2)$.

If $\pi_1(N_2) \neq 0$, simply iterate this construction:

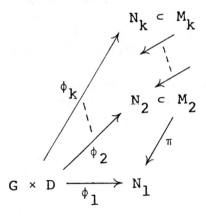

or

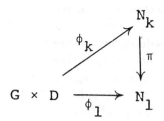

for short.

Just as in the proof of the ordinary Dehn Lemma, the tower eventually stops with $\pi_1 N_k = 0$ because at each step the singularity set of ϕ_i, appropriately measured, is reduced.

Now, again $\partial N_k \cong \cup S^2$, with standard action. As before there is a 2-disk $\Delta \subset \partial N_k$ such that $\partial \Delta = \phi_k(e \times S^1)$ and $\Delta \cap \phi_k(G \times S^1) = \partial \Delta$. Identifying Δ with a standard 2-disk D, define $\phi' : G \times D \to M$ to be the following composition:

$$G \times D \cong G(\Delta) \subset N_k \to N_1 \subset M$$

Clearly ϕ' is a G-immersion transverse to F. It remains to check that $\phi'^{-1}(F) \cap \Sigma(\phi'|e \times D) = \varnothing$ and $\#\phi'^{-1}(F) < \#\phi^{-1}(F)$. One must examine $\Delta \cap \pi^{-1}(F_1)$.

Now $F_1 = \cup A_j$, a disjoint union of arcs transverse to $\phi(G \times D)$, such that $A_j \cap \phi(G \times D)$ is just one point. The hypothesis that $\phi^{-1}(F) \cap \Sigma(\phi|e \times D) = \varnothing$ implies that $\pi^{-1}(A_j)$ consists of one arc fixed by G or p disjoint arcs cyclically permuted: $\pi^{-1}(A_1)$ consists of arcs transverse to $\phi_k(G \times D)$; a given $\phi_k(g \times D)$ meets $\pi^{-1}(A_j)$, and only once; otherwise either A_j meets $\phi(g \times D)$ more than once or at a singular point of $\phi|g \times D$.

A homology argument implies that $\pi^{-1}(A_j) \cap \Delta$ consists of at most one point. Pass to G-orbit spaces $N_k^* \supset \Delta^*$. Here Δ^* is a 2-disk in ∂N_k^* and N_k^* is a simply connected 3-manifold. If the arc $\pi^{-1}(A_j)^*$ had both end points in Δ^*, then one could construct a loop λ meeting the relative simplicial 2-cycle $\phi(G \times D)^*$ exactly once. This would mean $[\lambda] \neq 0$ in $H_1(N_k^*)$. Since $H_1(N_k^*) = 0$ this is a contradiction, and easily implies that $\pi^{-1}(A_j)$ meets Δ at most once.

From this one sees that $\phi'^{-1}(F) \cap \Sigma(\phi'|e \times D) = \varnothing$. Moreover, by construction, some $\pi^{-1}(A_{j_0})$ is a fixed arc. Then $\Delta \cap \pi^{-1}(A_{j_0}) = \varnothing$. Therefore $\#(\Delta \cap \pi^{-1}(F_1)) < \#(\phi^{-1}(F) \cap e \times D)$, so that $\#\phi'^{-1}(F) < \#\phi^{-1}(F)$. This completes the proof in Case I.

Proof in Case II. Here $gC = C$ for all $g \epsilon G$. First note that it suffices to produce a G-immersion $\phi : D \to M$ (where G acts by suitable rotations on the disk D) such that $\phi|\partial D$ parametrizes C, ϕ is transverse to F, and $\phi^{-1}(F) = \{0\}$. To see this pass to orbit spaces. The induced map $\phi^* : D^* \to M^*$ is transverse to F^* and $\phi^{*-1}(F^*) = \{0^*\}$. Excise a small neighborhood of F^* in M^* and of 0^* in D^*.

One may apply, for instance, the Shapiro-Whitehead version ([5], see also [2]) of the Dehn Lemma for planar domains to replace the corresponding singular annulus by an embedded one. Reinserting the neighborhood of 0^* and lifting to M provides the required embedded, invariant disk.

It is not even immediately obvious that there is any G-map $D \to M$ whose boundary parametrizes C. Therefore, begin by applying the ordinary Dehn Lemma to obtain a G-immersion $\phi : G \times D \to M$ where $\phi(g \times S^1) = C$ for all $g \epsilon G$, with ϕ transverse to F, and $\phi^{-1}(F) \cap \Sigma(\phi|e \times D) = \varnothing$.

Necessarily $\phi^{-1}(F) \neq \varnothing$, otherwise C could not be invariant. We show how either to reduce $\#\phi^{-1}(F)$ or to produce the required G-map $D \to M$.

Triangulate M and $G \times D$ very finely so that the actions and the map ϕ are simplicial, with $\phi(G \times D)$ a full subcomplex. Let $N_1 = N(\phi(G \times D), M)$ be the equivariant simplicial neighborhood of $\phi(G \times D)$ in M. The fixed set F_1 of G in N_1 consists of arcs A_j transverse to $\phi(G \times D)$.

If $\pi_1(N_1) = 0$, then C is an invariant curve on a 2-sphere $S^2 \subset \partial N_1$. Let Δ be the closure of one of the complementary domains of C in S^2. Then $D \cong \Delta$ is the desired equivariant disk.

When $\pi_1(N_1) \neq 0$ we again construct an equivariant tower of universal covering spaces. The elementary tower is constructed just as before:

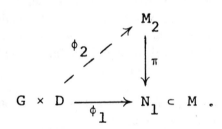

Here $\pi : M_2 \to N_1$ is the universal cover; $\phi_1|e \times D$ is lifted to some $\phi_2 : e \times D \to M_2$ arbitrarily; the action is lifted from N_1 to M_2 fixing some chosen point of $\phi_2(e \times D)$ over a fixed point of N_1. (This time this is a crucial choice.) Then ϕ_2 is extended to $G \times D \to M_2$ by equivariance. It may happen that $\phi_2(G \times \partial D)$ consists of p disjoint circles in ∂M_2.

LEMMA. *If $\pi_1(N_1) \neq \varnothing$, then there is an elementary tower with*
(i) $\phi_2(G \times S^1)$ consisting of p disjoint circles or with
(ii) $\phi_2(G \times S^1)$ consisting of one invariant circle and $\phi_2(G \times D) \cap \pi^{-1}(F_1)$
consisting entirely of fixed points.

PROOF OF LEMMA. Fix an arbitrary lift $\phi_2 : e \times D \to M_2$ of $\phi_1|e \times D$. Let $F_1 = F \cap \phi_1(e \times D) = \{x_1, \ldots, x_n\}$ and let $\pi^{-1}(F) \cap \phi_2(e \times D) = \{y_1, \ldots, y_n\}$ with

$y_i \in \pi^{-1}(x_i)$. There are at most n G actions on M_2 fixing one of the y_i, covering the G action on N_1. Suppose all these lifted actions leave $\phi_2(e \times S^1)$ invariant.

We show that all the lifted G actions coincide, fixing the entire set $\{y_1, \ldots, y_n\}$.

Choose paths $\lambda_1, \ldots, \lambda_n$ in $e \times D$ emanating from a base point z_0 as shown in Figure 1.

Here $\phi(z_i) = x_i$. (Note that $\#\phi^{-1}(x_i) \cap e \times D = 1$, since $\phi^{-1}(F) \cap \Sigma(\phi|e \times D) = \varnothing$.)

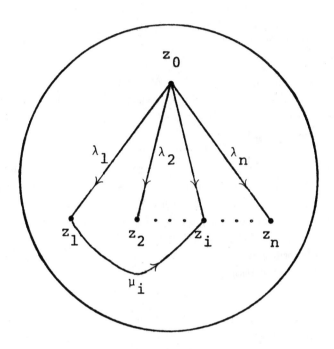

FIGURE 1

Now $[g\phi\lambda_i] = [\phi\lambda_i]$ in the set $[I, 0, 1; N_1, x_i, C]$ of homotopy classes, since the lifted G action fixing y_i leaves $\phi_2(e \times S^1)$ invariant.

Let μ_i be a path in $e \times D$ from z_1 to z_i, as indicated. We claim that $g\phi\mu_i \simeq \phi\mu_i$ rel end points

$$
\begin{aligned}
g\phi\mu_i &\simeq g(\phi\lambda_1^{-1} * \phi\lambda_i) \\
&\simeq g\phi\lambda_1^{-1} * g\phi\lambda_i \\
&\simeq \phi\lambda_1^{-1} * \tau * \phi\lambda_i \quad (\tau \text{ a loop in } C) \\
&\simeq \phi\lambda_1^{-1} * \phi\lambda_i \quad (C \simeq 0 \text{ in } N_1) \\
&\simeq \phi\mu_i .
\end{aligned}
$$

Covering space theory now implies that the action fixing y_1 also fixes an arbitrary y_i. This proves the lemma.

Now erect a complete tower.

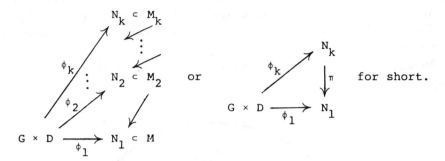

Let F_i denote the fixed set in N_i. At each stage $\phi_i^{-1}(F_i) \neq \varnothing$ and either $F_i = \pi^{-1}(F_1)$ and $\phi_i(G \times S^1)$ is a single circle or $\phi_i(G \times S^1)$ consists of p disjoint circles permuted by G. If at some stage $\phi_i(G \times S^1)$ consists of p circles, continue building the tower as in Case I.

As before, the tower eventually must terminate with $\pi_1 N_k = 0$.

Case (a). $\phi_k(G \times S^1)$ consists of p circles in $\partial N_k = \cup S^2$. Then there is a 2-disk $\Delta \subset \partial N_k$ such that $\partial \Delta = \phi_k(e \times S^1)$ and $\Delta \cap \phi_k(G \times S^1) = \partial \Delta$. As before, $F_1 = \cup A_j$, where each A_j is an arc transverse to $\phi(G \times D)$, and $\pi^{-1}(A_j)$ in N_k consists either of one arc fixed by G or of p arcs permuted by G. Again $\#\Delta \cap \pi^{-1}(A_j) \leq 1$ and for some $j_0 \Delta \cap \pi^{-1}(A_{j_0}) = \varnothing$, since some $pi^{-1}(A_{j_0})$ is a fixed arc by construction.

Let ϕ' be the composition

$$G \times D \cong G(\Delta) \subset N_k \to N_1 \subset M.$$

Then ϕ' is transverse to F, $\phi'^{-1}(F) \cap \Sigma(\phi'|e \times D) = \varnothing$, and, most important, $\#\phi'^{-1}(F) < \#\phi^{-1}(F)$.

Case (b). $\phi_k(G \times S^1)$ consists of a single invariant circle, and $\pi^{-1}(F_1) = F_k$. (In particular each $\pi^{-1}(A_j)$ is a single fixed arc.)

Now $\phi_k(G \times S^1)$ is an invariant simple closed curve in a 2-sphere S^2 in ∂N_k. Let Δ be the closure of one of the complementary domains of $\phi_k(G \times S^1)$ in S^2. Then $\Delta \cap F_k = \Delta \cap \pi^{-1}(F_1)$ is a single point! Identifying Δ with a standard disk D with G acting by rotations, we obtain the required equivariant disk meeting F just once as the composition

$$D \cong \Delta \subset N_k \to N_1 \subset M.$$

Concluding Remarks. It seems likely that the present techniques can be pushed further to yield the equivariant loop theorem. In any case, the argument of Meeks and Yau [3; Theorem 6] to produce a nontrivial loop in the boundary suitable for applying the Dehn Lemma is a reasonably straight forward argument about closed geodesics on a surface.*

---

*Although in the proof of Assertion 2 of Theorem 6 Meeks and Yau apply the Corollary to Theorem 5, which is proved via minimal surfaces, in fact the result actually required is a simple consequence of standard covering space theory.

It is possible that these techniques can be extended to yield some of the further results recently obtain by Meeks and Yau, Freedman, Haas, and Scott, et. al., using minimal surfaces. But clearly the present approach does not provide the powerful guiding principles found in the least area point of view.

REFERENCES

[1] A. L. Edmonds, *A topological proof of the equivariant Dehn lemma*, to appear.

[2] J. Hempel, 3-*Manifolds*, *Princeton University Press*, Princeton, 1976.

[3] W. Meeks III and S.-T. Yau, *The equivariant Dehn's lemma and loop theorem*, Comment. Math. Helv. **56**, (1981), 225–239.

[4] C. Papakyriakopolous, *On Dehn's lemma and the asphericity of knots*, Annals of Math. **66**, (1957), 1–26.

[5] A. Shapiro and J. H. C. Whitehead, *A proof and extension of Dehn's lemma*, Bull. Amer. Math. Soc. **64**, (1958), 174–178.

INDIANA UNIVERSITY

Contemporary Mathematics
Volume 44, 1985

VIRTUALLY HAKEN MANIFOLDS

John Hempel

1. INTRODUCTION. We are concerned with the question: which 3-manifolds are *virtually Haken?*—that is, have a finite sheeted covering space which is a *Haken manifold* (compact, orientable, irreducible, and sufficiently large). This is closely related to other problems in the structure theory for 3-manifolds, and in particular to the conjecture of Thurston [T] that every 3-manifold can be decomposed into "geometric pieces".

Our approach to this question is to ask which 3-manifolds have the stronger (assuming irreducibility) property that their fundamental group have a subgroup of finite index which maps epimorphically to the infinite cyclic group, $\mathbb{Z}$—or as we say is *virtually representable* to $\mathbb{Z}$. See [H-2] and [H-3] for more discussion of these properties. Note that for a finite complex X, $\pi_1(X)$ is virtually representable to $\mathbb{Z}$ if and only if there is a finite covering space $\tilde{X} \to X$ of X with $H_1(\tilde{X})$ infinite.

We conjecture, in fact, that these properties are equivalent: that is, that the fundamental group of every Haken manifold is virtually representable to $\mathbb{Z}$. In Section 2 we prove this for a certain class of 3-manifolds.

In [H-2] we showed that the fundamental group of a closed, orientable, irreducible 3-manifold $M \neq S^3, P^3$ is virtually representable to $\mathbb{Z}$ provided

(i) there is an orientation reversing involution $\tau : M \to M$ and

(ii) $\pi_1(M)$ has a subgroup of even index.

While the first assumption is clearly restrictive, this is not clear for the second condition (which is also necessary for the conclusion). Thus it is reasonable to

1.1. Conjecture. *If M is a closed 3-manifold with $\pi_1(M)$ infinite, then $\pi_1(M)$ has a subgroup of even index.*

In Section 3 we give some improvements to the results of [H-2]. We show the conclusion is valid without assumption (ii) unless some finite cover of the manifold gives a counterexample to

1.2. Conjecture. *If M is a closed 3-manifold with $\pi_1(M)$ infinite, then $\pi_1(M)$ has a proper subgroup of finite index.*

Finally, in Section 4 we provide numerous applications of these results in a setting (regular branched covers of S^3) in which conjecture 1.2 is known to be valid and the orientation reversing involutions exist for obvious geometric reasons.

We point out that there are examples (c.f. [Hg]) of finitely presented, infinite groups with no nontrivial finite quotient groups. Whether any such groups occur as fundamental groups of 3-manifolds remains open.

We work through in the piecewise-linear, PL, category, but note that the results remain valid in the smooth category. We refer to [H-1] for definitions.

2. UNIONS OF KNOT MANIFOLDS. Perhaps the simplest way of constructing Haken manifolds with finite first homology is by identifying the boundaries of two *knot manifolds* (complements, in S^3, of open regular neighborhoods of nontrivial knots). Such a manifold will have infinite first homology if and only if the identification of boundaries matches the two longitudes.

We say that a knot manifold, M, is *representable* if there is a representation

$$\rho : \pi_1(M) \to PSL_2(\mathbb{C})$$

which takes the longitude to a nontrivial parabolic element. We note that every simple (atoroidal) knot manifold, M, is representable. For then either M has a hyperbolic structure [T], or M is the complement of a torus knot. The discrete, faithful representation, in the first case, or a representation of $\pi_1(M)$ to a triangle group (c.f. [R; Theorem 7]) in the second, has the desired property.

Note also that if $f : (M_1, \partial M_1) \to (M_2, \partial M_2)$ is a map of knot manifolds with $f|\partial M_1$ a homeomorphism, then M_1 is representable provided M_2 is. Such maps of knot spaces arise when, for the corresponding knots, k_1 is a satellite about another knot k and k_2 is obtained from k_1 by trivializing k. So, for example, the space of any twisted double is representable.

The following generalizes 2.5 of [H-3].

2.1. THEOREM. *Let $M_i, i = 1, 2$, be representable knot spaces, $f : \partial M_1 \to \partial M_2$ be any homeomorphism, and $M = M_1 \cup_f M_2$. Then $\pi_1(M)$ is virtually representable to $\mathbb{Z}$.*

PROOF. It follows from Lemma 2.2 below that there are finite, nonabelian representations $\tau_i : \pi_1(M_i) \to G_i$ such that

$$(f|\partial M_1)_*(Ker(\tau_1 \circ i_{1*})) = Ker(\tau_2 \circ i_{2*}).$$

The proof then follows as in [H-3]. Briefly: for the corresponding finite, regular covers $p_i : \tilde{M}_i \to M_i$ it follows that $p_i^{-1}(\partial M_i)$ has $n_i > 1$ components. We take $[n_1, n_2]/n_i$ copies of $\tilde{M}_i, i = 1, 2$, and identify the copies of the components of $\partial \tilde{M}_1$ to those of $\partial \tilde{M}_2$ by homeomorphisms covering f in such a way that the result, $\tilde{M}$, is a connected covering of M. The copies of $\partial \tilde{M}_i$ become nonseparating in $\tilde{M}$; thus $\beta_1(\tilde{M}) > 0$.

2.2. LEMMA. *Let M be a representable knot space. Then for almost all primes, p, there is a finite, nonabelian representation $\tau : \pi_1(M) \to G$ such that $Ker\{\tau \circ i_* : \pi_1(\partial M) \to G\}$ is the characteristic subgroup of index p^2 in $\pi_1(\partial M)$.*

PROOF. We must find a finite representation $\tau : \pi_1(M) \to G$ such that $\tau(i_* \pi_1(\partial M)) \cong \mathbb{Z}_p \times \mathbb{Z}_p$. Then G, being noncyclic, is nonabelian and $ker(\tau \circ i_*)$ is the unique characteristic subgroup of index p^2.

By assumption there is a representation $\rho : \pi_1(M) \to PSL_2(\mathbb{C})$ such that $\rho(\lambda)$ is parabolic (μ, λ the meridional and longitudinal elements). After conjugating this representation we may assume that $\rho(\lambda)$ is represented by the matrix $\begin{pmatrix} 1 & 1 \\ 0 & 1 \end{pmatrix}$. Commutativity forces that $\rho(\mu)$ be represented by $\begin{pmatrix} 1 & w \\ 0 & 1 \end{pmatrix}$. for some $w \in \mathbb{C} - \{1\}$.

Now $\rho(\pi_1(M))$ is contained in $PSL_2(A)$ for some subring A of $\mathbb{C}$ which is finitely generated as a ring (by the entries of $\rho(g_1), \ldots, \rho(g_n)$ for a finite set of generators $g_1, \ldots, g_n$ of $\pi_1(M)$). The multiplicative group of units of A is known to be finitely generated [S] and so can contain only finitely many prime integers. If p is a prime which is not a unit of A then p is contained in some maximal ideal I of A. The field A/I is finitely generated as a ring and therefore finite (c.f. [A,M; pg. 84]) and clearly of characteristic p. We have the induced representation $\overline{\rho} : \pi_1(M) \to PSL_2(A/I)$ with $\overline{\rho}(\mu^p) = \overline{\rho}(\lambda^p) = 1$.

There is a unique representation $\psi : \pi_1(M) \to \mathbb{Z}_p$ and the representation

$$\tau = (\overline{\rho}, \psi) : \pi_1(M) \to PSL_2(A/I) \times \mathbb{Z}_p$$

is the desired one.

3. INVOLUTIONS. Let M be a closed, orientable 3-manifold and $\tau : M \to M$ an orientation reversing PL involution. The possibilities $M = S^3$, Fix $\tau = S^2$ and $M = S^3$, Fix $\tau = S^0$ occur. The second induces one with $M = P^3$ and Fix $\tau = P^2 \cup P^0$.

The following theorem shows that any other example has $\pi_1(M)$ virtually representable to $\mathbb{Z}$, or there is a counterexample to the Poincaré conjecture, or a counterexample to conjecture 1.2 which covers M in a specified manner.

3.1. THEOREM. *Let M be a closed, orientable 3-manifold with $o(\pi_1(M)) > 2$ and which admits an orientation reversing PL involution $\tau : M \to M$.*
Then either:
(i) *$\pi_1(M)$ is virtually representable to $\mathbb{Z}$, or*
(ii) *Fix τ is a 2-sphere and $\pi_1(M)$ is an infinite group with no subgroups of finite index, or*
(iii) *Fix τ contains a projective plane, P, and (ii) applies to the double cover $(\tilde{M}, \tilde{\tau})$ of (M, τ) obtained by untwisting at P, or,*
(iv) *#(Fix τ) = 2 and the commutator subgroup of $\pi_1(M)$ is an infinite group with no proper subgroup of finite index and which has odd index in $\pi_1(M)$.*

PROOF. In Theorem 2.2 of [H-2] we showed with these hypotheses that $\pi_1(M)$ is virtually representable to $\mathbb{Z}$ provided that M is irreducible (and different from S^3, P^3), and in case $dim($ Fix $\tau) = 0$, that $\pi_1(M)$ has a subgroup of even index.

Irreducibility is used in the argument of [H-2] only in the case Fix τ is a separating 2-sphere or contains a projective plane. In the first case Fix τ splits M into homeomorphic manifolds M_1 and M_2 which are interchanged by τ. Since $\pi_1(M) \cong \pi_1(M_1) * \pi_1(M_2)$ is nontrivial, it is infinite. If $\pi_1(M)$ has a proper subgroup of finite index, then intersecting this subgroup with its finitely many distinct conjugates gives a proper normal subgroup of finite index in $\pi_1(M)$; hence one such in both $\pi_1(M_1)$ and in $\pi_1(M_2)$. This allows the construction, just as in the proof of Theorem 2.1, of a finite cover $\tilde{M} \to M$ which contains a nonseparating 2-sphere and thus having $\beta_1(\tilde{M}) > 0$. The alternate gives conclusion (ii).

If Fix τ contains a projective plane we easily reduce to the above case on a double cover of M and obtain either conclusion (i) or (iii).

Thus we assume that Fix τ is a finite, nonempty set which must contain an even number, $2k$, of points. Let U be a τ-invariant regular neighborhood of Fix τ and $X = M - \text{Int } U$. Then τ acts freely on X and $Y = X/\tau$ is a nonorientable 3-manifold whose boundary consists of $2k$ projective planes.

Suppose $k > 1$. From

$$H_1(Y;\mathbb{Z}_2) \cong H_2(Y,\partial Y;\mathbb{Z}_2) \to H_1(\partial Y;\mathbb{Z}_2) \to H_1(Y;\mathbb{Z}_2)$$

we see that $\pi_1(Y)$ has a subgroup of index 4. Since $\pi_1(X)$ is a subgroup of index 2 in $\pi_1(Y)$, $\pi_1(X)$ has a subgroup of even index and conclusion (i) follows from [H-2].

Now suppose that $k = 1$. If $H_1(M)$ is infinite or has even order then we again have conclusion (i). So we assume that $[\pi_1(M),\pi_1(M)]$ has (finite) odd index in $\pi_1(M)$. We must have $[\pi_1(M),\pi_1(M)]$ infinite; otherwise, by [E], $\pi_1(Y) \cong \mathbb{Z}_2$ and we get the contradiction $\pi_1(M) \cong \pi_1(X) = \{1\}$.

If $[\pi_1(M),\pi_1(M)]$ has any proper subgroup of finite index then it contains a proper characteristic subgroup, N, of finite index. Then N is normal in $\pi_1(M) \cong \pi_1(X)$ and is invariant under $\tau_* : \pi_1(M) \to \pi_1(M)$; so N is also normal in $\pi_1(Y) \cong \pi_1(X) \rtimes \mathbb{Z}_2$. Note that $\pi_1(M)/N$ is finite and nonabelian. We assume it has an odd order.

Let

$$\tilde{M} \quad \to \quad M$$
$$p : \bigcup \qquad \bigcup$$
$$\tilde{X} \quad \to \quad X$$

be the regular coverings corresponding to N. Put Fix $\tau = \{x_1, x_2\}$ and let S_i be the boundary of the component of $M - X$ which contains x_i. Choose $\tilde{x}_1 \in p^{-1}(x_1)$. Since $\tau_*(N) = N$, there is an orientation reversing involution $\tilde{\tau} : (\tilde{M}, \tilde{x}_1) \to (\tilde{M}, \tilde{x}_1)$ covering τ.

The fixed points of $\tilde{\tau}$ are in one to one correspondence with the components of $\partial \tilde{X}$ which are invariant under $\tilde{\tau}$. One such component, $\tilde{S}_1$, of $p^{-1}(S_1)$ bounds a regular neighborhood of $\tilde{x}_1$. Since $\tilde{\tau}$ must permute the (odd number of) components of $p^{-1}(S_2)$ in pairs, some component $\tilde{S}_2$ of $p^{-1}(S_2)$ is τ-invariant. Let α be a loop in X based in S_1 and representing an element $a \in \pi_1(X) \backslash N$ and let $\tilde{\alpha}$ be the lifting of α to $\tilde{X}$ which begins in $\tilde{S}_1$. Then $\tilde{\alpha}$ ends in a component S_1^* of $p^{-1}(S_1)$ different from $\tilde{S}_1$. If $\tilde{\tau}(S_1^*) = S_1^*$ then $\tilde{\tau}$ has more than two fixed points and the previous case applies to show that $\pi_1(\tilde{M})$ (hence $\pi_1(M)$) is virtually representable to $\mathbb{Z}$.

If $\tilde{\tau}(S_1^* \neq S_1^*$, then $a\,t\,a^{-1}t^{-1} \notin N$ where

$$t \in \pi_1(Y) = \pi_1(X) \times \mathbb{Z}_2$$

is represented by the nontrivial loop in S_1/τ. Thus we assume that $a\,t\,a^{-1}t^{-1} \notin N$ for all $a \in \pi_1(X) \backslash N$, or, equivalently, that for all

$$1 \neq g \in \pi_1(X)/N \subset \pi_1(Y)/N \cong \pi_1(X)/N \times \mathbb{Z}_2$$

that $g\,t\,g^{-1}t^{-1} \neq 1$.

Thus the function $g \mapsto g\,t\,g^{-1}t^{-1}$ is one to one and so maps the finite set $\pi_1(X)/N$ onto itself. So for $h \in \pi_1(X)/N$, there is some $g \in \pi_1(X)/N$ with $h = g\,t\,g^{-1}t^{-1}$. Then (recalling $t^2 = 1$)

$$t\,h\,t^{-1} = t\,g\,t\,g^{-1}t^{-2} = t\,g\,t^{-1}g^{-1} = h^{-1}$$

But $h \mapsto t\,h\,t^{-1}(= h^{-1})$ is an automorpohism of $\pi_1(X)/N$ —which can only hold if $\pi_1(X)/N$ is abelian. This is the final contradiction which completes the proof.

3.2. COROLLARY. *If a closed, orientable 3-manifold, M, admits an orientation reversing involution and $o(\pi_1(M)) > 2$, then $\pi_1(M)$ is infinite.*

In [Sh] Shalen defines a group G to be *half-way residually finite* if either G is finite or G has subgroups of arbitrarily large finite index, and proved that the fundamental group of every regular branched cover of S_3 is half-way residually finite. This condition eliminates possibilities (ii), (iii), and (iv) of Theorem 3.1. Thus we have

3.3. COROLLARY. *Let M be a closed, orientable 3-manifold with $\pi_1(M)$ half-way residually finite and of order greater than two. If M admits an orientation reversing PL involution, then $\pi_1(M)$ is virtually representable to $\mathbb{Z}$.*

4. **EXAMPLES.** A link $L \subset S^3$ is *strongly amphicheiral* if there is a PL involution

$$\tau : (S^3, L) \to (S^3, L)$$

which reverses orientation on S^3. Examples are easily constructed by choosing L to be symmetric about the fixed point set of one of the orientation reversing PL involutions of S^3. Figure 1 illustrates, in this way, that the figure eight knot is strongly amphicheiral.

The strongly amphicheiral 2-bridge knots are classified in [H,K].

The double cover of S^3 branched over a two-bridge knot is always a lens space—illustrating that the orientation reversing involution $\tau : (S^3, L) \to (S^3, L)$

need not induce one on a branched cover. However, for odd sheeted branched covers we have:

4.1. THEOREM. *Let $p : M \to S^3$ be an odd sheeted regular branched covering whose branch set $L \subset S^3$ is a strongly amphicheiral link. If $o(\pi_1(M)) > 2$, then $\pi_1(M)$ is virtually representable to $\mathbb{Z}$.*

PROOF. There is a PL involution $\tau : (S^3, L) \to (S^3, L)$ which reverses orientation on S^3.

Let U be an invariant regular neighborhood of $L, X = S^3 - \mathrm{Int}\,U$ and $\hat{X} = p^{-1}(X)$. Then $p|\hat{X}; \hat{X} \to X$ is a regular (unbranched) covering with an odd number, λ, of sheets. Let

$$N = p_*\pi_1(\hat{X}) \cap \tau_* p_* \pi_1(\hat{X}) < \pi_1(X).$$

Then N is a normal, τ_*-invariant subgroup of $\pi_1(X)$ whose index (a divisor of λ^2) is also odd. Let $q : \tilde{X} \to X$ be the regular cover corresponding to N. If S is any component of $\partial \hat{X}$ then

$$p_* Ker(\pi_1(S) \to \pi_1(M)) < N.$$

Thus the regular cover $\tilde{X} \to \hat{X}$ extends to an unbranched regular cover $r : \tilde{M} \to M$.

Since $\tau_*(N) = N$, τ is covered by a map $\tilde{\tau} : \tilde{M} \to \tilde{M}$. $\tilde{\tau}^2$ is a covering transformation so must have odd order, say k. Then $\tilde{\tau}^k$ must reverse orientation and have order 2. It follows from Corollary 3.3 and Theorem 2.1 of [Sh] that $\pi_1(\tilde{M})$, hence $\pi_1(M)$, is virtually representable to $\mathbb{Z}$.

FIGURE 1. Knot invariant under the involution $x \to -x$ of $R^3 \cup \{\infty\}$

REFERENCES

[A,M] M. F. Atiyah and I. G. MacDonald, *Introduction to Commutative Algebra*, Addison Wesley (1969).

[E] D.B.A. Epstein, *Projective planes in 3-manifolds*, Proc. Lond. Math. Soc. *(3)* **11** (1961), 469–484.

[H,K] Richard Hartley and Akio Kawauchi, *Polynomials of Amphicheiral knots*, Math. Ann. **243** (1979), 63–70.

[H-1] John Hempel, *3-Manifolds*, Annals of Math Study No. *81*, Princeton Univ. Press, 1976.

[H-2] John Hempel, *Orientation reversing involutions and the first betti number for finite coverings of 3-manifolds*, Invent. Math. **67** (1982), 133–142.

[H-3] John Hempel, *Homology of coverings*, Pacific J. Math. **112** (1984), 83–113.

[Hg] G. Higman, *A finitely generated infinite simple group*, J. Lond. Math. Soc. **26** (1951), 61–64.

[R] Robert Riley, *Parabolic representations of knot groups, I*, Proc. Lond. Math. Soc. *(3)* **24** (1972), 217–242.

[S] P. Samuel, *A propos du théorème des unités*, Bull. Sci. Math., **90** (1966), 89–96.

[Sh] Peter B. Shalen, *A torus theorem for regular branched covers of S^3*, Mich. Math. J. **28** (1981), 347–358.

[T] William P. Thurston, *Three dimensional manifolds, kleinian groups and hyperbolic geometry*, Bull. Amer. Math. Soc. *(new series)* **6** (1982), 357–381.

RICE UNIVERSITY

Contemporary Mathematics
Volume **44**, 1985

LECTURES ON 3-FOLD SIMPLE COVERINGS AND 3-MANIFOLDS

José Maria Montesinos*

To Professor Arthur Stone

In these lectures we will essentially study 3-fold simple branched coverings of the 3-sphere S^3 as a procedure for representing closed, orientable 3-manifolds ("3-manifolds" along the paper). After motivating our theme by a look to the case of surfaces, we will concentrate on 3-manifolds.

1. Introduction: the case of surfaces.

1.1 A simplicial map $f : M^n \to N^n$ between two compact, triangulated n-manifolds M and N is a *branched covering* if it is an ordinary covering out of the $(n-2)$-skeleton of N. The points of N whose preimage has less points than the number m of sheets of f, form a subcomplex R^{n-2} of N^n which is called the *branching set*. We will say that "f is an m-fold covering of N branched over R."

Given R^{n-2}, sub-complex of N^n, each covering of $N - R$ has a unique completion (cfr. [6]) called the *associated branched covering*. If B is the preimage of R we have the following diagram

$$
\begin{array}{ccc}
M - B & \lhook\joinrel\longrightarrow & M \\
{\scriptstyle p'}\big\downarrow & & \big\downarrow{\scriptstyle p} \\
N - R & \lhook\joinrel\longrightarrow & N
\end{array}
$$

where p' is the original covering and p is the associated branched covering. Let $b \in B$ and let U be a neighborhood of $p(b)$ in N. Let V be the component of $p^{-1}U$ containing b. Then $V - B \to U - R$ is a u-fold covering. The minimal u for every U is the *branching index in b*. If this index is finite for every $b \in B$, then M is triangulated and B is a $(n-2)$-subcomplex. If R is a locally flat submanifold, then M is a manifold and B is a submanifold.

---

*Supported by "Comisión Asesora de investigación científica y técnica"

Thus in the branched covering $p: M \to N$, M is determined by $M - B$, and this manifold is determined by the monodromy

$$\omega : \pi_1(N - R) \to S_m$$

of the (unbranched) covering p', where S_m is the symmetric group of m indices.

1.2 Let K be the group generated by reflections on the sides of the triangle of $\mathbb{R}^2$ with vertices in $(0,0), (1,1)$ and $(1,0)$. The translations of vectors $(4,0)$ and $(0,4)$ generate a normal subgroup G of K. Let H be the normal subgroup of K which extends G by the reflection u through $(0,0)$. We have the commutative diagram

$$\mathbb{R}^2 \longrightarrow \mathbb{R}^2/H \cong S^2$$

$$\mathbb{R}^2/G = T^2 \qquad p$$

Thus p is a 2-fold covering of the 2-sphere S^2 branched over the 4 points A_i of Figure 1. Here T^2 is the 2-torus, and S^2 is the quotient by u, the 180° rotation around the axis passing through the points $\tilde{A}_i$.

1.3 The group K/G acts on T^2 and this action projects to the action of K/H on S^2. The group K/H, being the full group of symmetries of S^2 fixing the set $\{A_1, A_2, A_3, A_4\}$, coincides with $\mathbb{Z}_2 \times D_4$, where D_4 is the dihedral group of order 8. The element t of this group which is a 90° rotation of S^2 sending A_1 to A_4 followed by a reflection on the plane of the points A_i, is the projection of a 90° rotation of T^2 around $(1,1)$ followed by reflection on $\tilde{H}'$. We call this homeomorphism $\tilde{t}$ and we see that the homeomorphism $\tilde{t}_* : H_1(T^2; \mathbb{Z}) \circlearrowleft$ is given by the equations:

$$[\tilde{t}_* \tilde{Q}, \tilde{t}_* \tilde{H}] = [\tilde{Q}, \tilde{H}] \begin{bmatrix} 0 & 1 \\ 1 & 0 \end{bmatrix}$$

1.4 The homeomorphism of $\mathbb{R}^2$ given by $(x,y) \mapsto (x, y - x)$ is isotopic to the one of Figure 2, which is a suitable translation on the shadowed vertical bands. This homeomorphism projects to the homeomorphisms $\tilde{v}^{-1}: T^2 \to T^2$ and $v^{-1}: S^2 \to S^2$ shown in Figure 3. Thus $\tilde{v}^{-1}$ is what is called a *negative Dehn-twist along $\tilde{Q}'$* inducing the automorphism

$$[\tilde{v}_*^{-1} \tilde{Q}, \tilde{v}_*^{-1} \tilde{H}] = [\tilde{Q}, \tilde{H}] = \begin{bmatrix} 1 & -1 \\ 0 & 1 \end{bmatrix}$$

in homology. The homeomorphism v^{-1} is a *positive half-twist along Q'*.

1.5 The homeomorphism $v^{a_1} t v^{a_2} t \ldots v^{a_r} t$ of S^2, where a_i are integers, lifts to a homeomorphism of T^2 with matrix

$$\begin{bmatrix} a_1 & 1 \\ 1 & 0 \end{bmatrix} \begin{bmatrix} a_2 & 1 \\ 1 & 0 \end{bmatrix} \cdots \begin{bmatrix} a_r & 1 \\ 1 & 0 \end{bmatrix} = \begin{bmatrix} \alpha & \gamma \\ \beta & \delta \end{bmatrix},$$

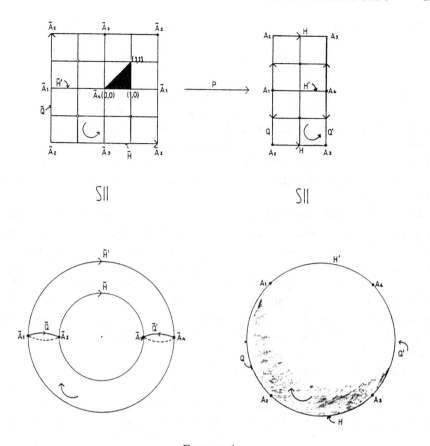

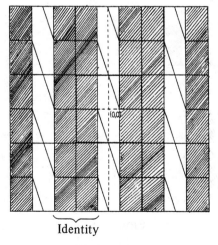

FIGURE 1

FIGURE 2

and is easy to check that

$$\frac{\alpha}{\beta} = a_1 + \cfrac{1}{a_2 + \cdots + \frac{1}{a_r}}.$$

The image of $\tilde{Q}$ by this homeomorphism $\tilde{f}(\alpha/\beta)$ is $\alpha\tilde{Q} + \beta\tilde{H}$. The projection of $\tilde{f}(\alpha/\beta)$ is called $f(\alpha/\beta)$. What we essentially have is that, up to isotopy, every homeomorphism of T^2 commutes with the involution u (see [4], [28]).

1.6 It is easy to see that every closed, orientable surface F_g of genus $g > 1$ is a 2-fold covering of S^3 branched over $2g+2$ points. But except for $g = 1, 2$, it is not true that every homeomorphism of F_g commutes with the covering involution (cfr. [2], [42], [7]).

1.7 We want to give another example of branched covering between surfaces, namely an irregular 3-fold covering, coming from the algebraic geometry of curves in $\mathbb{C}P^2$. Take a non singular cubic C in $\mathbb{C}P^2$ and a point p in its complement so that among the lines passing through p, exactly 6 are tangent to C (cfr. [3], p. 268). The set of lines through p is $\mathbb{C}P^1 \cong S^2$ and we have a map $g : C \to \mathbb{C}P^1$ which is 3 to 1 in general. But if $x \in \mathbb{C}P^1$ is tangent to C, $g^{-1}(x)$ consists of 2 points. The map g is an irregular covering, C is a torus and the preimage of a point in the branching set consists of a point of branching index 2 (the point of tangency) and a point of branching index 1.

2. Two-fold branched coverings of S^3.

2.1 The typical example is the 2-fold covering $P : S^3 \to S^3$ where the covering involution U is a 180° rotation around an axis of $S^3 = \mathbb{R}^3 + \infty$. Thus the branching set is the trivial knot T of S^3. More generally, given a link L of S^3 we consider the representation $\omega : \pi_1(S^3 - L) \to S_2$ which sends the meridians to

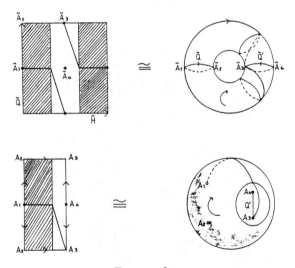

FIGURE 3

the transposition $(12) \in S_2$. The manifold $M(L, \omega)$ is the *2-fold covering of S^3 branched over L*.

2.2 We consider the torus T^2 of Figure 1 embedded in $S^3 = \mathbb{R}^3 + \infty$ so that the axis of the involution U passes through the points $\tilde{A}_i$, thus inducing the involution u in T^2 as in Figure 4.

The core of the solid torus V bounded by T^2 is a *strongly-invertible knot* K(or *link*, if we use a number of tori T^2, which are left invariant by U). The image of T^2 under $P : S^3 \to S^3$ is also shown in Figure 4. We select the curve $\tilde{H}$ on T^2 so that it is homologous to zero in $S^3 - \text{Int} V$. The manifold $M(K, \alpha/\beta)$ obtained by Dehn-surgery (with new meridian $\alpha\tilde{Q} + \beta\tilde{H}$) on K is $V \bigcup_{\tilde{f}(\alpha/\beta)}(S^3 - \text{Int} V)$ and this is a 2-fold covering of $S^3 \cong B^3 \bigcup_{f(\alpha/\beta)}(S^3 - \text{Int} B^3)$, the branching set is obtained replacing the arcs $A_1 A_2, A_3 A_4$ by the

rational tangle ⊗ defined as in Figure 5 (for $\alpha/\beta = 38/7 = 5 + \frac{1}{2 + \frac{1}{3}}$).

This shows that *every 3-manifold obtained by Dehn-surgery on a strongly invertible link is a 2-fold branched covering of S^3* [31], [43].

In particular, the lens space $L(\alpha, \beta)$ is the result of Dehn-surgery α/β on the trivial knot in S^3 which is strongly invertible. Thus $L(\alpha, \beta)$ is a 2-fold covering

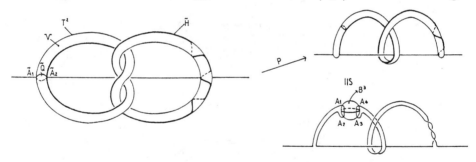

FIGURE 4

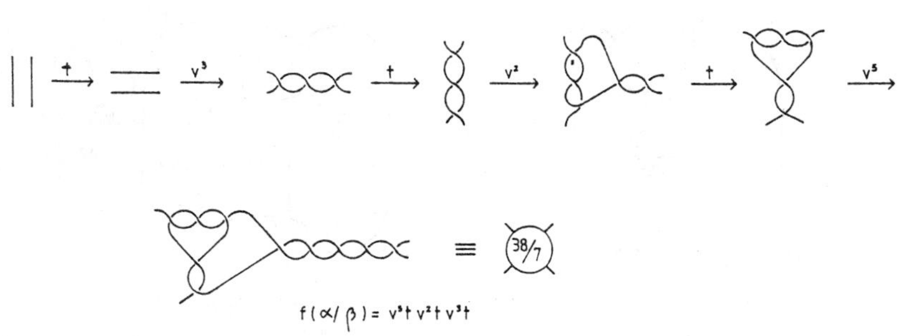

$$\| \xrightarrow{t} = \xrightarrow{v^3} \text{)○○(} \xrightarrow{t} \text{)(} \xrightarrow{v^2} \text{)(} \xrightarrow{t} \text{)(} \xrightarrow{v^5}$$

$$\text{)(} \text{○○○○○} \equiv \text{⊗} \,38/7$$

$$f(\alpha/\beta) = v^3 t\, v^2 t\, v^3 t$$

FIGURE 5

of S^3 branched over the *rational link* $\left(\begin{smallmatrix} \alpha/\beta \end{smallmatrix}\right) = R(\alpha/\beta)$. This is a knot when α is odd [38].

2.3 It is easy to see (cfr. [35], p.114) that the orientable Seifert manifolds have the Dehn-surgery description shown in Figures 6 and 7. The links of Figure 6 and 7 are symmetric with respect to an axis. Then from 2.2 it follows that these manifolds are branched coverings as depicted in Figure 8. Thus the manifolds with base S^2 or a non orientable surface are 2-fold branched coverings of S^3. The ones with orientable base of genus g are 2-fold branched coverings of $g\#S^1 \times S^2$. Similar results hold for the graph-manifolds of Waldhausen (cfr. [28], [32]).

2.4 We remark that there are 3-manifolds obtained by Dehn-surgery on a non-invertible link which are 2-fold branched coverings of S^3 (cfr. [48], [9]). On

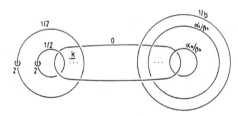

$$(\, O \, o \, g / b \, ; \quad \alpha_1/\beta_1 \, \cdots \cdots \, \alpha_s/\beta_s)$$

FIGURE 6

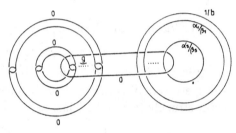

$$(\, O \, n \, k / b \, ; \quad \alpha_1/\beta_1 , \ldots , \, \alpha_s/\beta_s)$$

FIGURE 7

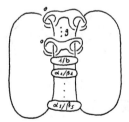 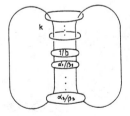

FIGURE 8

the other hand, there are 3-manifolds which are not 2-fold branch coverings of S^3, for instance $S^1 \times S^1 \times S^1$ [7] (cfr. [27], [28], [19], [24], [9], [37]).

2.5 Though S^3 is a 2-fold branched covering of S^3 in essentially one way [45], and the same happens with some other manifolds ($n\#S^1 \times S^2$ [45]; lens spaces [20]), there are manifolds with different presentations as 2-fold branch coverings spaces (see [42], [25], [49], [44], [32], [8], [50], [36]); for instance, the torus knot $\{3,7\}$ and the pretzel knot $\{-2,3,7\}$ have the same associated 2-fold covering (cfr. also [32], [21; problem 3.41], [22]).

2.6 Since there are 3-manifolds without periodic homeomorphisms [41], it is impossible to represent the class of all 3-manifolds as regular coverings of S^3. This is why we look to irregular coverings.

3. Irregular 3-fold coverings.

3.1 The irregular 3-fold coverings were considered long ago by Heegaard [10] (cfr. [52],[40]) in relation with algebraic surfaces. If we project $\mathbb{C}^3$ onto $\mathbb{C}^2$ by $(x,y,z) \mapsto (y,z)$, and we restrict this map to the algebraic affine surface F given by the equation $x^3 + yx - z = 0$, we obtain a map $p : F \to \mathbb{C}^2$ with the following properties: the preimage of each point $(y,z) \in \mathbb{C}^2$ consists of the points $(x_i, y, z), 1 \leq i \leq 3$, where x_i is a solution of $x^3 + yx - z = 0$ (see Figure 9; cfr. the beautiful book [3, p.233]).

This map is a 3-fold branched covering, the ramification being the discriminant $D \equiv 27z^2 + 4y^3 = 0$. The preimage of the points of D consists of 2 points except for $(0,0)$ which has only one. Thus this covering is irregular. The restriction of p to $p^{-1}S^3$, where $S^3 = \{(y,z) \in \mathbb{C}^2 | y\bar{y} + z\bar{z} = 1\}$, defines a 3-fold covering $g : S^3 \to S^3$ branched over the trefoil 3_1 (Figure 10), because $p^{-1}S^3$, being the boundary of a regular neighborhood of $p^{-1}(0,0)$ in F, must be S^3. The monodromy $\omega : \pi_1(S^3 - 3_1) \to S_3$ sends meridians to transpositions, because the branching indices of the two points which form the preimage of $P \in 3_1$ must be 2 and 1. Since ω must be onto, we have, up to conjugation in S_3, $\omega(x) = (12)$, $\omega(y) = (13)$ [and $\omega(z) = (23)$]. The covering $p^{-1} : D^4 \to D^4$ where $D^4 = \{(y,z) \in \mathbb{C}^2 | y\bar{y} + z\bar{z} \leq 1\}$ is equivalent to the cone of $g : S^3 \to S^3$.

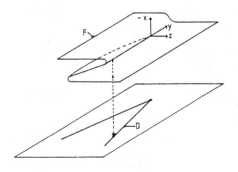

FIGURE 9

3.2 We are going to study 3-fold coverings $p : M^3 \to S^3$ branched over a link $L \subset S^3$, given by a transitive representation

$$\omega : \pi_1(S^3 - L) \to S_3$$

which send meridians to transpositions. We call this sort of coverings *3-fold simple coverings*. The example of 3.1 is typical. The monodromy group of the covering is the dihedral group of 6 elements S_3 and the group of covering transformations is trivial. A nice way for defining ω, given L, is due to Ralph Fox [5]. Three colors (say G=green, R=red, B=blue) are used to color the bridges of the link. This is done in such a way that three colors that meet at an overcrossing are either all the same, or all distinct. Then a representation is defined on the meridians, $G \to (12)$, $R \to (13)$, $B \to (23)$. The color condition guarantees that the defining relations are sent to the identity in the Wirtinger presentation. If at least two colors are used, the representation is transitive. A representation defined in this way is called a *colored knot or link*. Since $p : M \to S^3$ is not the quotient by a group acting on M, it is a little difficult to visualize p. We describe now a 3-fold simple covering $p : S^3 \to S^3$ as the starting step for proving that every 3-manifold is a 3-fold simple covering.

3.3 We describe $p : S^3 \to S^3$ as the double $2p'$ of a map $p' : D^3 \to D^3$. The map p' is depicted in Figure 11, where the ball D^3 is the result of performing identifications in the sides of the cylinder as depicted. The map p' is the result of folding D^3 along the axis B_{12} and B'_{13}, exactly as one does when folding a letter.

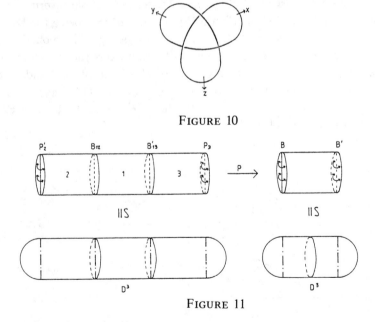

FIGURE 10

FIGURE 11

Then $p'^{-1}B$ is the union of B_{12} (branching index 2) and P_3 (branching index 1). We will call $B_{12} \cup B'_{13}$ the *branching cover* and $P'_2 \cup P_3$ the *pseudobranching cover*. Note the important fact that p' near B_{12} (and B'_{13}) works like a 2-fold branched covering. We use this and the facts about 2-fold covering, displayed in section 2, to represent any 3-manifold as a 3-fold simple covering.

Now the map $p = 2p : 2D^3 \to 2D^3$ is very easy to understand. The branching set is composed of two unlinked trivial knots and its preimage consists of four unlinked trivial knots.

The coverings q (of 3.1) and p are clearly different in as much as they have different branching sets. But they are closely related as we shall see later. The multiplicity of representations of a manifold as a simple 3-fold covering of S^3 is typical as we shall see.

3.4 Consider some 3-manifold M obtained by surgery on a link L of S^3. To fix the ideas, assume L is the non invertible knot 8_{17} of Figure 12, and M is obtained by surgery on it. Take 8_{17} and place it between the axis P'_2 and B_{12} of Figure 11. We then reflect 8_{17} through B_{12} to get a 180° rotated copy $8'_{17}$ of 8_{17} between B_{12} and B'_{13}. Finally we make the connected sum of 8_{17} and $8'_{17}$ to get the strongly invertible knot $8_{17}\#8'_{17}$ with respect to the axis B_{12}, as depicted in Figure 13.

FIGURE 12

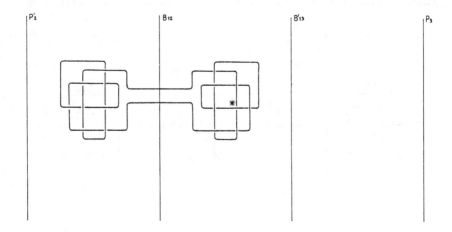

FIGURE 13

We now look for crossings of L such that local changes of underpasses by overpasses replace L by a trivial split link (in $L = 8_{17}$, the crossing marked with a star). We now introduce around those crossings a set T of unlinked trivial knots (Figure 14) which we make symmetric with respect to B'_{13}. Now the manifold M can be obtained by surgery on $8_{17} \# 8'_{17} \cup T$ because doing the Hempel's trick [11] on T we perform a local change of an underpass by an overpass and thus $8_{17} \# 8'_{17}$ becomes the original knot 8_{17}. Thus we can assume that M is obtained by α/β-surgery on $8_{17} \# 8'_{17}$ and ± 1 surgery on T.

Projecting Figure 14 through p we get Figure 15 and the image of $8_{17} \# 8'_{17} \cup T$ is the union of the arcs $p(8_{17} \# 8'_{17}), p(T)$ with their endpoints on the branching set $B \cup B'$. Replacing the rational tangles $N(p(8_{17} \# 8'_{17})) \cap B$, $N(p(T)) \cap B'$ [where $N(\cdots)$ stands for regular neighborhoods] by the tangles α/β and ± 1, the branching set B change to $\overline{B}$ and the corresponding 3-fold covering changes by α/β-surgery on $8_{17} \# 8'_{17}$, (± 1)-surgery on T, and two surgeries on the regular neighborhood of the two arcs which are contained in the preimage of $p^{-1}(p(8_{17} \# 8'_{17} \cup T))$. Thus the covering space changes from S^3 to M and this shows that M is a simple 3-fold covering of S^3 branched over $\overline{B}$.

Since every 3-manifold can be obtained by surgery on a link of S^3 [23] [46], the above argument shows that *every 3-manifold is a 3-fold simple cover of S^3* [12] [29] (cfr. [13], [30], [18], [33], [1]) *and thus it corresponds to a colored link.*

3.5 As an example, consider the 3-torus T^3 which is the result of 0-surgery on the borromean rings. Figure 16 depicts the borromean rings. Figure 17 shows the branching set.

3.6 The last example is interesting because the branching set is the boundary of two orientable surfaces. This suggests a method to obtaining many examples of colored links (and of branched coverings). Thus let G and R be a pair of bounded disjoint surfaces in S^3, which we can think of as discs with bands in "general position with respect to a plane" so that all the singularities of the

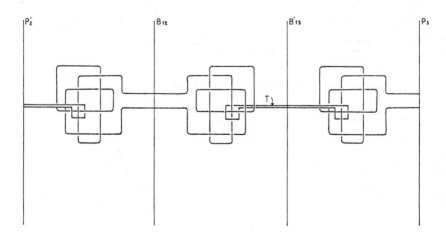

FIGURE 14

projection are the type shown in Figure 18. If the boundary of G^2 is colored green except for those parts that are under R^2, and the boundary of R^2 is colored red except for those parts under G^2, we obtain a colored link which we call *natural*. Then the proof of 3.4 shows that *every 3-manifold is a 3-fold covering of S^3 branched over the natural colored link associated to connected surfaces G, R*(cfr. [14]), moreover, the surfaces *can be supposed to be orientable*[17] (cfr. [16], [21]). The proof consists in showing that we can obtain an analogous situation to the one of Figure 14 but with *even-integer* framings, but we will not give the details here.

4. Moves.

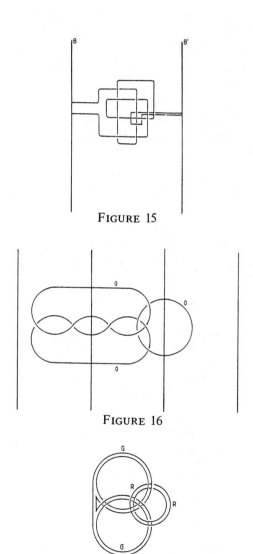

FIGURE 15

FIGURE 16

FIGURE 17

4.1 Every link L in S^3 has a *n-bridge presentation*, i.e. a triad (S^3, L, S^2), where the 2-sphere S^2, which separates S^3 into two balls D_1^3, D_2^3, is such that $D_i \cap L$ is a collection of n unknotted and unlinked arcs properly embedded in D_i, for $i = 1, 2$. If $p : M^3 \to S^3$ is a covering branched over L, then the sphere S^2 lifts to a closed orientable surface which defines a Heegard splitting of M^3.

4.2 If $p : M^3 \to S^3$ is a simple 3-fold covering, the genus g of $F_g = p^{-1}S^2$ is $n - 2$. Hence if L is a 2-bridge link, the manifold M^3, having genus zero, must be S^3. This is what happens with the example of 3.1 because the trefoil knot has two bridges.

4.3 The class of 2-bridge links coincides with the class of rational links $R(\alpha/\beta)$ (see 2.2) and $R(\alpha/\beta)$ can be colored exactly when 3 divides α. Hence if in the projection of a colored knot we perform the move of Figure 19 the covering manifold does not change because we are cutting out a ball upstairs and regluing it back again in a different way. This follows from 4.2 [7], [25], [26], [18].

4.4 Using these moves one can represent a 3-manifold M as a simple 3-fold covering of S^3 in many different ways (for instance, if $M = S^3$ in at least as many as rational links $R(3\alpha/\beta)$). The moves $M(3\alpha/\beta)$ are a consequence of the moves $M(\pm 3)$ (see Figure 20), as it is very easy to check. We have long ago posed the following

Problem. *Find a set of moves which do not change the covering and such that if two colored links have the same cover, they are related by a finite sequence of moves.*

One would like that $M(\pm 3)$ be the required set of moves [25] [7], but this is not the case [51]. The interest of the problem is that if that set does exist, we can translate the classification problem of 3-manifolds in the problem of classification of colored links under some combinatorial moves.

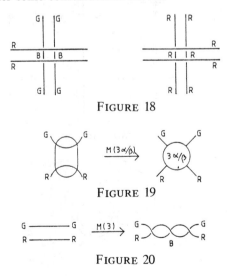

FIGURE 18

FIGURE 19

FIGURE 20

4.5 We invite the reader to apply moves $M(\pm 3)$ in the places of Figure 21 and to realize how the knots split in the components shown under them. We may say that a colored link L is *separable* if, using moves $M(\pm 3)$, it is possible to obtain a link $L_1 \cup L_2$, where L_1 is colored in green and L_2 in red (cfr. [25], [7], [34]).

It is easy to see that every colored closed braid with 2 or 3 strings is separable [25]. However, there are non-separable links, because *if the colored link L is separable into $L_1 \cup L_2$, the corresponding covering manifold M is a 2-fold covering of S^3 (branched over any of the connected sums $L_1 \# L_2$[25], [7]).* The proof of this statement is easy, and its converse is not true in general [51]. Thus, the colored link of Figure 17 [7] (see [27], [28]) is non-separable, while the manifolds which are associated to the knots of Figure 21 are 2-fold coverings of S^3; namely $S^3, \mathbb{R}P^3, L(3,1), L(4,1), L(5,2), S^1 \times S^2$, and the homology 3-sphere discovered by Poincaré (cfr. [47]).

Problem. *If the covering manifold associated to the colored link L is S^3, is L separable?*

Fox Conjecture. [7] *A colored link, having a simply connected associated covering manifold, is separable.*

This conjecture implies the Poincaré conjecture, because by the strong Smith conjecture, the associated covering manifold, being a 2-fold branched covering, must be S^3. On the other hand, the Poincaré conjecture implies the Fox conjecture exactly when the answer to the last problem is "yes".

4.6 It is easy to see that every colored link can be converted in a knot by a number of applications of moves $M(\pm 3)$. Hence we have that *every 3-manifold is a 3-fold simple covering of S^3 branched over a knot* [12], [29]. The simpler knot for S^3 is the trefoil. The branching set can also be converted in a pure closed braid [13] or even in a fibred link (using moves $M(\pm 3)$ to get the conditions of Stallings [39] for fibred closed braids). This fibration can be lifted to an open book decomposition of the associated covering 3-manifold.

5. Some applications.

5.1 We want to give a simpler and constructive proof, due to Hilden and the author, of the result [14] that *every 3-manifold is a simple 3-fold covering of S^3 branched over a knot so that the branching cover bounds an embedded disc* (the pseudo branching cover, though, can be knotted). The starting point is a 3-manifold M represented as a simple 3-fold covering of S^3 branched over the natural colored link associated to the connected surfaces G and B. The manifold M is constructed taking three copies $(S_K^3, G_K, B_K), K = 1, 2, 3$ of the triple (S^3, G, B), splitting along $G_K \cup B_K$ and gluing $G_1^{\pm}$ with $G_2^{\mp}$ and $B_2^{\pm}$ with $B_3^{\mp}$. Thus the branching cover bounds $G_2^+ \cup B_2^-$. We want to convert $G_2^+ \cup B_2^-$ in two discs which we will connect later. This is obtained using the moves which depend on $M(\pm 3)$ (see Figures 22 and 23):

To understand this, take a ball containing the clasp of Fig. 23 (see Figure 24) in the copy (S_2^3, G_2', B_2'). Cutting open this ball along $B_2' \cup G_2' \cup C_2'$, we obtain the ball depicted in Figure 25, where $a_2, b_2, c_2, b_2', c_2'$, are are the preimages of $\bar{a}_2$

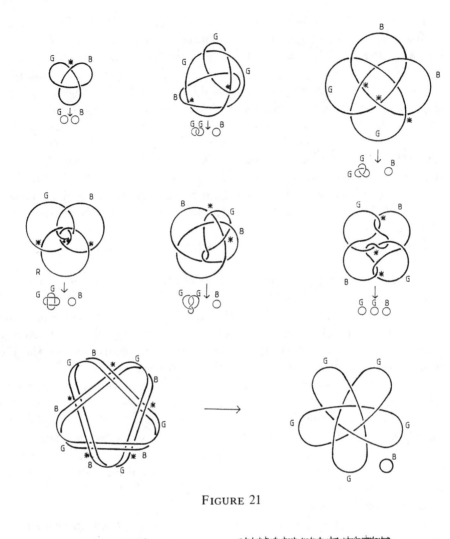

FIGURE 21

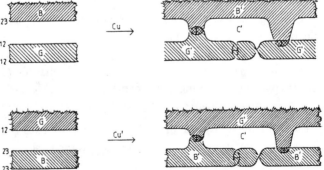

FIGURE 22

(for details of this process see [25], [26]). The arcs marked b (resp. p) belong to the branching (pseudo branching) cover.

Note that $a_3 = c_3$ because $G_3'^-$ and $G_3'^+$ must be identified. Also $a_3 = b_2$, $c_3 = b_2'$ because $B_3'^+$ is identified with $B_2'^-$. Hence $b_2 = b_2'$. By analogous reason $c_2 = c_2'$. Thus $(B_2'^-/b_1 = b_2') \cap (G_2'^+/c_2 = c_2') = \varnothing$ and $(B_2'^-/b_2 = b_2') \cup (G_2'^+/c_2 = c_2')$ bound the branching cover. In the preimage of the ball containing the clasp, we have the situation depicted in Figure 26. Thus the clasp downstairs is an unclasp in the branching cover.

Take now the ball containing the crossing of Figure 23 in the copy (S_2^3, G_2', B_2'), depicted in Figure 27. Cutting open this ball along $B_2' \cup G_2' \cup C_2'$ we obtain the ball depicted in Fig. 28, where $B_2'^-$ and $G_2'^+$ touch along an arc and thus they are connected.

FIGURE 23

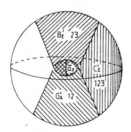

FIGURE 24

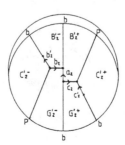

FIGURE 25

Using move C_u, it is possible to cut one by one all the bands of $G_2'^+$. The similar move C_u' would cut all the bands of $B_2'^-$. Finally move C_o would connect $B_2'^-$ with $G_2'^+$. This proves our theorem.

A nice consequence of the last theorem is that *every 3-manifold is parallelizable* because so it is out of the disc D bounded by the branching cover since here the projection is a local homeomorphism. On the other hand, since $\pi_2(SO(3)) = 0$ there is no obstruction to extend the parallelization to a ball containing the disc D [14].

5.2 We now show that *any 3-manifold is the pullback of any 3-fold simple covering* $p : S^3 \to S^3$ *and some smooth map* $\Omega : S^3 \to S^3$ *transversal to the*

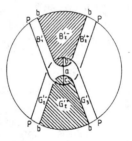

FIGURE 26

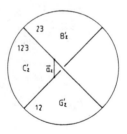

FIGURE 27

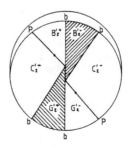

FIGURE 28

branching set of p. [17] . The starting point is a 3-manifold M represented as a simple 3-fold covering of S^3 branched over the natural colored link L associated with the *orientable* surfaces G and R. If we send green (red) meridians to the letter $g(r)$, we have a homeomorphism $\omega_* : \pi_1(S^3 - L) \to F_2$, where F_2 is the free group in the letters g, r. Note that to prove that ω_* is a homomorophism, we need both G and R to be orientable. The composition $\lambda\omega_*$, where $\lambda(g) = (12)$, $\lambda(r) = 13$ defines the representation giving rise to the covering $g : M \to S^3$ branched over L.

Let $S^1 \vee S^1$ be two circles identified along a point p. Since $S^1 \vee S^1$ is aspherical, there is a continuous map $\omega : S^3 - L \to S^1 \vee S^1$, realizing ω_*. We extend ω to $\Omega' : S^3 \to D^2 \vee D^2$ in the most natural way, where $D^2 \vee D^2$ are two discs identified along p.

We now take $p : S^3 \to S^3$ a 3-fold simple covering branched along the colored link K, and we embedded $D^2 \vee D^2$ in such a way that it cuts K transversaly in the centers of the two discs (Figure 29). The composition of Ω' and this embedding can be approximated by a smooth map. $\Omega : S^3 \to S^3$ transversal to K. Then clearly $\Omega^{-1}(K) = L$. Thus the pullback of p and Ω is a 3-fold simple covering of S^3 branched over L, corresponding to the representation

$$\pi_1(S^3 - L) \overset{\omega_*}{\to} \pi_1(S^1 \vee S^1) = F_2 \overset{i_*}{\to} \pi_1(S^3 - K) \overset{\eta}{\to} S_3$$

where i is the natural embedding and η is the representation corresponding to p. But $\eta i_* = \lambda$ hence that pullback is $g : M \to S^3$. This proves the theorem. An interesting question is to find the relationship between two $\Omega, \tilde{\Omega}$ giving rise to the same 3-manifolds.

5.3 A consequence of the last Theorem is the following partial sharpening of a beautiful theorem of Hilden [15] (cfr. [17]). *It is possible to embed any 3-manifold M in $S^3 \times D^2$ so that the composition with the projection in the first factor is a 3-fold simple covering.*Thus M is a sort of Riemann space as in the initial example of 3.1 with which we started our study.

In fact, during the proof of the last theorem we found a map $\Omega' : S^3 \to D^2 \vee D^2$. We have a 3-fold covering $p : E^2 \to D^2 \vee D^2$ branched over the centers of the discs (see Figure 30, compare with Figure 11).

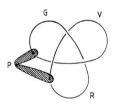

FIGURE 29

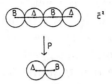

$$\text{FIGURE 30}$$

Clearly the pullback of p and Ω' is the covering $g : M \to S^3$. We then have the commutative diagram

$$
\begin{array}{ccc}
M & \overset{\tilde{\Omega}^{1'}}{\to} & E^2 \\
g\downarrow & & \downarrow p \\
S^3 & \overset{\Omega^{1'}}{\to} & D^2 \vee D^2
\end{array}
$$

The required embedding $M \to (S^3 \times D^2)$ is given by $(g, i\tilde{\Omega}')$, where $i : E^2 \to D^2$ is an embedding. Note that the number of points in $g^{-1}x$ coincides with the number of points of $\tilde{\Omega}' g^{-1}x$. This proves the theorem.

Note that $S^3 \times D^2$, being part of the boundary of D^6, embeds in S^5, hence we obtain the classical result of Morris Hirsch that every 3-manifold embeds in S^5.

REFERENCES

[1] I. Berstein and A. Edmonds, *On the construction of branched coverings of low-dimensional manifolds*, Trans. Amer. Math. Soc. **247** (1979), 87–124.

[2] J. S. Birman and H. M. Hilden, *The homeomorphism problem for S^3*. Bull. Amer. Math. Soc. **79** (1973), 1006–1010.

[3] E. Brieskorn and H. Knörrer, *Ebene Algebraische Kurven*, Birkhäuser (1981) Basel, Boston, Stuttgart.

[4] J. H. Conway, *An enumeration of knots and links, and some of their algebraic properties.* Computational Problems in Abstract Algebra, John Leech, Ed. Pergamon Press, New York (1970), 329–59.

[5] R. H. Crowell and R. H. Fox, *Introduction to Knot Theory.* Springer (1977).

[6] R. H. Fox, *Covering spaces with singularities.* Algebraic Geometry and Topology, A symposium in honour of S. Lefschetz. Princeton, 1957.

[7] R. H. Fox, *A note on branched cyclic coverings of spheres*, Revista Mat. Hisp.-Amer. **32** (1972), 158–166.

[8] C. Giller, *A family of links and the Conway calculus*, Trans. Amer. Math. Soc. **270** (1982), 75–107.

[9] M. Boileau, F. González and J. M. Montesinos, *Surgery on doubles and symmetries.*

[10] P. Heegaard, *Sur l'Analysis situs*, Soc. Math. France Bull. **49** (1916), 161–242.

[11] Hempel, J., *Construction of orientable 3-manifolds.* Topology of 3-manifolds, ed. M. K. Fort, Prentice Hall, 1962.

[12] H. M. Hilden, *Every closed, orientable 3-manifold is a 3-fold branched covering space of S^3*, Bull. Amer. Math. Soc. **80** (1974), 1243–4.

[13] H. M. Hilden, *Three-fold branched coverings of S^3*, Amer. J. Math. **98** (1976), 989–997.

[14] H. M. Hilden, J. M. Montesinos and T. Thickstun, *Closed orientable 3-manifolds as 3-fold branched coverings of S^3 of a special type*, Pacific J. Math **65** (1976), 65–75.

[15] H. M. Hilden, *Embeddings and branched coverings spaces for three and four dimensional manifolds*, Pacific J. Math. **78** (1978), 139–147.

[16] H. M. Hilden and J. M. Montesinos, *A method of constructing 3-manifolds and its application to the computation of the μ-invariant*, Proc. Sympos. in Pure Math. **32**, Amer. Math. Soc., Providence, RI (1977), p. 477–485.

[17] H. M. Hilden, M. T. Lozano and J. M. Montesinos, *All three-manifolds are pullbacks of a branched covering S^3 to S^3*, Trans. Amer. Math. Soc. **279** (1983), 729–735.

[18] U. Hirsch, *Über offene Abbildungen auf die 3-sphäre*, Math. Z. **140** (1974), 203–230.

[19] U. Hirsch and W. D. Neumann, *On cyclic branched coverings of spheres*, Math. Ann. **215** (1975), 289–291.

[20] C. D. Hodgson, *Involutions and isotopies of lens spaces*, M.S. Thesis, Melbourne (1981).

[21] R. Kirby, *Problems in low dimensional manifold theory*, Proc. of Symp. in Pure Math. **32** (1978), 273–312.

[22] S. Kojima, *Finiteness of symmetrics on 3-manifolds*, (preprint).

[23] W.B.R. Lickorish, *A representation of orientable combinatorial 3-manifolds*, Annals of Math. **76** (1962), 531–540.

[24] R. Myers, *Homology 3-spheres which admit no PL involutions*, Pacific J. Math. **94** (1981), 379–384.

[25] J. M. Montesinos, *Sobre la conjetura de Poincaré y los recubridores ramificados sobre un nudo*, Doctoral Thesis, Madrid 1971. (Departamento de publicaciones de la Facultad de Ciencias de la Universidad Complutense de Madrid).

[26] J. M. Montesinos, *Una nota a un teorema de Alexander*, Rev. Mat. Hisp.-Amer. **32** (1972), 167–87.

[27] J. M. Montesinos, *Una familia infinita de nudos representados no separables*, Rev. Mat. Hisp. Amer. (4) **33** (1973), 32–35.

[28] J. M.Montesinos, *Variedades de Seifert que son recubridores cíclicos ramificados de dos hojas*, Bol. Soc. Mat. Mejicana **18** (1973), 1-32.

[29] J. M. Montesinos, *A representation of closed, orientable 3-manifolds as 3-fold branched coverings of S^3*, Bull. Amer. Math. Soc. **80** (1974), 845–846.

[30] J. M. Montesinos, *Three-manifolds as 3-fold branched covers of S^3*, Quarterly J. Math. Oxford (2), **27** (1976), 85–94.

[31] J. M. Montesinos, *Surgery on links and double branched covers of S^3*, Ann. of Math. Studies (The R. H. Fox Memorial Volume) **84**, 1975.

[32] J. M. Montesinos, *Revêtement ramifiés de noeuds, Espaces fibres de Seifert et Scindement de Heegaard*, Orsay (1978).

[33] J. M. Montesinos, *A note on 3-fold branched covering of S^3*, Math. Proc. Camb. Phi. Soc. (1980).

[34] J. M. Montesinos, *Algunos aspectos de la teoría de cubiertas ramificadas*, Pub. Mat. UAB (1980), 109–126.

[35] J. M. Montesinos, *Variedades de Mosaicos*, Instituto Politécnico Nacional, México (1982).

[36] J. M. Montesinos and W. Whitten, *Constructions of two-fold branched covering spaces* (preprint).

[37] M. Sakuma, *Surface bundles over S^3 which are 2-fold cyclic coverings of S^3*, Math. Sem. Notes **9** (1981), 159–180.

[38] H. Schubert, *Knoten mit zwei Brücken*, Math. Z. **65** (1956), 133–170.

[39] J. R. Stallings, *Constructions of fibered knots and links*, Proc. Sympos. in Pure Math. **32**, A.M.S., Providence, RI (1977), 55–60.

[40] H. Tietze, *Uber die topologischen Invarianten mehrdimensionaler Mannigfaltigkeiten*, Monatsh. Math. **19** (1908), 1–118.

[41] J.L. Tollefson, *A 3-manifold with no PL involutions*, Notices of the Amer. Math. Soc. **22** (1975), A–231.

[42] O. Ja. Viro, *Linking, 2-sheeted branched coverings and braids*, Math. USSR, Sbornik **16** (1972), 223–36 (English translation).

[43] O. Ja. Viro, *Two-sheeted branched coverings of a three-dimensional sphere*, Zap. Nau. Sem. Leningrad Inst. **36** (1973), 6–39.

[44] O. Ja. Viro, *Non projecting isotopies and knots with homeomorphic coverings*, J. Soviet Math. **12** (1979), 86–96.

[45] F. Waldhausen, Über Involutionen der 3-Sphäre, *Topology* **8** (1969), 81–91.

[46] A. D. Wallace, *Modifications and cobounding manifolds*, Canad. J. Math. *(1960), 503–528.*

[47] *C. Weber and H. Seifert, Die beiden Dodekaederräume, Math. Z. **37** (1933), 237–253.*

[48] *W. Whitten, Inverting double knots, Pacific J. Math. **97** (1981), 209–216.*

[49] *J. S. Birman, F. González-Acuña and J. M. Montesinos, Heegaard splittings of prime 3-manifolds are not unique, Michigan Math. J. (1976), 97–103.*

[50] *C. Livingstone, More 3-manifolds with multiple knot-surgery and branched-cover description, Math. Proc. Camb. Phil. Soc. **91** (1981), 473–475.*

[51] *J. M. Montesinos, A note on moves and irregular coverings of S^4 (A. Stone proceedings, 1982).*

[52] *K. Brauner, Das Verhalten der Funktionen in der Umgebung ihrer Verzweigungsstellen, Abh. Hamburg **6** (1928), 1–55.*

FACULTAD DE CIENCIAS, UNIVERSIDAD DE ZARAGOZA

Contemporary Mathematics
Volume 44, 1985

WITT CLASSES OF TORSION LINKING FORMS

HAE SOO OH*

§1. Introduction.

By a *Seifert manifold*, we mean an oriented closed 3-manifold admitting a fixed point free action of the circle group S^1. Such a manifold is equivariantly classified by its *Seifert invariant* ([6], [8]). In this paper we shall use non-normalized Seifert invariants. They are described in §2. Our orientation convention for Seifert manifolds and lens spaces are the same as those adopted in [3], [6], and [7]. If M is an oriented manifold with an S^1-action and M/S^1 is an orientable manifold, orient M/S^1 so that the orientation of M/S^1 followed by the natural orientation of the orbits gives the orientation of M. Orient the boundary ∂M of an oriented manifold M so that an orientation of ∂M followed by an inward normal vector gives the orientation of M.

Let $\mathbb{Z}/a$ be a cyclic group of order a and let b be an integer relatively prime to a. Then $w(b/a)$ denotes an element of the Witt group $W(\mathbb{Q}/\mathbb{Z})$ of finite forms represented by the finite form $(\mathbb{Z}/a, f)$, where $f(1,1,) = b/a \epsilon \mathbb{Q}/\mathbb{Z}$. The purpose of this paper is to prove the following:

(1.1) MAIN THEOREM. *Let M be a Seifert manifold with associated Seifert invariant $(g : (\alpha_1, \beta_1), ..., (\alpha_n, \beta_n))$ and suppose p/q is $e(M) = -\sum_{i=1}^{n} (\beta_i/\alpha_i)$ expressed in lowest term. Then the Witt class of a torsion linking form of M is*

$$\begin{cases} -w(\beta_1/\alpha_1) & -\cdots - & w(\beta_n/\alpha_n) - w(1/pq) \text{ if } e(M) \neq 0, \\ -w(\beta_1/\alpha_1) & -\cdots - & w(\beta_n/\alpha_n) \text{ if } e(M) = 0. \end{cases}$$

Since reversing the orientation of M replaces the Seifert invariant $(g : (\alpha_i, \beta_i))$ by $(g : (\alpha_i, -\beta_i))$ and hence replaces $1/pq$ by $-1/pq$, the Witt class of a torsion linking form of $-M$ is $w(\beta_1/\alpha_1) + \ldots + w(\beta_n/\alpha_n) + w(1/pq)$ or $w(\beta_1/\alpha_1) + \ldots +_w (\beta_n/\alpha_n)$ according as $e(M) \neq 0$ or $e(M) = 0$.

The *torsion linking form* of a closed oriented $(4n-1)$-manifold N is a symmetric non-singular finite form defined by $\lambda(x,y) = < z \bigcup x, [N] > \epsilon \mathbb{Q}/\mathbb{Z}$, where $x, y \epsilon$ Tor $H^{2n}(N; \mathbb{Z})$ and z is a preimage of y under the Bockstein homomorphism

*The author is grateful to Professor P.E. Conner for his valuable suggestions and his encouragement.

$H^{2n-1}(N;\mathbb{Q}/\mathbb{Z}) \to H^{2n}(N;\mathbb{Z})$. The Witt class represented by $(Tor\, H^{2n}(N;\mathbb{Z}),\lambda)$ is denoted by $Lk(N)\epsilon W(\mathbb{Q}/\mathbb{Z})$.

Suppose B^{4n} is a closed oriented $4n$-manifold with $\partial B = N$ and let V be the image of the homomorphism $i^* : H^{2n}(B,\partial B;\mathbb{Q}) \to H^{2n}(B;\mathbb{Q})$. Then there exists a symmetric non-singular bilinear form $\mu : V \times V \to \mathbb{Q}$ defined by $\mu(i^*x, i^*y) =< x \bigcup i^*y, [B,\partial B] > \epsilon \mathbb{Q}$. This form represents an element $w(B)$ of the Witt group $W(\mathbb{Q})$ of rational forms. It follows from [1] that

(1.2) $$Lk(N) = \pm\partial(w(B)).$$

Here ∂ is the map in the Knebush sequence,

(1.3) $$0 \to W(\mathbb{Z}) \to W(\mathbb{Q}) \xrightarrow{\partial} W(\mathbb{Q}/\mathbb{Z}) \to 0.$$

Let $\pi : M \to M/S^1$ be the Seifert fibration. Then the mapping cylinder $Z(\pi)$ of π may not be a manifold in general. Cutting out the small neighborhoods (these are actually cones over lens spaces $L(\alpha_1,\beta_1),\ldots,L(\alpha_n,\beta_n)$) of the bad points corresponding to singular orbits from $Z(\pi)$, we obtain an oriented 4-manifold B. (See figure 1.) In §2, it will be shown that the boundary of B is a disjoint union of M and several lens spaces.

In §3, we prove that the Witt class $w(B)$ is represented by the rank one rational form f defined by $f(1_B,1_B) = e(M)$, the Euler number of M (defined in §2). By assembling these results and by using (1.2), we prove our main theorem in §4. Finally we give some examples supporting the main theorem.

Throughout this paper, M always denotes a Seifert manifold with associated Seifert invariant $(g : (\alpha_1,\beta_1),\ldots,(\alpha_n,\beta_n))$.

§2. The 4-manifold B.

In order to determine the boundary of the 4-manifold B described in §1, we reproduce a standard construction of a Seifert manifold which can be found in

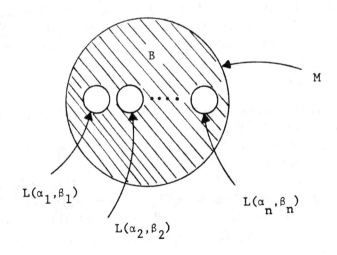

FIGURE 1

([3], [6], [8]): let $(g : (\alpha_1, \beta_1), \ldots, (\alpha_n, \beta_n))$ be a tuple of integers with $g \geq 0, \alpha_i \geq 1$ for each i, and $gcd(\alpha_i, \beta_i) = 1$ for each i. We permit β_i to be positive, negative, or zero (if $\alpha_i = 1$). Let M^* be an oriented surface of genus g. Choose n points $x_1^*, \ldots, x_n^*$ in M^*. Let $D_1, \ldots, D_n$ be disjoint closed disc neighborhoods of $x_1^*, \ldots, x_n^*$ and $M_0^* = M^* - int(D_1 \cup D_2 \cup \ldots \cup D_n)$. Let $M_0 = (M_0^*) \times S^1$ with the obvious S^1-action. Then the boundary of M_0 consists of n-tori $S_1^* \times S^1, \ldots, S_n^* \times S^1$. Let T_j be a copy of the solid torus $D^2 \times S^1 = \{(re^{i\tau}, e^{i\phi}) | 0 \leq r \leq 1, 0 \leq \tau, \phi < 2\pi\}$.

Define an S^1-action on T_j by

$$z(re^{i\tau}, e^{i\phi}) = (z^{\nu_j} re^{i\tau}, z^{\alpha_j} e^{i\phi}),$$

where z ranges over the complex numbers of norm 1, $0 < \nu_j \leq \alpha_j$, and $\beta_j \nu_j \equiv 1 \bmod \alpha_j$. Parametrize $S_j^* \times S^1$ by $(e^{i\tau}, e^{i\phi})$, where $0 \leq \tau < 2\pi$ and $0 \leq \phi < 2\pi$, increasing τ orients S_j^* as a boundary component of M_0^*.

Define an equivariant homeomorphism $f_j : S_j^* \times S^1 \to \partial T_j$ by

(2.1) $$f_j(e^{i\tau}, e^{i\phi}) = (e^{i\overline{\tau}}, e^{i\overline{\phi}}),$$

where

$$\begin{pmatrix} \overline{\theta} \\ \overline{\phi} \end{pmatrix} = \begin{pmatrix} \rho_j & \nu_j \\ \beta_j & \alpha_j \end{pmatrix} \begin{pmatrix} \theta \\ \phi \end{pmatrix} \text{ and } \rho_j = \beta_j \nu_j - 1)/\alpha_j.$$

The action on ∂T_j admits a cross-section Q_j (Say $Q_j = f_j(S_j^* \times \{1\})$. Let H be any orbit in ∂T_j, oriented by the circle group. Orient Q_j so that the ordered pair (Q_j, H) gives the same orientation of the ordered pair (M_j, L_j), where $M_j = \{(e^{i\tau}, e^{i\phi_0}) | 0 \leq \tau < 2\pi\}$ and $L_j = \{(e^{i\tau_0}, e^{i\phi}) | 0 \leq \phi < 2\pi\}$ are a meridian and a longitude of $T_j = D_j^2 \times S^1$ respectively. Then we have homology relations,

(2.2) $$M \sim \alpha_j Q_j + \beta_j H$$
$$L \sim -\nu_j Q_j - \rho_j H$$

Hence $Q_j = \{(e^{i\rho_j \phi}, e^{i\beta_j \phi}) | 0 \leq \phi < \pi\} \subset \partial T_j$ oriented by decreasing ϕ will satisfy the above condition.

The manifold $M = M_0 \bigcup_{f_1} T_1 \bigcup, \ldots, \bigcup_{f_n} T_n$ with the assigned S^1-action is a Seifert manifold and its non-normalized Seifert invariant is $(g : (\alpha_1, \beta_1), \ldots, (\alpha_n, \beta_n))$. The following theorem shows that any two Seifert manifolds with the same associated Seifert invariants are orientation preservingly homeomorphic. Hence we may consider a Seifert manifold as an S^1-manifold constructed from its Seifert invariant by the method described above.

(2.3) THEOREM. ([3],[4]). Let M and M' be two Seifert manifolds with associated Seifert invariants $(g : (\alpha_1, \beta_1), \ldots, (\alpha_n, \beta_n))$ and $(g : (\alpha_1', \beta_1'), \ldots, (\alpha_m', \beta_m'))$ respectively. Then M and M' are orientation preserving homeomorphic by a fiber preserving homeomorphism if and only if, after reindexing the Seifert pairs if necessary, there exists an integer k such that

(i) $\alpha_i = \alpha_1'$ for $i \leq k$ and $\alpha_i = \alpha_j' = 1$ for $i, j, > k$,

(ii) $\beta_i \equiv \beta_i'$ (mod α_i) for $i \leq k$,

(iii) $\sum_{i=1}^{n}(\beta_i/\alpha_i) = \sum_{i=1}^{m}(\beta_i'/\alpha_i')$. $\square$

   (2.4) Definition ([4]). The number $-\sum_{i=1}^{n}(\beta_i/\alpha_i)$ is called the *Euler number of M* and is denoted by $e(M)$. If M is a genuine S^1-bundle then $e(M)$ is the usual Euler number.

   (2.5) THEOREM. *([3],[4]). Let $\mathbb{Z}/m$ be a cycle group of order m and suppose M is a Seifert manifold. Then $M/(\mathbb{Z}/m)$, the orbit space by the $(\mathbb{Z}/m)$-action inside the S^1-action, has the Euler number $me(M)$.* $\square$

   Suppose M is a Seifert manifold with associated Seifert invariant $(g:(\alpha_1,\beta_1),\ldots,(\alpha_n,\beta_n))$ and $\pi: M \to M^*$ is the Seifert fibration. Then from the foregoing construction of M, we have the following commutative diagram:

$$
\begin{array}{ccccc}
\partial M_0 \supset S_j^* \times S^1 & \xrightarrow{f_j} & \partial T_j & \subset & T_j \\
\downarrow{\scriptstyle \pi_0} \quad \downarrow{\scriptstyle (\partial\pi_0)_j} & & \downarrow{\scriptstyle \partial\pi_j} & & \downarrow{\scriptstyle \pi_j} \\
\partial(M_0)^* \supset S_j^* & \xrightarrow{f_j^*} & \partial(T_j/S^1) & \subset & T_j/S^1
\end{array}
$$

   It is obvious that f_j^* is a homeomorphism. Hence $(f_j \times id) \bot f_j^* : (S_j^* \times S^1) \times [0,1] \bot S_j^* \to (\partial T_j \times [0,1]) \bot (\partial T_j)^*$ induces a homeomorphism F_j from the mapping cylinder $Z((\partial\pi_0)_j)$ of $(\partial\pi_0)_j$ onto the mapping cylinder $Z(\partial\pi_j)$ of $\partial\pi_j$.

   (2.6) LEMMA. *Let $\pi_0: M_0^* \times S^1 \to M_0^*$ be the natural projection and suppose $Z(\pi)$ is the mapping cylinder of the Seifert fibration π. Then $Z(\pi) = Z(\pi_0) \bigcup_{F_1} Z(\pi_1) \bigcup_{F_2}, \ldots, \bigcup_{F_n} Z(\pi_n)$.* $\square$

   Let E_j be a subset $[(T_j \times [0,1/2]) \bot T_j^*]/\approx$ of the mapping cylinder $Z(\pi)$. Then E_j is actually a neighborhood of the jth (singular) orbit in $Z(\pi)$.

   (2.7) LEMMA. *If $B = Z(\pi) - \bigcup\{int E_j \mid 1 \leq j \leq n\}$, then we have*

   (i) B is a compact 4-manifold whose boundary is a disjoint union of lens spaces $L(\alpha_1,\beta_1),\ldots,L(\alpha_n,\beta_n)$ and M.

   (ii) $H_1(B;\mathbb{Q}) \approx \mathbb{Q}^{2g}, H_2(B;\mathbb{Z}) \approx \mathbb{Z}, H_3(B;\mathbb{Z}) \approx Z^n$, and $H_4(B;\mathbb{Z}) \approx 0$.

   Moreover, $H_4(B,\partial B;\mathbb{Z}) \approx \mathbb{Z}$ and hence B is orientable.

   PROOF. (i) We can see that the mapping cylinder $Z(\eta)$ of an orbit map $\eta: \partial T_j \to \partial T_j^*$ is a solid torus $D^2 \times S^1$ with a cross section Q_j and an orbit H as a longitude and a meridian respectively. Since ∂E_j is a space constructed by gluing two solid tori T_j and $(1/2)Z(\eta)$ together along their boundaries, it follows from the homology relations (2.2) that ∂E_j is a lens space $L(\alpha_j,\beta_j)$.

   (ii) Let $Z^k = Z(\pi) - \bigcup\{int E_j \mid 1 \leq j \leq k\}$, then from the reduced Mayer-Vietoris sequence with integer coefficients for (Z^1, E_1), we have

$$0 \to H_4(Z^1) \to H_4(Z(\pi)) \to \mathbb{Z} \to H_3(Z^1) \to H_3(Z(\pi)) \to 0$$
$$\to H_2(Z^1) \to H_2(Z(\pi)) \to \mathbb{Z}/\alpha_1 \to H_1(Z^1) \to H_1(Z(\pi)) \approx \mathbb{Z}^{2g} \to 0.$$

Since $Z(\pi)$ is homotopy equivalent to an oriented surface M^* of genus g, we have $H_4(Z^1) \approx 0, H_3(Z^1) \approx \mathbb{Z}, H_2(Z^1) \approx \mathbb{Z}$ and $H_1(Z^1; \mathbb{Q}) \approx \mathbb{Q}^{2g}$. Successive applications of the Mayer-Vietoris sequence for $(Z^k, E_k), k = 1, \ldots, n$ give rise to the conclusion. Furthermore, from the homology exact sequence for $(B, \partial B)$, $0 \approx H_4(B; \mathbb{Z}) \to H_4(B, \partial B; \mathbb{Z}) \to H_3(\partial B; \mathbb{Z}) \approx \mathbb{Z}^{n+1} \to H_3(B; \mathbb{Z}) \approx \mathbb{Z}^n \to$, we have $H_4(B, \partial B; \mathbb{Z}) \approx \mathbb{Z}$ and hence B is a compact oriented 4-manifold. $\square$

(2.8) LEMMA. $H^2(B, \partial B; \mathbb{Q}) \approx \mathbb{Q}$ and the inclusion map $i : (B, \varnothing) \to (B, \partial B)$ induces an isomorphism $i^* : H^2(B, \partial B; \mathbb{Q}) \to H^2(B; \mathbb{Q})$.

PROOF. It follows from the duality and (2.7) that $H^2(B, \partial B; \mathbb{Q}) \approx H_2(B; \mathbb{Q}) \approx H_2(Z(\pi); \mathbb{Q}) \approx \mathbb{Q}$, and $H^3(B, \partial B; \mathbb{Q}) \approx H_1(B; \mathbb{Q}) \approx \mathbb{Q}^{2g}$. By the universal coefficient theorem and (2.7), we have $H^2(B; \mathbb{Q}) \approx Hom(H_2(B; \mathbb{Q}), \mathbb{Q}) H^3(B; \mathbb{Q}) \approx \mathbb{Q}^n$, and $H^4(B; \mathbb{Q}) \approx 0$. Since ∂B consists of $n+1$ closed oriented 3-manifolds, $H^3(\partial B; \mathbb{Q}) \approx \mathbb{Q}^{n+1}$. From the cohomology exact sequence with rational coefficients for $(B, \partial B)$, we have

$$\to H^2(B, \partial B) \approx \mathbb{Q} \to H^2(B) \approx \mathbb{Q} \to H^2(\partial B) \approx \mathbb{Q}^{2g} \to H^3(B, \partial B) \approx \mathbb{Q}^{2g}$$
$$\to H^3(B) \approx \mathbb{Q}^n \to H^3(\partial B) \approx \mathbb{Q}^{n+1} \to H^4(B, \partial B) \approx \mathbb{Q} \to H^4(B) \approx 0.$$

Hence i^* must be an isomorphism. $\square$

(2.9) Remark. In this section we tacitly assumed that the Euler number $e(M)$ was not zero. If $e(M)$ is zero, then the map i^* in (2.8) is a trivial homomorphism.

§3. The Witt class $w(B)$.

Suppose M is a Seifert manifold with associated Seifert invariant $(g : (\alpha_1, \beta_1), \ldots, (\alpha_n, \beta_n))$ and let m be the least common multiple of $\alpha_1, \alpha_2, \ldots,$ and α_n.

Factoring by the $(\mathbb{Z}/m)$-action inside the S^1-action, we have a commutative diagram,

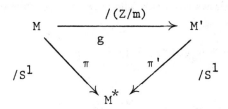

Now M' is a genuine S^1-bundle over M^*. It follows from (2.5) that the Euler number of M' is $me(M) = -m \sum_{i=1}^n (\beta_i/\alpha_i)$. Furthermore, it is known that the intersection paring $\mu : H_2(Z(\pi'); \mathbb{Z}) \otimes H_2(Z(\pi'); \mathbb{Z}) \to \mathbb{Z}$ is the rank 1 form defined by $\mu(1, 1) = me(M)$, where $Z(\pi')$ is the mapping cylinder of π'.

By the commutative diagram, a map $(g \times id) \perp id : (M \times [0, 1]) \parallel M^* \to (M' \times [0, 1]) \parallel M^*$ induces a well defined continuous map $\tilde{G} : Z(\pi) \to Z(\pi')$. Let E'_j be the image of E_j (see the construction followed by (2.7)) under $\tilde{G}$. Then E'_j is a 4-ball and $\tilde{G}^{-1}(E'_j) = E_j$. Define $W = Z(\pi') - \bigcup \{int\ E'_j \mid 1 \le j \le n\}$. Then by an argument similar to (2.7), we have $H_2(W; \mathbb{Z}) \approx H_2(Z(\pi'); \mathbb{Z}) \approx \mathbb{Z}$.

(3.1) LEMMA.   *If 1_W is a generator of $H^2(W, \partial W; \mathbb{Z})$, then we have* $< 1_W, 1_W > = me(M)$. *Here $<, >$ denotes the rank 1 bilinear form $H^2(W, \partial W; \mathbb{Z}) \otimes$*

$$H^2(W, \partial W; \mathbb{Z}) \xrightarrow{\cup} H^4(W, \partial W; \mathbb{Z}) \xrightarrow{\cap [W, \partial W]} H_0(W; \mathbb{Z}) \xrightarrow{\epsilon} \mathbb{Z}.$$

PROOF.   Let $E' = \bigcup\{E'_j \mid 1 \leq j \leq n\}$. Then it follows from the cohomology sequence of the triple $(Z(\pi'), M' \bigcup E', M')$ that $j^* : H^*(Z(\pi'), M' \bigcup E'; \mathbb{Z}) \to H^*(Z(\pi'), M'; \mathbb{Z})$ is an isomorphism. Suppose $1, 1'$, and 1_W are generators of $H^2(Z(\pi'), M'; \mathbb{Z}), H^2(Z(\pi'), M' \bigcup E'; \mathbb{Z})$ and $H^2(W, \partial W; \mathbb{Z})$ respectively such that $j^*(1') = 1$ and $e^*(1') = 1_W$, were e^* is an excision isomorphism. Then we have $me(M) = < 1, 1 > = \epsilon''((1 \cup 1) \cap [Z(\pi'), M']) = \epsilon''(j^*(1' \cup 1') \cap [Z(\pi'), M']) = \epsilon' j_*(j^*(1' \cup 1') \cap [Z(\pi'), M']) = \epsilon'((1 \cup 1) \cap j_*([Z(\pi'), M'])) = < 1', 1' >$, where ϵ' and ϵ'' are augmentations.

Let $e_* : H_4(W, \partial W; Z) \to H_4(Z(\pi'), M' \bigcup E'; \mathbb{Z})$ be an excision isomorphism and suppose $e_*^{-1}(j_*[Z(\pi'), M'])$ is a fundamental class. Then by an argument similar to the preceding one, we have $< 1_W, 1_W > = \epsilon'(1' \cup 1') \cap e_*[W, \partial W]) = \epsilon'((1' \cup 1') \cap j_* Z([\pi'), M']) = me(M)$.

Since g is a map of degree m, it follows from the homology ladder of $\tilde{G}$: $(Z(\pi), M) \to (Z(\pi'), M')$ with integer coefficients,

$$\begin{array}{ccccccccc}
0 = & H_4(M) & \to & H_4(Z(\pi), M) & \to & H_3(M) & \to & H_3(Z(\pi)) & = 0 \\
& \downarrow & & \downarrow \tilde{G}_* & & \downarrow g_* & & \downarrow & \\
0 = & H_4(M') & \to & H_4(Z(\pi'), M') & \to & H_3(M') & \to & H_3(Z(\pi')) & = 0,
\end{array}$$

that an induced homomorphism $\tilde{G}_*$ is a multiplication by m.

(3.2) LEMMA.   *If G is the restriction of $\tilde{G}$ to B then G is a map of degree m.*

Proof.   The lemma follows from the following commutative diagram:

$$\begin{array}{ccccc}
H_4(Z(\pi), M) & \xrightarrow[\approx]{i_*} & H_4(Z(\pi), M \cup E) & \xrightarrow[\approx]{e_*} & H_4(B, \partial B) \\
\downarrow \tilde{G}_* & & \downarrow \tilde{G}_* & & \downarrow G_* \\
H_4(Z(\pi'), M') & \xrightarrow[\approx]{i'_*} & H_4(Z(\pi'), M' \cup E') & \xleftarrow[\approx]{e_*'} & H_4(W, \partial W),
\end{array}$$

where $E = \bigcup_{i=1}^n E_i, E' = \bigcup_{i=1}^n E'_i, e'_*$ are excision isomorphisms.   □

It follows from (2.8) that $H^2(B, \partial B; \mathbb{Q})$ is a vector space of dimension one and $H^2(B, \partial B; \mathbb{Q}) \xrightarrow{i^*} H^2(B; \mathbb{Q})$ is an isomorphism. Recall $w(B)$ is an element of $W(\mathbb{Q})$ represented by a symmetric bilinear form $\mu : V \times V \to \mathbb{Q}$, where V is the image of i^*.

(3.3) LEMMA.   *$w(B)$ is a Witt class $w(e(M))$ represented by the rank one form f, where $f(1, 1) = e(M)$.*

PROOF. Since $Z(\pi) \simeq M^* \simeq Z(\pi')$ and G is an identity map on M^*, G^* : $H^*(Z(\pi'); \mathbb{Q}) \to H^*(Z(\pi); \mathbb{Q})$ is an isomorphism. It is known that there exists a homomorphism $\lambda : H^*(M; \mathbb{Q}) \to H^*(M'; \mathbb{Q})$ such that λg^* is a multiplication by m, where g is an orbit map $M \to M' = M/(Z/m)$. Hence $g^* : H^*(M'; \mathbb{Q}) \to H^*(M; \mathbb{Q})$ is an isomorphism, since two vector spaces $H^p(M; \mathbb{Q})$ and $H^p(M'; \mathbb{Q})$ have the same dimension for each p. Therefore it follows from the five lemma that $G^* : H^2(Z(\pi'), M'; \mathbb{Q}) \to H^2(Z(\pi), M; \mathbb{Q})$ is an isomorphism. The following diagram implies $G^* : H^2(W, \partial W; \mathbb{Q}) \to H^2(B, \partial B; \mathbb{Q})$ is also an isomorphism.

$$
\begin{array}{ccccc}
H^2(Z(\pi'),M';\ \mathbb{Q}) & \longrightarrow & H^2(Z(\ '),M'\cup E';\ \mathbb{Q}) & \xrightarrow{\ exci\ } & H^2(W,\partial W;\ \mathbb{Q}) \\
\approx \;\downarrow \tilde{G}* & & \downarrow \tilde{G}* & & \downarrow G* \\
H^2(Z(\pi),M;\ \mathbb{Q}) & \xrightarrow{\approx} & H^2(Z(\pi),M\cup E;\ \mathbb{Q}) & \xrightarrow[\approx]{\ exci\ } & H^2(B,\partial B;\ \mathbb{Q})
\end{array}
$$

Let 1_W and 1_B be generators of $H^2(W, \partial W; \mathbb{Q}) \approx \mathbb{Q}$ and $H^2(B, \partial B; \mathbb{Q}) \approx \mathbb{Q}$ respectively such that $G^*(1_W) = 1_B$. Then we have

$$
\begin{aligned}
\mu(1_B, 1_B) &= < 1_B, 1_B > = \epsilon((1_B \cup 1_B) \cap [B, \partial B]) \\
&= \epsilon(G^*(1_W \cup 1_W) \cap [B, \partial B]) \\
&= \epsilon' G_*(G^*(1_W \cup 1_W) \cap [B, \partial B]) \\
&= \epsilon'((1_W \cup 1_W) \cap G_*[B, \partial B]) \\
&= \epsilon'((1_W \cup 1_W) \cap m[W, \partial W]) \text{ by (3.2)} \\
&= m < 1_W, 1_W > = m(me(M)) \text{ by (3.1)}.
\end{aligned}
$$

Thus we have $w(B) = w(m^2 e(M)) = w(e(M))$. $\square$

Suppose (V, f) is a rational form and $L \subset V$ is a free $\mathbb{Z}$-module such that $L \otimes \mathbb{Q} = V$ and $f(x, y) \epsilon \mathbb{Z}$ for all $x, y \epsilon L$. If $L^\#$ denotes a $\mathbb{Z}$-module $\{x \epsilon V | f(x, y) \epsilon \mathbb{Z}$ for all $y \epsilon L\}$, then the map ∂ of (1.3) actually is a homomorphism defined by $\partial(V, f) = (L^\#/L, \overline{f})$, where $\overline{f}(x + L, y + L) = \rho f(x, y)$ and ρ is reduction (see [1]).

If p/q is $e(M) = -\sum_{i=1}^{n} (\beta_i/\alpha_i)$ expressed in lowest term and ξ is $(1/m)1_B$, then it follows from (3.3) that $< \xi, \xi >$ is p/q or 0 according as $e(M) \neq 0$ or $e(M) = 0$. Let $\varsigma = q\xi$ and $L = \{k\varsigma | k \epsilon \mathbb{Z}\}$ and suppose $e(M)$ is not zero. Then $L^\#$ is a $\mathbb{Z}$-module generated by $\{(k/pq)\varsigma$ or $k'\varsigma | k, k' \epsilon \mathbb{Z}\}$ and hence $L^\#/L \approx \{(k/pq)\varsigma | k \equiv pq \pmod{1}, k \epsilon \mathbb{Z}\}$. On the other hand, we have $< (1/pq)\varsigma, (1/pq)\varsigma > = 1/pq$. Thus we have the following.

(3.4) LEMMA.

$$
\partial w(B) = w(1/pq) \text{ if } e(M) \neq 0,
$$
$$
\partial w(B) = 0 \text{ if } e(M) = 0. \quad \square
$$

§4. Proof of Main Theorem

We now assemble the above lemmas to prove our main theorem.

PROOF OF (1.1). By (2.7), the boundary ∂B of B is a disjoint union of $L(\alpha_1, \beta_1), \ldots, L(\alpha_n, \beta_n)$ and M, and hence $Lk(\partial B) = Lk(L(\alpha_1, \beta_1)) + \ldots + Lk(L(\alpha_n, \beta_n)) + Lk(M)$. By (1.2) and (3.4), $w(1/pq) = \partial w(B) = \pm Lk(\partial B)$. It is known [9] that the linking form of a lens space $L(\alpha, \beta)$ is $\pm \beta / \alpha$, where the positive or negative sign depends upon the orientation of the lens space. We thus have

$$(4.1) \quad \pm \{w(\beta_1/\alpha_1) + \ldots + w(\beta_n/\alpha_n) + Lk(M)\} = \begin{cases} w(1/pq) & \text{if } e(M) \neq 0, \\ 0 & \text{if } e(M) = 0. \end{cases}$$

In the proof of (2.7), ∂E_j was regarded as a lens space constructed by gluing two solid tori T_j and $(1/2)Z(\partial \pi_j)$ together along their boundaries by a gluing map $h : \partial T_j \rightarrow \partial(Z(\partial \pi_j))$, where

$$h = \begin{pmatrix} \beta_j, & -\rho_j \\ \alpha_j, & -\nu_j \end{pmatrix}.$$

We may regard that, as subsets of ∂B, T_j and $(1/2)Z(\partial \pi_j)$ are oriented in the standard way (that is, the ordered pairs $(H \sim (1,0), Q_j \sim (0,1))$ and $(M \sim (1,0), L \sim (0,1))$ give orientations on $(1/2)Z(\partial \pi_j)$ and ∂T_j respectively). Since $det(h) = -1$, the orientation of B induces the positive orientation on $L(\alpha_j, \beta_j) \subset \partial B$. It follows from [1] that if an orientation of B induces the positive orientation of ∂B then we have the negative sign in (1.2). Hence $Lk(M)$ is $-w(\beta_1/\alpha_1) - \ldots - w(\beta_n/\alpha_n) - w(1/pq)$ or $-w(\beta_1/\alpha_1) - \ldots - w(\beta_n/\alpha_n)$ according as $e(M) \neq 0$ or $e(M) = 0$. □

(4.2) Example. Suppose M is a Seifert manifold with associated Seifert invariant $(g : (\alpha_1, \beta_1), \ldots, (\alpha_n, \beta_n))$. Then by [8], we have $H_1(M; \mathbb{Z}) = \mathbb{Z}^{2g} \oplus < Q_j, H | Q_1 + \ldots + Q_n = 0, \alpha_j Q_j + \beta_j H = 0 >$. Furthermore, suppose M is a homology sphere then it was shown in [4] that the determinant of the presentation matrix of $H_1(M; Z)$ is $\pm(\alpha_1 \alpha_2 \ldots \alpha_n)(\sum_{i=1}^{n} (\beta_i / \alpha_i)) = \pm 1$ and hence $\alpha_1, \alpha_2, \ldots, \alpha_n$ are pairwise coprime.

Suppose det (the presentation matrix) $= +1$. Then $e(M) = -1/(\alpha_1 \alpha_2 \ldots \alpha_n)$. Hence we have

(4.2.a) $Lk(M) = -\{w(\beta_1/\alpha_1) + \ldots + w(\beta_n/\alpha_n)\} - \{w(-1/(\alpha_1 \ldots \alpha_n))\}$.

Since $\alpha_1, \alpha_2, \ldots, \alpha_n$ are pairwise coprime, we obtain

(4.2.b)

$$w(1/(\alpha_1 \ldots \alpha_n)) = w(\alpha_2 \ldots \alpha_n / \alpha_1) + w(\alpha_1 \alpha_3 \ldots \alpha_n / \alpha_2) + \ldots + w(\alpha_1 \alpha_2 \ldots \alpha_{n-1} / \alpha_n).$$

Since

$(\alpha_1 \ldots \alpha_n) \sum_{i=1}^{n} (\beta_i / \alpha_i) = \beta_1 (\alpha_2 \ldots \alpha_n) + \beta_2 (\alpha_1 \alpha_3 \ldots \alpha_n) + \ldots + \beta_n (\alpha_1 \ldots \alpha_{n-1}) = 1$,

we have

(4.2.c)

$$\beta_i(\alpha_1 \ldots \check{\alpha}_i \ldots \alpha_n) \equiv 1 \pmod{\alpha_i} \text{ for each } i.$$

For each i, we have

$$-w(\beta_i/\alpha_i) + w(\alpha_1\ldots\check{\alpha}_i\ldots\ldots\alpha_n/\alpha_i)$$
$$= w(\{-\beta_i + (\alpha_1\ldots\check{\alpha}_i\ldots\alpha_n)\}/\alpha_i)$$
$$\quad + w(\{-\beta_i(\alpha_1\ldots\check{\alpha}_i\ldots\alpha_n)/\alpha_i\}\{-\beta_i + (\alpha_1\ldots\check{\alpha}_i\ldots\alpha_n)\}/\alpha_i) \text{ (by p.85 of [2])}$$
$$= w(\{-\beta_i + (\alpha_1\ldots\check{\alpha}_i\ldots\alpha_n)\}/\alpha_i) + w(-\{-\beta_i + (\alpha_1\ldots\check{\alpha}_i\ldots\alpha_n)\}/\alpha_i)$$
$$= 0 \text{ (by (4.2.c))}$$

and hence $Lk(M) = 0$. Similarly we also have $Lk(M) = 0$ for the case of $det = -1$. □

Next example shows that if we pick the positive sign in (4.1), the Witt class of torsion linking form of a homology sphere may not be zero. Hence it is absurd to choose the positive sign in (4.1).

(4.3) Example. Suppose M is a Seifert manifold with associated Seifert invariant $(0 : (4,3),(7,-5))$. Then M is a homology sphere and $e(M) = -(3/4 - 5/7)) = -1/(4 \times 7)$.

Since $w(-5/7) = w(2/7) = w(2 \times 3^2/7) = w(4/7)$,

$$+ \{w(3/4) + w(-5/7)\} - w(-1/28)$$
$$= \{w(3/4) + w(4/7)\} + \{w(4/7) + w(7/4)\} \text{ by (4.2.b)}$$
$$= w(4/7) + w(4/7), \text{ since } w(\mathbb{Z}/m, f) = 0$$

if and only if m is a square.

By (p.87 of [2]), $W(F_7)$ is a cyclic group of order 4. Hence $w(4/7) + w(4/7) \neq 0$, since $w(4/7)$ is actually a generator of $W(F_7)$.   □

Since a Seifert manifold with associated Seifert invariant $(0 : (1,b))$ is a lens space $L(b,1)$ and hence $Lk(M) = w(1/b)$, $\partial w(B)$ can not be $w(-1/pq)$ in (3.4) and (1.1).

(4.4) Remark. A Seifert manifold M is a boundary of a plumbed manifold P. Furthermore, the intersection form of P is diagonalizable (see p176 of [4]) and hence we can also prove the main theorem in a purely algebraic way, which will be sketched in [10]. But our direct geometric proof is interesting and will be useful in further development.

As an application of the main theorem, we prove in [10] that any Witt class of a finite form can be realized by a Seifert manifold.

REFERENCES

[1] J. Alexander, G. Hamrick, and J. Vick: *Linking Forms and Maps of Odd Prime Order.* Trans. of Amer. Math. Soc. **221**, (1976), 169–185.

[2] D. Husemoller and J. Milnor: *Symmetric Bilinear Forms.* Springer-Verlag, (1973).

[3] W. D. Neumann: S^1-*Actions and the α-invariant of their involutions.* Bonner Math. Schriften **44**, Bonn (1970).

[4] W. D. Neumann and F. Raymond: *Seifert Manifolds, Plumbing, μ-invariant and Orientation reversing maps.* Algebraic and Geometric Topology, Proceedings, Santa Barbara 1977, Lecture Note Math. No. **664**, Springer-Verlag (1978), 161–196.

[5] P. Orlik: *Seifert Manifolds.* Lecture Notes Math. No. **291**, (1972).

[6] P. Orlik and F. Raymond: *Actions of S0(2) on 3-Manifolds.* Proc. of the Conf. on Transformation Groups, New Orleans 1967, Springer-Verlag, (1968), 297–318.

[7] F. Raymond: *Classification of the actions on 3-Manifolds.* Trans. of Amer. Math. Soc. **131**, (1968), 51–78.

[8] H. Seifert: *Topologie Dreidimensionaler Gefaserter Räume.* Acta Math. **60**, (1933), 147–238.

[9] H. Seifert and W. Threlfall: *A Textbook of Topology,* Academic Press (1980).

[10] H. S. Oh: *The Witt Classes of Torsion Linking Forms of (4n − 1)-Manifolds with Pseudo-Free Circle Actions,* preprint.

DEPARTMENT OF MATHEMATICS

UNIVERSITY OF NORTH CAROLINA AT CHARLOTTE

Contemporary Mathematics
Volume **44**, 1985

OUTERMOST FORKS AND A THEOREM OF JACO

Martin Scharlemann*

In [J, Theorem 2] Jaco proves the following

THEOREM 0.1. *Let M be an orientable compact 3-manifold and let J be a simple closed curve in ∂M such that $\partial M - J$ is incompressible. If M has compressible boundary, then the 3-manifold obtained by adding a 2-handle to M along J has incompressible boundary.*

This generalizes the thesis of J. Przytycki [P].

As a corollary, Jaco shows that a knot in S^3 has property R if it has an orientable, incompressible spanning surface S such that $\pi_1(S^3 - S)$ is a free product.

The purpose of this paper is to reprove Jaco's theorem, using the method of outermost forks developed in [Sc]. The purpose is not so much to improve Jaco's proof (the argument here is no easier) as to give an exposition of the outermost fork argument in a relatively straightforward application. Like Jaco's use of pre-disks and coplanar curves, it seems to be useful in some of those cases in which the standard innermost disk, outermost arc argument does not suffice. For the benefit of the author, as well as of those familiar with Jaco's proof, the arguments are as parallel as possible.

§1. *Main Lemma and its Preliminaries*

Let M be a compact 3-manifold and J a two-sided simple closed curve in ∂M. A planar surface P in M is a *disc with J-holes* if all but at most one component s (denoted $\overline{\partial P}$ and called the pre-boundary) of ∂P are parallel in ∂M to J. A simple closed curve s in $\partial M - J$ is called *J-essential* if there is no disk with J-holes P in ∂M such that $\overline{\partial P} = s$. If N is the 3-manifold obtained from M by adding a 2-handle to M along J, then $s \subset \partial M - J$ is J-essential if and only if it is essential in ∂N, and a disk with J-holes in M with pre-boundary s can be completed to a disk in N with boundary s. Thus 0.1 is an immediate corollary of

LEMMA 1.1.. *Let M be a compact orientable 3-manifold and J be a simple closed curve in ∂M such that $\partial M - J$ is incompressible. If M has compressible*

*Supported in part by an NSF grant.

boundary, then there is no disk with J-holes in M whose pre-boundary is J-essential in ∂M.

PROOF. Of all disks with J-holes in M whose pre-boundaries are J-essential in ∂M, let P have minimal number of boundary components $\{s, p_1, \ldots, p_n\}$. Label the simple closed curves $\{p_1, \ldots, p_n\}$, which are all parallel to J, in order, so that p_i and p_{i+1} cobound an annulus A_i in $\partial M - \partial P$, $i = 1, \ldots, n-1$.

Let D be a boundary reducing disk for M, transverse to P, chosen to minimize the number of components of $D \cap M$. This ensures, in particular that each arc of $\partial D \cap A_i$ runs from one end of the annulus to the other.

(1) ASSERTION: *No component of $D \cap P$ is a simple closed curve or an inessential spanning arc on P.*

The proof is a standard innermost disk, outermost arc argument; cf [J, Assertions 1,2].

Now let α be a component of $D \cap P$ that is outermost on D; i.e., there is an arc $\beta \subset \partial D$ such that $\partial \alpha = \partial \beta$ and the curve $\alpha \cup \beta$ bounds a disk Δ in D with $\Delta \cap P = \phi$.

(2) ASSERTION: *The component α of $D \cap P$ does not have both ends in the same component of ∂P.*

Since ∂D intersects each A_i in an arc running from one end to the other either both ends of β (hence α) lie in s or $n = 1$ and the assertion follows from (1). If both ends lie in s, then a boundary compression of P at α using the disk Δ results in two new planar surfaces P_1 and P_2, each with fewer boundary components than P, and each a J-holed disk. Furthermore, at least one must have J-essential pre-boundary. This contradicts the choice of P.

(3) ASSERTION: *The component α of $D \cap P$ does not have one end in s and one end in another component of ∂P.*

If it did, boundary compression of P at α via Δ results in a new planar surface P' with one fewer boundary components. Furthermore, P' is a J-holed disk whose pre-boundary is J-essential.

(4) ASSERTION: *The component α of $D \cap P$ must have one end in p_1 and the other in $p_n, n > 1$.*

From the previous assertions, we know that the ends of α must lie in distinct components of $\{p_1, \ldots, p_n\}$. Since $\beta \subset M - \partial P$, the only other possibility is that β is an arc connecting one end of an annulus A_i to the other. But then a boundary compression of P at α via Δ results in a new planar surface with one of its boundary components an inessential disk in A_i. Capping it off would produce a J-holed disk P' with the same pre-boundary as P, but two fewer boundary components, a contradiction.

§**2.** *Bilabelled trees and outermost forks.*

Following assertion (1), all components of $P \cap D$ are arcs. Construct a planar tree T (in D, to be concrete) as follows. The vertices of T consist of a single point v_i in the interior of each disk component C_i of $D - P$. Each arc component γ of $D \cap P$ is contiguous to two disks components C_i and C_j of $D - P$; for each such γ an edge of T will be an arc e_γ connecting v_i and v_j and an intersecting γ in a single point. The edge e_γ of T divides γ into two arcs γ_1 and γ_2, on opposite sides of e_γ. One end of each γ_i is a point lying in ∂P; label the corresponding side of e_γ by i if the point lies in p_i and by ∞ if it lies in s. The resulting planar tree, with both sides of each edge given a label is called a *bilabelled tree*.

In general, for T a planar tree, define an *outermost vertex* to be a vertex which is in the boundary of a single edge, which is called an *outermost edge*. A *fork* is a vertex in the boundary of three or more edges. Let F be the collection of forks of T and remove from T all components of $T - F$ which contain an outermost vertex. An outermost vertex of the resulting tree is an *outermost fork*. (If T has no forks, then let any vertex *not* an outermost vertex be called an outermost fork.) If v is an outermost fork, then the components of $T - \{v\}$ which contain no forks are called *outermost lines* of v. If v is any fork then two components of $T - \{v\}$ are *adjacent* if a small circle around v in the plane contains an arc intersecting only those components of $T - v$.

Now assertion 4 translates to

(4') ASSERTION: *Each outermost edge is labelled* $\{1, n\}, n > 1$.

Furthermore, certain patterns are not possible anywhere in T.

(5) ASSERTION: *For* $2 \le i \le n - 1$, *the pattern*

$$\frac{i \, . i + 1}{ii = 1}$$

never occurs in T.

Consider the corresponding arc of $D \cap P$ with ends labeled $\{i, i\}$. The dotted arc shown lies in a small neighborhood of P in M and its ends can be connected by a small arc in the interior of A_i. The result is a circle in a neighborhood of P in M intersecting P in exactly one point. Since P is 2-sided, this is impossible. (See Figure 1.)

(6) ASSERTION: *If $n \ge 3$, then for no $1 \le i \le n - 1$ is there a vertex in T all of whose p adjacent edges, $p \ge 2$, are labelled as shown.*

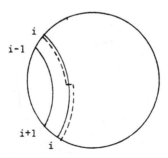

FIGURE 1

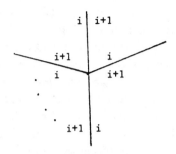

If all but one of the p edges is outermost, the same is true for $n = 2$.

(The case $p = 1$ is contained in (4').)

The corresponding 2-cell F of $D - P$ has boundary (oriented clockwise in Figure 2) consisting of arcs running from p_i to p_{i+1} in P and from p_{i+1} to p_i in A_i. Consider the genus one punctured surface Q obtained by attaching A_i to P along $p_i \cup p_{i+1}$. (Q is orientable since it is 2-sided and M is orientable. This is the only point at which the orientability of M is used.) Surgery on Q using F produces a disk with J-holes with fewer boundary components than P, a contradiction.

(7) ASSERTION: *The number of edges in two adjacent outermost lines of T must total at least n.*

By (4) all outermost edges are labelled $\{1, n\}$. Since an arc in A_i of $\partial D - P$ must run from one end of A_i to the other, it follows that any edge adjacent to an outermost edge in an outermost line must be labelled $\{2, n - 1\}$. Continuing in this manner, the arc in ∂E corrsponding to the path T from one outermost vertex to another must pass through all of $\{p_1, \ldots, p_n\}$.

(8) ASSERTION: *No outermost line has more than $n/2$ edges.*

If it does, since the outermost vertex is labelled $\{1, n\}$, the outermost line contains one of the following patterns, depending on whether n is odd or even.

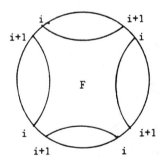

FIGURE 2

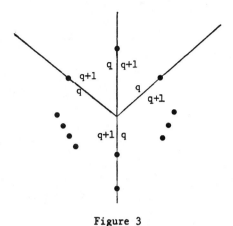

Figure 3

$$\cdots\frac{q\quad q+1}{q\quad q-1}\cdots\frac{2q-2\quad 2q-1}{2\quad 1}\quad n=2q-1$$

$$\cdots\frac{q\quad q+1}{q+1\quad q}\cdots\frac{2q-1\quad 2q}{2\quad 1}\quad n=2q.$$

One case contradicts (5), the other (6).

(9) ASSERTION: *T has no outermost forks.*

Combining claims 7 and 8, each outermost line of an outermost fork must have length $n/2 = q$. Hence the labelling around the fork is as in Figure 3.

This contradicts (6).

Following (9), T is either a single edge or a single point. The former case contradicts (4′) and the latter the incompressibility of $\partial M - J$.

A final remark: Jaco only requires that ∂M be orientable while the above proof requires M to be orientable. It is a pleasant but not a trivial exercise, using the orientable double cover, to show that the theorem in the case when M is orientable implies the theorem when only ∂M is orientable.

Added in proof: It now appears that Jaco's theorem can be generalized further, using a generalization of a Haken algorithm cf. A. Casson and C. Gordon, *Reducing Heegaard Splittings of 3-Manifolds*, T.A.M.S. **4** (1983), 182.

BIBLIOGRAPHY

[J]   W. Jaco, *Adding a 2-handle to a 3-manifold: an application to property R* (to appear).

[Sc] M. Scharlemann, *Tunnel number one knots satisfy the Poenaru conjecture*, Topology and its Applications (to appear).

[P]   J. Przytycki, *Incompressible surfaces in 3-manifolds*, Thesis: Columbia University (1981).

Contemporary Mathematics
Volume 44, 1985

SURFACES IN 3-MANIFOLDS*

John R. Stallings

Introduction

This is an outline of part [5]. The major theorem is:

Let A, B be compact oriented 3-manifolds with boundary such that B is contained in the interior of A and such that $H_2(A, B) = 0$. Suppose that T is a compact oriented surface contained in A, such that the boundary $Bd\, T$ is contained in the interior of B. Then there is a compact oriented surface $T_1 \subset B$ with oriented boundary $Bd\, T_1 = Bd\, T$, such that genus $T_1 \le$ genus T.

The case of this theorem when A and B are handlebodies suggests, but does not prove, some group-theoretic conjectures about free groups. One of these conjectures would prove Kervaire's conjecture that a free product $G * Z$ cannot be killed by adding a single relation unless $G = \{1\}$.

1. Proof of theorem

1.1 We do not assume that surfaces are connected unless this is made explicit. The *genus* of a surface T is defined to be the sum of the genera of its connected components.

The hypothesis $H_2(A, B) = 0$, with integer coefficient group, implies that there is a prime number p such that $H_2(A, B; Z_p) = 0$, since A and B are compact. Fix this prime number from now on; all homology groups are construed to have coefficient group Z_p; the ith Betti number $\beta_i(X)$ is the rank over Z_p of $H_i(X; Z_p)$.

The genus of T is then related to its Euler characteristic by this formula:

$$\chi(T) + \beta_0(Bd\, T) = 2(\beta_0(T) - \text{ genus } T).$$

1.2 An explicit elementary case of the theorem is:

LEMMA. *Let M be a compact orientable connected 3-manifold whose boundary is the union of two 2-manifolds E and D intersecting along their boundaries. $Bd\, M = E \cup D$, $E \cap D = Bd\, E = Bd\, D$. Suppose $H_2(M, E) = 0$. Then genus $E \le$ genus D.*

PROOF. Add 2-handles to M along $Bd\, D$ to obtain M'. Then $Bd\, M'$ is the disjoint union of E' and D', which are basically E and D with their boundaries

*This work was partly supported by NSF Grant MCS 80-2858..

closed off by 2-cells. By excision and homotopy, $H_2(M', E') \approx H_2(M, E) = 0$. By Lefschetz duality, $H_1(M', D') = 0$, and the exact sequence of (M', D') implies $\beta_1(M') \leq \beta_1(D')$. Standard exercise in algebraic topology shows the genus of $Bd\,N$ is $\leq \beta_1(N)$. And so,

$$\text{genus } E' + \text{ genus } D' \leq \beta_1(M') \leq \beta_1(D') = 2 \text{ genus } D'.$$

1.3. Now suppose we consider the situation in the theorem, where $H_2(A, B) = 0$ and T is a 2-manifold in A whose boundary lies in B. We can make T transversal to $Bd\,B$, and consider a component D of $T - B$. Then D represents a 2-chain in (A, B) and so bounds a 3-chain which is supported on a certain 3-manifold M_D whose boundary is $D \cup E$, where $E \subset Bd\,B$. Inside M_D may be other components of $T - B$, but there is an innermost one. Thus we can suppose that $M_D \cap T = D$.

Now $H_2(M_D, E) \approx H_2(M_D \cup B, B)$. This fits into the exact sequence of the triple $(A, M_D \cup B, B)$, where

$$H_3(A, B) \to H_3(A, M_D \cup B)$$

is surjective (in the interesting cases these groups are both 0), and where, by assumption, $H_2(A, B) = 0$. Therefore $H_2(M_D, E) = 0$, and **1.2** then implies that genus $E \leq$ genus D.

Replace D in T by E to get T'. A computation shows that, even though E may not be connected, since genus $E \leq$ genus D, we get genus $T' \leq$ genus T.

Push T' towards the interior of B in a neighborhood of E. This produces a situation where the number of components of $T - B$ has been reduced by one. After a finite number of such changes T is replaced by $T_1 \subset B$, thus proving the theorem.

2. Some corollaries

2.1. *Suppose that $A \supset B$ is a pair of compact oriented 3-manifolds such that $H_2(A, B) = 0$. Then if k is a curve in B which is unknotted in A, it follows that k is unknotted in B. More generally, if k is null-homologous in A, then the genus of k in B is equal to the genus of k in A.*

This is simply the theorem for the case that $Bd\,T$ has only one component k, and we look at the component of T_1 containing k.

I am surprised that this fact has not been previously noticed, and rather annoyed at not being able to find a significant consequence.

2.2. *As before, suppose $H_2(A, B) = 0$. Suppose k_1 and k_2 are knots in B bounding an annulus in A. Then they bound an annulus in B.*

3. Some conjectures

3.1. Suppose F and G are finitely generated free groups, the fundamental groups of finite 1-complexes. By thickening the 1-complex of F we get a 3-dimensional handlebody A. Given a homomorphism $\alpha : G \to F$, we can find a map of the 1-complex of G into A, which can be taken to be an embedding; then thicken the 1-complex of G to obtain $B \subset A$. Then $H_2(A, B) = 0$ if and only if the abelianization of α is injective.

If the major theorem were about singular, or even immersed, 2-manifolds T and T_1, we could prove the following. But since the theorem is only about *embedded* manifolds, these are only *conjectures*.

3.2. Conjectures:

Suppose $G \subset F$ are finitely generated free groups, and that, abelianized,

$$H_1(G) \to H_1(F)$$

is injective. Then:

(i) *If $\beta, \alpha_1, \ldots, \alpha_n \in G$ and there exist $w_1, \ldots, w_n \in F$ such that*

$$\beta = \prod_1^n w_i \alpha_i w_i^{-1} ,$$

then there exist $u_1, \ldots, u_n \in G$ such that

$$\beta = \prod_1^n u_i \alpha_i u_i^{-1}$$

More grandiosely,

(ii) *If $\beta, \alpha_1, \ldots, \alpha_n \in G$, and there exist $w_1, \ldots w_n \in F$, $r_1, \ldots, r_k$, $s_1, \ldots, s_k \in F$ such that*

$$\beta = \prod_1^n w_i \alpha_i w_i^{-1} \cdot \prod_1^k r_j s_j r_j^{-1} s_j^{-1} ,$$

then we can find elements w_i, r_j, s_j of G satisfying the same equation.

3.3. Comment. The case $n = 1$ of conjecture (i) is true [5]. That is, if $H_1(G) \to H_1(F)$ is injective, then two elements of G which are conjugate in F are conjugate in G. The proof of this is of fair difficulty and does not seem to generalize to $n > 1$.

Explicit cases of these conjectures seem quite difficult. For example, let F be free on x, y, and G be the subgroup generated by yxy^{-1} and xyx^{-1}. I do not know any result related to any particular case of these conjectures for this specific pair of groups, except for conjecture (i), $n = 1$.

4. How conjecture (i) implies Kervaire's conjecture

4.1. Let G be an arbitrary group, Z the infinite cyclic group generated by t, and $w \in G * Z$. If we suppose that $w = 1$ kills $G * Z$, it is easy to show that the exponent-sum of t in w is ± 1. Thus Kervaire's conjecture (quoted in [3], page 403) will follow from the following (which is true for locally residually finite [1] and locally indicible [2] groups G):

If the exponent-sum of t in w is non-zero, then the composition

$$G \to G * Z \to G * Z / < w = 1 >$$

is injective.

Trying to prove this, we can assume that G is finitely generated, $G = F/R$, where R is a normal subgroup of a finitely generated free group F.

Let α belong to the kernel of $G \to G * Z / < w = 1 >$. Let α_1 be a representative of α in F, and w_1 a representative of w in $F * Z$; the exponent-sum of t in w_1 is non-zero. Let S be the free product of F and an infinite cyclic group generated by w_2; map $S \to F * Z$ by the identity on F and $w_2 \to w_1$; this is injective on homology because of the exponent sum condition, and so, by [4], is itself injective, and so we consider S to be a subgroup of $F * Z, w_2 = w_1$.

Now α_1 belongs to the smallest normal subgroup of $F * Z$ containing $R \cup \{w_1\}$, and so there is an equation:

$$(*) \qquad\qquad \alpha_1 = \prod_1^n u_i \beta_i u_i^{-1}$$

where $u_i \in F * Z$, and each β_i either belongs to R, or is of the form w_1 or w_1^{-1}. In particular, α_1, β_i belong to S, and all these belong to F except for the β_i of the form $w_1^{\pm 1}$.

If conjecture (i) should happen to be true, then the "coefficients" u_i in $(*)$ can be taken to belong to S. Now, the interesting thing about S is that it retracts to F, sending $w_1 \to 1$. Apply this retraction to $(*)$; we see that each term $u_i \beta_i u_i^{-1}$ is mapped into R; if $\beta_i = w_1^{\pm 1}$ it is completely killed,otherwise it goes into a conjugate in F of an element of R. On the other hand, α_1 is left fixed by the retraction. Hence $\alpha_1 \in R$, and so $\alpha = 1$.

REFERENCES

[1] Gerstenhaber, M., Rothaus, O.S., *The solution of sets of equations in groups*. Proc. Nat. Acad. Sci. USA **48** (1962), 1531–1533.

[2] Howie, J., *On pairs of 2-complexes and systems of equations over groups*. J. Reine Angew. Math. **324** (1981), 165–174.

[3] Magnus, W., Karass, A., Solitar, D., *Combinatorial Group Theory* (New York: Wiley, 1966).

[4] Stallings, J., *Homology and central series of groups*. J. Alg. **2** (1965), 170–181.

[5] Stallings, J., *Surfaces in three-manifolds and non-singular equations in groups*. Math. Zeit. **184** (1983), 1–17.

UNIVERSITY OF CALIFORNIA, BERKELEY

HOMOTOPY THEORY AND INFINITE DIMENSIONAL TOPOLOGY

Contemporary Mathematics
Volume 44, 1985

TAMING HOPF INVARIANTS

M. G. Barratt

(Dedicated to Dorothy and Arthur Stone)

PREAMBLE

Some generalizations of the Hopf Invariant, which appear in a number of contexts, have been obtained from the Hilton-Milnor Theorem splitting the loop space of a bouquet of suspensions (cf. J. Milnor, "The Construction FK, Princeton 1955/6). Here contemplation of this theorem, and generalizations of binomial coefficients and of Lie algebras, produces some relations between the invariants.

THE MILNOR-HILTON INVARIANTS

Suppose $\{w\}$ is a Hall basis for the free Lie algebra on the ordered set $\{x_2, x_1\}$ and that A_2, A_1 are pointed C.W. complexes—for the moment supposed for simplicity to be both countable and suspensions. Each w generates a space A_w by interpreting x_i as A_i and the Lie product as a smash product. Then any version of the Milnor-Hilton Theorem—and here Milnor is followed except that factors are split off on the right—express $\Omega\Sigma(A_1 \vee A_2)$ as a (weak) product of the $\Omega\Sigma A_w$.

Suppose $A = A_1 = A_2$. By looping the co-multiplication on ΣA and using the projections defined by the theorem there are obtained maps

$$\overline{H}_w : \Omega\Sigma \to \Omega\Sigma A_w$$

Suppose P is a category 1 space; by adjoining there arise the *Hilton-Milnor Invariant* homomorphisms

$$H_w : [\Sigma P, \Sigma A] \to [\Sigma P, \Sigma A_w].$$

There follows at once the classic Hilton-Milnor Theorem that

THEOREM 0. *If $\alpha : \Sigma P \to \Sigma A$ and $x_1, x_2 : \Sigma A \to X$, then $(x_1 + x_2 \circ \alpha)$ is the sum of $W \circ H_w(\alpha)$, where the Lie product is replaced by the Whitehead Product.*

However, contemplation of Milnor's proof proves more. First, the nth *James-Milnor Invariant* H_n is to mean the $H_{w[n]}$ associated with the unique $w = w[n]$ which has weight n and is of degree 1 in the (second) variable x_1. It is easy to relate this with the nth classical James Invariant, but this is not done here.)

Suppose B is a suspension (or category 1 space) and $\beta : \Sigma B \to X$. Let α, x_2 be as in Theorem 0, and let x_1 be re-interpreted as the Whitehead Product $[\beta, x_2]$. The following generalizes a Theorem of Barcus and Barratt, and is proved by the following Milnor's argument with an appropriate substitution of commutator for one variable.

THEOREM 1.   *The Whitehead Product $[\beta, x_2 \circ \alpha]$ is the sum for all $n \geq 1$ of $w[n]_\circ \underline{B} \wedge H_n(\alpha)$, where $\underline{B}$ is the identity map of B.*

(Here, and below, these infinite sums are defined by limiting processes if $\dim(P)$ is not finite).

There is a more general result for arbitrary B, proved in the same way. When $B = S^o$ this becomes

THEOREM 2.   *If $\beta \in \pi_1(X)$ and x_1 is again $[\beta, x_2]$, then $[\beta, x_2 \circ \alpha]$ is the sum of all $w_\circ H_w(\alpha)$ for $w \neq x_2$. (In fact, this is equivalent to Theorem 0).*

SOME CONSEQUENCES

If W is of weight n, $A_w = A^{(n)}$, on which the symmetric group $\$_n$ acts by permuting factors; the action is on the left and suspends to an action of $\$_n$ on $[\Sigma A^{(n)}, \Sigma A^{(n)}]$ by acting on the target. Since A is supposed to be a suspension, this abelian group is a ring and a ring homomorphism $\mathbb{Z}(\$_n) = \Lambda_n \to [\Sigma A^{(n)}, \Sigma A^{(n)}]$ is induced. The following is deduced from Theorem 1 with $B = S^1$ by means of some algebra disclosed below.

THEOREM 3.   *For each w of weight n and each integer $m \geq 1$ there is a $\sigma_m \in \mathbb{Z}(\$_{nm})$ such that*
(i) $\underline{S}^1 \wedge H_w(\alpha) = \sigma_n \cdot \underline{S}^1 \wedge H_n(\alpha)$,
(ii) $\underline{S}^1 \wedge H_m(H_w(\alpha)) = \sigma_m \cdot \underline{S}^1 \wedge H_{nm}(\alpha)$.

A recipe for computing the "coefficients" σ_m is given below. This will imply the following: Let the *twisted binomial coefficient* $\mathbb{C}_q^n \in \mathbb{Z}(\$_n)$ be defined to be zero unless $0 \leq q \leq n$, and otherwise is to be the sum of all $(q, n - q)$ shuffles (that is, the permutations $\varnothing$ in which the relative orders of $\phi^{-1}(1) \ldots \phi^{-1}(q)$, and of $\phi^{-1}(q + 1) \ldots \phi^{-1}(n)$, are preserved).

THEOREM 4. $\mathbb{C}_q^n \cdot \underline{S}^1 \alpha) = 0$ *for* $1 \leq q < n$.

It follows from this, the recipe, and the relations (in which $\tau_m = (1, 2, \ldots, m)$)

$$ \mathbb{C}_q^n = 1 \otimes \mathbb{C}_{q-1}^{n-1} + \tau_{q+1}^{-1}(1 \otimes \mathbb{C}_q^{n-1}), $$

that, if $\not{Z}_m$ reverses the order of the first m factors,

COROLLARY 5. $\underline{S}_\wedge^1 H_{2n}(\alpha) = (-1)^{n-1} \not{Z}_n \cdot \underline{S}_\wedge^1 H_2(H_n(\alpha))$.
(*Remark:* $(-1)^{n-1} \not{Z}_n$ *is invertible in* Λ_{2n}.)

SOME ALGEBRA

Let $\Lambda_n = \mathbb{Z}(\$_n)$. The obvious embedding $\$_p \times \$_q \subset \$_{p+q}$ induces an imbedding $\phi : \Lambda_p \times \Lambda_q \to \Lambda_{p+q}$. Following [Barratt, Twisted Lie Algebras, Springer Lectures Notes v.658] a *twisted algebra* $A = \{A_n\}$ is a graded algebra such that A_n is a right Λ_n module and the multiplication $\alpha : A_p \times A_q \to A_{p+q}$ makes the diagram

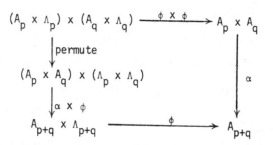

commute. A *twisted Lie algebra* is a twisted algebra in which the multiplication (written $[,]$) satisfies, for all $\alpha \in A_p$, $\beta \in A_q$, $\gamma \in A_r$,

(i) $0 = [\alpha, \beta] + [\beta, \alpha]\tau_n^q \quad (n = p + q)$

(ii) $0 = [[\alpha, \beta], \gamma] + [[\beta, \gamma], \alpha]\tau_m^{-p} + [[\gamma, \alpha], \beta]\tau_m^r \; (m = p + q + r)$.

It is shown (*loc. cit.*) that there is a free twisted Lie algebra $L(S) = \{L_n\}$ on a (finite) set S with L_1 the free $\mathbb{Z}$-module on S, and a version of the Birkhoff-Poincare-Witt theorem applies. $L(S)$ can be embedded in the free twisted associative algebra $A(S) = \{A_n\}$ on S, with $A_1 = L_1$. Moreover, if the recipe for a Hall basis be followed with the twisted Lie product $[\alpha, \beta] = \alpha\beta - (\beta\alpha)\tau_n^q$ for $\alpha \in A_p$, $\beta \in A_q$, then $A(S)$ is additively generated in the usual way by monotone monomials in the basis. However, the additive structure of $L(S)$ is more fun.

In $L(S)$ let $\xi_k(y)$ mean y if $k = 1$ and $[\xi_{k-1}(y), y]$ if $k > 1$. It has been shown (*loc. cit.*) that L_n is not a free Λ_n module if $n > 1$, but nevertheless is spanned by all $\xi_k(w)$ such that $n = k(\text{ Weight }(w))$, subject only to the relations implied by

$$0 = \xi_k(w) \otimes_{\Lambda_n} ((p - \beta_{p,q}) \otimes u)$$

where $q = \text{ weight }(w)$, $kq = n, 2 \le p \le k$, u is the unit of $\Lambda_{q(k-p)}$ and the (generalized) Dynkin-Specht-Wever element $\beta_{p,q}$ is defined for $q = 1$ by

$$\beta_{1,1} = 1, \beta_{p+1,1} = (1 - \tau_{p+1})(\beta_p \times |)\,(\tau_{p+1} = (1, 2, \ldots, p+1),$$

and $\beta_{p,q}$ is the image of $\beta_{p,1}$ under the ring-homomorphism $\Lambda_p \to \Lambda_{pq}$ induced by making each permutation of $\$_p$ permute blocks of length q.

EXPLICIT FORM OF THEOREM 3

The (left) action of Λ_n on $\Sigma(A^{(n)})$ induces a left action on the image K_n of (left) suspension

$$S_\Lambda^1 : [\Sigma P, \Sigma A^{(n)}] \to [S_\Lambda^1 \Sigma P, S_\Lambda^1 \Sigma A^{(n)}].$$

Let $S = \{y_2, y_1\}$ and let $A(S)$ denote the free twisted associative algebra on S, so that A_n is a free right Λ_n-module. Set $G_n = A_n \times_{\Lambda_n} K_n$.

THEOREM 6. *In* $G_n, ((y_1 + y_2)^n - y_1^n - y_2^n) \otimes \underline{S}^1 \wedge H_n(\alpha)$ *is the sum of all* $(w)^m \times \underline{S}^1 \wedge H_n(H_w(\alpha)$ *such that* $m \cdot$ weight $(w) = n$.

Theorem 3 follows by expressing $(y_1 + y_2)^m - y_1^n - y_2^n$ in terms of monotone monomials in a twisted Hall basis $\{w\}$. The coefficient of $y_2^q y_1^{n-q}$ is C_q^n; this term is not a $w^m (m > 1)$ if $1 \le q < n$, so Theorem 4 follows. From this one can prove by induction on p that

LEMMA 7. $(1 \times C_p^{n-1}) \cdot \underline{S}^1 \wedge H_n(\alpha) = (-1)^p \not{C}_{p+1} \cdot \underline{S}^1 \wedge H_n(\alpha)$
where $\not{C}_{p+1} = \tau_{p+1} \tau_p \cdots \tau_2$ *is the permutation reversing* $\{1, 2, \cdots, p+1\}$.

Suppose $n = 2K$, then the coefficient of $(w[k])^2$ is $1 \otimes C_{k-1}^{n-1}$, and Corollary 5 follows.

PROOF OF THEOREM 6

Let $X = S^2 \vee \Sigma A_1 \vee \Sigma A_2$, where $A = A_1 = A_2$ is a suspension; let β, y_1, y_2 represent the three inclusions. Here S will be $\{y_2, y_1\}$. We construct maps

$$\phi_n : G_n = A_n \times K_n \to [S^1 \wedge P, X]$$

which will turn out to be monomorphisms with disjoint images (save for 0). The first step is to construct a twisted version of the ad-operation. Let $L^* = L(y_2, y_1, b)$ be the free twisted Lie algebra on the ordered set $\{y_2, y_1, b\}$. L_{n+1}^* is Λ_{n+1} module which is turned into a Λ_n module by the morphism

$$\Lambda_n \to 1 \otimes \Lambda_n \subset \Lambda_{n+1}.$$

The homomorphism $ad_b : A_n \to_{n+1}^*$ is defined recursively by

$$ad_b(z) = [b, z] \text{ if } n = 1$$
$$ad_b(z \cdot y) = [ad_b(z), y] \text{ if } n > 1 \text{ and } y = y_1 \text{ or } y_2.$$

It is easy to verify that this makes $ad_b : A_n \to L_{n+1}^*$ into a Λ_n-homomorphism such that $ad_b(z) = [b, z]$ if z is a Lie element (i.e. $z, \epsilon i, m(L(S) \subset A(S))$. Furthermore, ad_b is a monomorphism, for a type of Hall basis for L^* can be concocted using a Hall basis $\{w\}$ for $L(S)$ and the prescription that b follows all w's: then ad_b carries a monotone monomial $w_1 \ldots w_k$ into the left normalized basis element $[b, w_1, \ldots, w_k]$.

On interpreting b, y_1, y_2 as maps and Lie products in L^* as Whitehead Products, we get a linear map

$$\psi_n : A_n \to [S^1 \wedge \Sigma A^{(n)}, X]$$

which carries each monotone monomial in the Hall basis for $L(S)$ into the images of distinct Hall bases for L^*. ψ_n extends to

$$\psi_n^1 : A_n \otimes_{\underline{Z}} K_n \to [S^1 \wedge \Sigma P, X] \text{ by } \psi_n'(z \otimes \underline{S}^1 \wedge \tau) = \psi_n(z) \circ (\underline{S}^1 \wedge \tau),$$

and it is easily verified that this factors through $A_n \otimes_{\Lambda_n} K_n$, yielding ϕ_n. The remark about monotone monomials mapping to basis elements in L^*, with the Hilton-Milnor Theorem, proves

LEMMA 8. *Each* $\phi_n : G_n \to [S' \wedge \Sigma P, X]$ *is a monomorphism and if* $n \neq m$, $Im(\phi_n) \cap Im(\phi_m) = \{0\}$.

Now by Theorem 1,

$$(8.1) \qquad [b, (y_1 + y_2)_0 \alpha] = \sum_n w[n]_0 \underline{S}^1 \Lambda H_n(\alpha)$$

where $w[n]$ is the Whitehead product defined by $w[n]$ with $x_1 = [b, y_1 + y_2]$ and $x_2 = y_1 + y_2$; thus

$$w[n]_0 \underline{S}^1 \wedge H_n(\alpha) = \phi_n(y_1 + y_2)^n \times \underline{S}^1 \wedge H_n(\alpha).$$

On the other hand, if $\{v\}$ is a Hall basis for $L(S)$, Theorem 0 gives

$$[b, (y_1 + y_2) \circ \alpha] = [b, \sum_v v \circ H_v(\alpha)]$$

$$= \sum_v [b, v \circ H_v(\alpha)]$$

(since A is a suspension). On expanding by Theorem 1, this is

$$(8.2) \qquad \sum_v \sum_m w(n, v) \circ \underline{S}^1 \wedge H_m(H_v(\alpha))$$

where $w(m, v)$ is $w[m]$ using $x_1 = [b, v], x_2 = v$. If $q = \dim(v)$, $w(m, v) = \phi_{mq}(v^q)$, so

$$q(m, v) \circ \underline{S}^1 \wedge H_m(H_v(\alpha)) = \phi_{mq}(v^q \times \underline{S}^1 \wedge H_m(H_v(\alpha)).$$

On applying ϕ_n^{-1} to (8.1) and (8.2) Theorem 6 follows.

NORTHWESTERN UNIVERSITY

and

THE INSTITUTE FOR ADVANCED STUDIES, JERUSALEM

Contemporary Mathematics
Volume 44, 1985

ARTIN'S BRAID GROUPS AND CLASSICAL HOMOTOPY THEORY

F. R. Cohen*

Let B_k denote Artin's braid group on k strands [1,3]. These groups are intimately related to classical homotopy theory in several seemingly diverse ways. In this expository note we consider just a few of these connections to topics such as Steenrod representability and the Hilton-Milnor theorem.

Sections 1 and 2 of this paper give some general algebraic and homological properties of the braid groups. Connections to Steenrod representability and bundle theory are given in Section 3. The homology of the braid groups is tied to the homology of the general linear groups and to the automorphism groups of free groups. A related natural homomorphism of the qth stable homotopy group of $\Omega^2 S^3$ to $K_{q-1}(\mathbb{Z})$ is introduced. Direct consequences are listed in Section 5. The remaining sections give proofs of these results.

We would like to thank the University of Rochester and especially John Harper and Richard Mandelbaum for organizing this conference in honor of the Stones, Arthur and Dorothy.

1. PRESENTATIONS AND OTHER PRELIMINARIES. The kth braid group B_k is generated by elements σ_i, $1 \leq i \leq k-1$, with relations

$$\sigma_i \sigma_j = \sigma_j \sigma_i \text{ if } |i-j| \geq 2, \text{ and } \sigma_i \sigma_{i+1} \sigma_i = \sigma_{i+1} \sigma_i \sigma_{i+1} \text{ for all } i.$$

By imposing the relation $\sigma_i^2 = 1$ for all i, one obtains a presentation for the symmetric group on k letters Σ_k. Thus there is a natural map

$$\pi : B_k \to \Sigma_k$$

which is an epimorphism. Denote the kernel of π by P_k; P_k is called the pure braid group. References for these facts are [1,2,3].

One may visualize the braid σ_i as

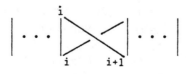

*The author was partially supported by the NSF and was an Alfred P. Sloan fellow during the preparation of this paper.

and $\pi(\sigma_i)$ is the transposition $(i, i+1)$. Embed B_k in B_{k+1} by "adding a trivial strand to the right." In particular, send σ_i to σ_i. Denote the limit, $\lim_{\to k} B_k$, by B_∞.

The abelianization homomorphism

$$h : B_k \to \mathbb{Z}$$

sends a word to its "length"; that is, $h(\sigma_{i_1}^{\epsilon_1} \cdots \sigma_{i_j}^{\epsilon_j}) = \sum_{i=1}^{j} \epsilon_i$. Write $[B_k, B_k]$ for the kernel of h in case $k \leq \infty$.

The center of B_k, Z_k, is isomorphic to the integers [8]. Z_k is generated by $(\sigma_1 \sigma_2 \cdots \sigma_{k-1})^k$ in case $k > 2$.

The braid groups support external sums and "tensor products" [10]

$$\bigoplus : B_j \times B_k \to B_{j+k}, \text{ and } \bigotimes : B_j \times B_k \to B_{jk} .$$

The sum $a \oplus b$ is defined on $a \times b$ by juxtaposition of braids a and b. The tensor product $a \otimes b$ is defined visually by thickening each strand of b and then embedding a in each tube [10]. For example

To be precise, we record the behavior of $\bigoplus$ and $\bigotimes$ on generators in $B_j \times B_k$:

1. $1 \oplus \sigma_i = \sigma_{i+j}$ and $\sigma_i \oplus 1 = \sigma_i$.
2. $\sigma_i \otimes 1 = \sigma_i \sigma_{i+j} \sigma_{i+2j} \cdots \sigma_{i+j(k-1)}$, and

$$1 \otimes \sigma_i = (\sigma_{(i-1)j+1} \cdots \sigma_{ij}) \cdots (\sigma_{ij-1} \sigma_{ij} \cdots \sigma_{ij+j-2})(\sigma_{ij} \sigma_{ij+1} \cdots \sigma_{ij+1(j-1)}) .$$

The pairings $\bigoplus$ and $\bigotimes$ satisfy a right distributivity law, however left distributivity fails. For example, let $[p]$ denote the identity element of B_p and observe that $[p] = [1] \underset{\leftarrow p \to}{\oplus \cdots \oplus} [1]$. Notice that

$$a \otimes [p] = a \underset{\leftarrow p \to}{\oplus \cdots \oplus} a$$

but left distributivity fails because

$$[p] \otimes a \neq a \underset{\leftarrow p \to}{\oplus \cdots \oplus} a$$

in general although $[p] \otimes a = ([1] \oplus \cdots \oplus [1]) \otimes a$.

The braid groups appear in the study of configuration spaces [13,14]. Let M be a topological space and let $F(M, k)$ denote the subspace of M^k given by $\{(m_1, \ldots, m_k) | m_i \neq m_j \text{ if } i \neq j\}$. The symmetric group Σ_k acts on $F(M, k)$ by permuting coordinates. Denote the orbit space $F(M, k)/\Sigma_k$ by $B(M, k)$.

It known that $B(R^2, j)$, $F(R^2, j)$ and $B(\lim_{\to n} \mathbb{R}^n, j)$ are Eilenberg-MacLane spaces of type $K(B_j, 1)$, $K(P_j, 1)$, and $K(\Sigma_j, 1)$, respectively [13,14]. Since $B(R^2, j)$ is finite dimensional, B_j has no elements of finite orders by P. A. Smith's theorem [13]. The inclusion $\mathbb{R}^2 \to \lim_{\to n} \mathbb{R}^n = \mathbb{R}^\infty$ induces a map $B(R^2, j) \to B(\mathbb{R}^\infty, j)$. On the level of fundamental groups this map is given by $\pi : B_j \to \Sigma_j$.

2. THE HOMOLOGY OF THE BRAID GROUPS. One of the lynchpins of some theorems described here is a computation of the homology of the B_k. Complete information is given in [9,11,28]; we record the following theorem where $\Omega^n X$ denotes the n-fold based loop space of a space X.

THEOREM 2.1. *There is a map* $\Theta : K(B_\infty, 1) \to \Omega^2 S^3$ *which induces a homology isomorphism for any trivial coefficient system. Furthermore, with field coefficients, Θ_* is an isomorphism of Hopf algebras.*

Let $S^3 < 3 >$ denote the 3-connected cover of S^3.

COROLLARY 2.2. *There is a map* $\Theta' : K([B_\infty, B_\infty], 1) \to \Omega^2 S^3 < 3 >$ *which induces a homology isomorphism for any trivial coefficient system.*

The inclusions $B_k \to B_\infty$ induce monomorphisms in homology and their images are given in [11]. The computations of the homology of $[B_k, B_k]$, $k < \infty$, seem more complex. Indeed, we have

COROLLARY 2.3. *The inclusion maps* $[B_k, B_k] \to [B_\infty, B_\infty]$ *have a kernel in homology.*

The pairings $\bigoplus$ and $\bigotimes$ induce pairings on the level of homology groups. The map induced by $\bigoplus$ evidently induces a commutative pairing in homology while explicit calculations with $\bigotimes$ [11-IV] also give a commutative pairing. Write $H_*(B_\infty^\otimes; \mathbb{Z}/p)$ for the ring structure induced by $\bigotimes$ where $\mathbb{Z}/p$ denotes the integers modulo p.

THEOREM 2.4. $H_*(B_\infty^\otimes; \mathbb{Z}/p)$ *is isomorphic to* $H_*(S^1 \times \Omega^2 S^3 < 3 >; \mathbb{Z}/p)$ *as a Hopf algebra over the Steenrod algebra.*

The point of this theorem is that $\Omega^2 S^3$ is homotopy equivalent to $S^1 \times \Omega^2 S^3 < 3 >$ but not multiplicatively. The ring structure of $H_*(\Omega^2 S^3; \mathbb{Z}/p)$ is isomorphic to the natural ring structure of $H_*(S^1 \times \Omega^2 S^3 < 3 >; \mathbb{Z}/p)$ if and only if $p > 2$.

3. BRAID GROUPS AND STEENROD REPRESENTABILITY. A classical theorem of Thom gives that every mod-2 homology class is Steenrod representable [24]. That is, if $x \in H_q(X; \mathbb{Z}/2)$, there is a q-dimensional compact manifold M without boundary and a map $f : M \to X$ with $f_*[M] = x$ where $[M]$ denotes the fundamental class of M. This result is closely related to the natural real representation of B_k discussed below.

Define $\rho_k : B_k \to 0(k)$ by sending a braid to the permutation in the orthogonal group determined by the ends of the braid; in particular, ρ_k is the composite

$$B_k \xrightarrow{\pi} \Sigma_k \xrightarrow{i} 0(k)$$

where i is the natural inclusion. Let $0 = \lim_{\to_n} 0(n)$. The maps ρ_k induce a map which commutes with $\bigoplus$ and $\bigotimes$; ρ induces a map on the level of classifying spaces

$$B\rho : BB_\infty \to BO$$

with $BB_\infty = K(B_\infty, 1)$. M is said to have a *braid orientation* if there is a homotopy commutative diagram

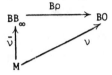

where ν classifies the stable normal bundle of M [9,7].

Let X be a topological space and let $\Omega_*^{Br}(X)$ denote the braid bordism classes of maps of braid oriented manifolds to X. The Hurewicz homomorphism

$$h : \Omega_*^{Br}(X) \to H_*(X, \mathbb{Z}/2)$$

is defined by $h[M, f] = f_*[M]$.

Observe that $2\Omega_*^{Br}(X) = 0$.

THEOREM 3.1.   *The Hurewicz homomorphism $h : \Omega_*^{Br}(X) \to H_*(X; \mathbb{Z}/2)$ is an isomorphism. Hence any element of $H_*(X; \mathbb{Z}/2)$ is represented by a braid oriented manifold which is unique up to braid bordism.*

COROLLARY 3.2.   *If M is a positive dimensional braid oriented manifold without boundary, then M is the boundary of a braid oriented manifold.*

Let η denote the vector bundle specified by $B\rho$.

LEMMA 3.3. $2\eta = 0$. *Consequently the stable normal bundle and stable tangent bundle of a braid oriented manifold M are isomorphic, the Pontrjagin classes of M vanish, the square of every Stiefel-Whitney class of M is zero. Furthermore, $w_1(M)$ is the mod-2 reduction of an integral class.*

The map $B\rho_k : BB_k \to BO(k)$ admits a nice description which follows from Ed Brown's elegant proof given in [12]. BB_k is the quotient space $F(R^2, k)/\Sigma_k = B(R^2, k)$. Let $\mathbb{R}^k \subset \mathbb{C}^k$ be the real points. Define $\eta_k : B(R^2, k) \to G_{k,2k}$, the Grassmanian of k planes in $\mathbb{R}^{2k}$ by the formula

$$\hat{\eta}_k\{\alpha_1, \ldots, \alpha_k\} = V(\mathbb{R}^k)$$

where V is the linear transformation in $\mathbb{C}^k$ given by the Vandermonde matrix

$$\begin{pmatrix} 1 & \cdots & 1 \\ \alpha_1 & & \alpha_k \\ \vdots & & \\ \alpha_1^{k-1} & & \alpha_k^{k-1} \end{pmatrix}.$$

The map $B\rho_k$ may be replaced by η_k up to homotopy.

It is natural to ask for some large class of braid oriented manifolds [9]. The surfaces which are braid oriented are connected sums of tori and Klein bottles. Other examples are solvmanifolds and some, but not all Bieberbach manifolds.

Ed Brown and Frank Peterson have studied related questions [5]. Shaun Bullett has pursued these directions in a more geometric direction [6,7] where he tried to give an explicit construction of a 4-dimensional braid oriented manifold with nonzero third Stiefel-Whitney class.

Theorem 3.1 was strongly suggested by work of Mahowald on Thom spectra [19] and also some elegant work of Madsen and Milgram [17] where 3.1 was proven by using manifolds with a "primitive Mahowald orientation" rather than a braid orientation.

4. BRAID GROUPS AND THE HILTON-MILNOR THEOREM. Let X denote a connected space having the homotopy type of a CW-complex and let $Y = \Sigma^2 X = S^2 \wedge X$. Pinching the equator in S^2 to a point, we obtain a map $\nabla: Y \to Y \nabla Y$ called the comultiplication. Iterating $\vee$ (say on the right-hand factor), we obtain a map $\nabla^j: Y \to \vee_j Y$. Let p be a fixed prime. Fold $\vee_{p^k} Y$ to Y to obtain a map

$$[p^k] : Y \to Y$$

which is the composite of ∇^{p^k} with the folding map. The map $[p^k]$ does not generally induce multiplication by the integer p^k on homotopy groups. For example, if $[2]: S^2 \to S^2$ is the map of degree 2, then $[2] * (\eta) = 4\eta$ where η is the generator of $\pi_3 S^2$. If Y is an H-space (or a p-local H-space), then $[p^k]$ induces multiplication by the integer p^k on homotopy groups.

The Hilton-Milnor theorem gives a method for studying the difference between multiplication by p^k on homotopy and the map induced by $[p^k]$. Here one considers $\Omega[p^k]$ which factors as

$$\Omega Y \overset{\Omega \nabla^{p^k}}{\to} \Omega(\vee_{p^k} Y) \overset{\Omega(fold)}{\to} \Omega Y.$$

Since Y is a suspension, $\Omega(\vee_{p^k} Y)$ is homotopy equivalent to a product $(\Omega Y)^{p^k} \times Z$ where Z is a weak direct product of loop spaces $\Omega\Sigma(Z_\alpha)$ with Z_α given by explicit (but complicated) smash products of ΣX [15,21,27]. This "factorization" of $\Omega[p^k]$ gives the formula

$$[p^k]_* = p^k + \Delta$$

on homotopy groups where p^k denotes multiplication by p^k and Δ is a sum of maps given by compositions of Whitehead products and Hopf invariants induced by the Hilton-Milnor decomposition of $\Omega(\vee_{p^k} Y)$ [27].

The main observation of this section is that $\Omega^2[p^k]$ and the p^k-th power map on $\Omega^2\Sigma^2 X$, denoted by p^k, are induced by the tensor product map for braid groups. Consider the family of homomorphisms

$$\hat{\phi}, \hat{\theta} : B_j \to B_{jp^k}$$

given by $\hat{\phi}(\alpha) = [p^k] \otimes \alpha$ and $\hat{\theta}(a) = \alpha \otimes [p^k]$.

PROPOSITION 4.1. $\hat{\phi}$ and $\hat{\theta}$ induce self-maps ϕ and θ of $\Omega^2\Sigma^2 X$ where ϕ is homotopic to $\Omega^2[p^k]$, and θ is homotopic to p^k.

Thus the failure of a two-sided distributivity law for $\oplus$ and $\otimes$ gives another description for the difference of $\Omega^2[p^k]$ and p^k in the abelian group $[\Omega^2\Sigma^2 X, \Omega^2\Sigma^2 X]$. We remark that M.G. Barratt has conjectured that if $[p^k]$ is null-homotopic, then p^{k+1} is null-homotopic.

Let C_*X denote the mod-p singular chains of a path-connected space X. We record an approximation for $C_*(\Omega^2\Sigma^2 X)$ which is related to the above remarks.

Let C denote a connected, homotopy commutative, and homotopy associative differential coalgebra over $\mathbb{Z}/p$ with diagonal $\Delta(C)$. Let $\overline{B}(G)$ denote the reduced bar resolution for the discrete group G. Notice that the symmetric group Σ_k acts on the pure braid group P_k and thus on $\overline{B}(P_k)$ by conjugation. Σ_k acts in the natural way on $C^{\otimes k} = C \otimes \ldots \otimes C$ by permutation of factors with the usual
$$\overset{\leftarrow k \rightarrow}{}$$
sign conventions. Denote the chain complex $\overline{B}(P_k) \otimes_{\Sigma_k} C^{\otimes k}$ by $B(k)$.

There are natural maps

$$\pi_i : \overline{B}(P_k) \otimes C^{\otimes(k-1)} \to B(k-1) \quad \text{and} \quad s_i : \overline{B}(P_k) \otimes C^{\otimes(k-1)} \to B(k)$$

for $1 \le i \le k$: Define $p_i : P_k \to P_{k-1}$ by "forgetting the ith string" and set $\pi_i = \overline{B}(P_i) \otimes 1$. Define $\sigma_i : C^{\otimes(k-1)} \to C^{\otimes k}$ by $\sigma_i(c_1 \otimes \ldots \otimes c_{k-1}) = c_1 \otimes \ldots \otimes e \otimes \ldots \otimes c_{k-1}$ the map which inserts the unit e in the ith coordinate and set $s_i = 1 \otimes \sigma_i$. Define $\Omega^2\Sigma^2(C)$ to be the pushout obtained from the s_i and the π_i for all i and k. In the following proposition, $H_*(X; \mathbb{Z}/p)$ has trivial differential.

PROPOSITION 4.2. *There is a map of chain complexes*

$$\alpha_* : \Omega^2\Sigma^2(H_*(X; \mathbb{Z}/p)) \to C_*(\Omega^2\Sigma^2 X)$$

which induces a homology isomorphism for each prime p.

Since C is a connected coalgebra, $C^{\otimes j}$ is a connected coalgebra with diagonal given by $\Delta(C^{\otimes j})$. Let $f : B_j \to B_{jp^k}$ denote a choice of the homomorphisms $\hat{\phi}$ or $\hat{\theta}$ which is fixed for all j. Notice that f restricts to give a homomorphism $f : P_j \to P_{jp^k}$. Consider the maps

$$\overline{B}(f) \otimes \Delta(C^{\otimes j}) : \overline{B}(P_j) \otimes C^{\otimes j} \to \overline{B}(P_{jp^k}) \otimes C^{\otimes jp^k}.$$

They induce maps $\overline{\phi}$ and $\overline{\theta}$ of $\Omega^2\Sigma^2(C)$; these maps correspond to maps induced by $\Omega^2[p^k]$ and p^k respectively.

5. SUNDRY PROPERTIES OF B_∞.

In this section we consider connections between B_∞, the automorphism group Φ_n of the free group on n generators F_n, and the general linear groups over $\mathbb{Z}$ and $\mathbb{Z}[t, t^{-1}]$. A classical representation of B_k induces a map $\pi_*^s\Omega^2 S^3 \to K_{*-1}\mathbb{Z}$ which is touched upon. Other curious properties are that B_3/center is infinite and has the homology of $K(\mathbb{Z}/6, 1)$.

To begin, let $f : BB_\infty \to X$ be a continuous map. We want to give a criterion for deciding whether f_* is monic on homology. Throughout this section, all homology groups are taken with $\mathbb{Z}/p$-coefficients for p a prime unless otherwise specified. Then $H_1 BB_\infty = \mathbb{Z}/2$ if $p = 2$ and $H_{2p-2}BB_\infty = \mathbb{Z}/p$ if $p > 2$. Let z denote a generator of this group for p a fixed prime. Finally, let y denote a generator of $H_1 BB_\infty$ if $p > 2$. By Theorem 2.1, $z^i \ne 0$ for all i.

LEMMA 5.1. *If $f_*(z^{p^j}) \neq 0$ for all j and when $p > 2, f_*(y) \neq 0$, then f_* is a monomorphism in homology.*

Let $i : \Sigma_n \to \Phi_n$ be the homomorphism which permutes the (fixed) generators of F_n. There is a well-known epimorphism [18]

$$\varsigma : \Phi_n \to GL_n(\mathbb{Z})$$

given by the formula

$$\varsigma(f)(e_\alpha) = \Sigma n_\alpha e_\alpha$$

where e_α runs over a basis for $\mathbb{Z}^n$ and g_α runs over a set of generators for F_n with $f(g_\alpha) = \pi(g_\alpha)^{n_\alpha}$. The groups Φ_n support "Whitney sums" $\oplus : \Phi_k \times \Phi_j \to \Phi_{k+j}$ which "commute" with ς and the Whitney sums for the $GL_n(\mathbb{Z})$. Set $\Phi_\infty = \lim_{\to n} \Phi_n$. Let $G = B_\infty, \Sigma_\infty, \Phi_\infty$, or $GL(\mathbb{Z})$. The homology of G is a ring with multiplication induced by Whitney sum. The usual map from $\pi_*^s S^o$ to $K_*(\mathbb{Z})$ factors through $\pi_*(B\Phi_\infty)^+$.

PROPOSITION 5.2. *With $\mathbb{Z}/2$-coefficients H_*G is isomorphic to $H_*B_\infty \otimes M_G$ as an H_*B_∞-module for some module M_G.*

We intend to return to the study of $H_*\Phi_\infty$.

There is yet another classical interesting representation of B_k namely the Burau representation [3]

$$b : B_k \to GL_k(\mathbb{Z}[t, t^{-1}]) .$$

Define b on generators σ_i by the formula

$$b(\sigma_i) = \begin{pmatrix} \begin{array}{c|cc|c} 1 & 0 & & 0 \\ \hline & 1\text{-}t & t & \\ 0 & 1 & 0 & 0 \\ \hline 0 & 0 & & 1 \end{array} \end{pmatrix}$$

where the middle entries are in the i and $(i+1)$-st rows and columns. Notice that b preserves the relations in B_k and that b commutes with $\oplus$ but not $\otimes$. Also, the following diagram commutes

$$\begin{array}{ccc} B_k & \xrightarrow{b_k} & GL_k(\mathbb{Z}[t, t^{-1}]) \\ \pi \downarrow & & \downarrow \text{evaluate at } t=1 \\ \Sigma_k & \xrightarrow{j} & GL_k(\mathbb{Z}) \end{array}$$

where j is the standard inclusion. We obtain a map

$$b : \Omega^2 S^2 \to BGL(\mathbb{Z}[t, t^{-1}])^+$$

by passage to group completions and thus a map

$$\hat{b} : \Omega^\infty \Sigma^\infty \Omega^2 S^2 \to BGL(\sigma\mathbb{Z})^+ \times BGL(\mathbb{Z})^+$$

where $\sigma \mathbb{Z}$ is the suspension of $\mathbb{Z}$ [26]. Thus there is an induced non-trivial map

$$\hat{b}_* : \pi_*^s \Omega^2 S^3 \to K_{*-1}(\mathbb{Z}) \oplus K_*(\mathbb{Z}).$$

The following proposition was proven by Don Anderson and the author in 1976 (unpublished).

PROPOSITION 5.3. b *deloops once, but not twice.*

We complete this section by listing some homological properties of the groups B_k/Z_k. Consider the morphism of short exact sequences

$$
\begin{array}{ccccccccc}
1 & \to & Z_k & \xrightarrow{\ i\ } & B_k & \xrightarrow{\ j\ } & B_k/Z_k & \to & 1 \\
& & \downarrow{\cong} & & \downarrow{h} & & \downarrow{\lambda} & & \\
1 & \to & \mathbb{Z} & \xrightarrow{k(k-1)} & \mathbb{Z} & \to & \mathbb{Z}/k(k-1) & \to & 1
\end{array}
$$

and denote B_k/Z_k by G_k. Analyzing the Lyndon spectral sequence, we have

PROPOSITION 5.4. *The map* j_* *on mod-2 homology is a split monomorphism and* $H_{q+2}(G_k; \mathbb{Z}/2) \cong H_{q+2}(B_k; \mathbb{Z}/2) \oplus H_q(G_k; \mathbb{Z}/2)$. *If* p *is prime to* $k(k-1)$, *then* $H_*(B_k; \mathbb{Z}/p) \cong H_*(S^1; \mathbb{Z}/p) \otimes H_*(G_k; \mathbb{Z}/p)$. *The map* $\lambda : G_3 \to \mathbb{Z}/6$ *induces a homology isomorphism for all trivial local coefficients although* G_3 *is infinite.*

6. PROOFS FOR SECTION 3. The maps $\rho_k : B_k \to 0(k)$ induce a compatible family of maps $\eta_k : K(B_k, 1) \to BO(k)$ and thus determine a bordism theory which we denote by $\Omega_*^{Br}(X)$. Recall that the suspension of the Thom complex obtained from the bundle η_k maps naturally to the Thom complex obtained from η_{k+1}. The resulting spectrum is denoted MBr here. Classically, $\Omega_*^{Br}(X)$ is isomorphic to $\pi_*^s(X \wedge MBr)$ and so it suffices to prove

LEMMA 6.1. MBr *is the Eilenberg-MacLane spectrum* $K(\mathbb{Z}/2, 0)$.

Notice that 3.1 follows directly from 6.1.

Next notice that $\Omega_*^{Br}(\text{point})=0$ if $* > 0$. Thus any positive dimensional braid oriented manifold without boundary is itself a braid oriented boundary and 3.2 follows.

To prove 3.3 we consider the vector bundle $\eta_{k,n}$ given by the projection

$$F(R^2, k) X_{\Sigma_k} (R^n)^k$$
$$\downarrow$$
$$F(R^2, k)/\Sigma_k$$

where Σ_k acts diagonally by permuting factors. By definition, $\eta_{k,n}$ is the n-fold Whitney sum of $\eta = \eta_{k,1}$. To prove 3.3, we must show that $\eta_{k,2}$ is a trivial bundle. That $\eta_{k,2}$ is trivial was first done using [9], but Ed Brown gave an elegant proof which is also given here [12].

Define $\hat{f} : F(R^2, k) \times (R^2)^k \to F(R^2, k) \times (R^2)^k$ by first regarding points of $\mathbb{R}^2$ as complex numbers and then defining

$$\hat{f}((\alpha_1, \ldots, \alpha_k), (x_1, \ldots, x_k)) = ((\alpha_1, \ldots, \alpha_k), (\Sigma_i x_i, \Sigma \alpha_i x_i, \Sigma \alpha_i^2 x_i, \ldots, \Sigma \alpha_i^{k-1} x_i)).$$

Notice that $\hat{f}$ induces a map f of $(2k)$-plane bundles

$$
\begin{array}{ccc}
F(\mathbb{R}^2,k)\times_{\Sigma_k}(\mathbb{R}^2)^k & \xrightarrow{\ f\ } & F(\mathbb{R}^2,k)/\Sigma_k\times(\mathbb{R}^2)^k \\
\Big\downarrow{\scriptstyle\Sigma_k} & & \Big\downarrow \\
F(\mathbb{R}^2,k)/\Sigma_k & \xrightarrow{\ \ 1\ \ } & F(\mathbb{R}^2,k)/\Sigma_k \ .
\end{array}
$$

The map f restricted to each fibre is given by the Vandermonde matrix of $(\alpha_1,\ldots,\alpha_k)$. The matrix is non-singular because $\alpha_i-\alpha_j$ is a unit in $\mathbb{C}$. Thus f is a bundle isomorphism and so $\eta_{k,2}$ is trivial. Proposition 3.3 follows directly except for the statement that w_1 is the reduction of an integral class. This last statement follows from the fact that $H^1(B_\infty;\mathbb{Z})\cong\mathbb{Z}$ and that ρ^* is an isomorphism on $H^1(;\mathbb{Z}/2)$.

We now finish the proof of 5.1. There is a natural map

$$\phi:\mathcal{A}\to H^*(MBr;\mathbb{Z}/2)$$

given by $\phi(\alpha)=\alpha(\bigcup)$ where $\bigcup$ denotes the Thom class of the bundle over BB_∞ and where $\mathcal{A}$ is the mod-2 Steenrod algebra. We shall show that ϕ induces an isomorphism.

Here, first notice that $B\rho_*:H_*(K(B_\infty,1);\mathbb{Z}/2)\to H_*(BO;\mathbb{Z}/2)$ is multiplicative in homology because ρ commutes with Whitney sums and the multiplication in homology is induced by Whitney sums. Thus ϕ is a coalgebra map. To show that ϕ is a monomorphism it thus suffices by classical results of Milnor and Moore [22] to show that ϕ is a monomorphism when restricted to the module of primitives. We remark that the proof of 5.1 is then finished because $\mathcal{A}$ and $H_*(B_\infty;\mathbb{Z}/2)$ have the same (finite) dimension over $\mathbb{Z}/2$ in any fixed degree.

The primitives in $\mathcal{A}$ are given by the Milnor Q_i where $Q_0=Sq^1$ and inductively $Q_{i+1}=[Q_i,Sq^{2^i}]=Q_iSq^{2^i}+Sq^{2^i}Q_i$.

It is known by work of Peterson and Toda [23] that

$$Q_k\bigcup=\bigcup\cup(w_{2^k-1}+\Gamma)$$

in $H^*(MO;\mathbb{Z}/2)$ where $\bigcup$ is the Thom class and Γ is decomposable. To show that $\rho^*(Q_i\cdot\bigcup)\neq0$, it suffices to show that $\rho^*(w_{2^k-1}+\Gamma)\neq0$ by the Thom isomorphism.

Notice that $H_*(B_\infty;\mathbb{Z}/2)$ is isomorphic as a Hopf algebra to $\bigotimes_{k\geq1}P[x_{2^k-1}]$ where x_{2^k-1} is of degree 2^k-1 and the Steenrod operations are specified by $Sq^1_*x_{2^k-1}=(x_{2^{k-1}-1})^2$ if $k>1$ and $Sq^{2^r}_*x_{2^k-1}=0$ if $r>0$ [9]. Let $<,>$ denote the evaluation map

$$<,>:H^q(X;\mathbb{Z}/2)\otimes H_q(X;\mathbb{Z}/2)\to\mathbb{Z}/2$$

obtained by regarding $H^q(X;\mathbb{Z}/2)\cong\mathrm{Hom}(H_qX;\mathbb{Z}/2)$. We shall show that $\rho^*(Q_k\bullet\bigcup)\neq0$ by showing $<\rho^*w_{2^k-1},x_{2^k-1}>=1$ and $<\rho^*\Gamma,x_{2^k-1}>=0$. Thus 5.1 follows.

Observe that $< \rho^*\Gamma, x_{2^k-1} >= 0$ because Γ is decomposable and x_{2^k-1} is primitive. Next, observe that

$$Sq^1 Sq^2 \cdots Sq^{2^{k-2}} w_{2^k-1} = w_{2^k-1} + \text{ decomposables}$$

by the Wu formulas. Compute $< \rho^* w_{2^k-1}, x_{2^k-1} >$:

$$
\begin{aligned}
< \rho^* w_{2^k-1}, x_{2^k-1} > &= < \rho^* Sq^1 Sq^2 \cdots Sq^{2^{k-2}} w_{2^k-1}, x_{2^k-1} > \\
&= < w_{2^k-1}, \rho_*(Sq_*^{2^{k-2}} Sq_*^{2^{k-3}} \cdots Sq_*^2 Sq_*^1 x_{2^k-1}) > \\
&= < w_{2^k-1}, \rho_* x_1^{2^{k-1}} > \\
&= < w_1 \otimes \cdots \otimes w_1, \rho_* x_1 \otimes \cdots \otimes \rho_* x_1 > \\
&\quad\quad \overleftarrow{2^{k-1}} \qquad\qquad \overleftarrow{2^{k-1}} \\
&= 1 .
\end{aligned}
$$

7. PROOFS FOR SECTION 4. To prove the propositions in Section 4, we first recall the space of little n-cubes invented by Boardman and Vogt [4]: $C_n(j)$ is the space of maps $< c_1, \ldots, c_j >: \amalg_j I^n \to I^n$ where each c_i is a product of maps f_{is} of the form $f_{is} : I \to I$ given by $f_{is}(t) = a + tb$ for $0 \le a < b$, $a + b \le 1$, and the interiors of $c_i(I^n)$ and $c_k(I^n)$ are disjoint if $i \ne k$. Boardman and Vogt show that $C_n(j)$ is equivariantly homotopy equivalent to $F(\mathbb{R}^n, j)$ [4]. Composition of maps gives a pairing $c : C_n(j)/\Sigma_j \times C_n(k)/\Sigma_k \to C_n(jk)/\Sigma_{jk}$ which is considered in [20,10]. In particular, when $n = 2$, this pairing is induced by the homomorphism $\otimes : B_j \times B_k \to B_{jk}$ which gives c on the level of fundamental groups [10].

Boardman and Vogt give a map $\rho : C_n(j) \times (\Omega^n X)^j \to \Omega^n X$ defined by

$$\rho(< c_1, \ldots, c_j >, (f_1, \ldots, f_j))(t) = \begin{cases} f_i(u), & \text{if } c_i(u) = t \\ *, & \text{if } t \notin \cup_i \text{ image } (c_i). \end{cases}$$

A check of the definition now gives that $\hat{\theta}$ induces the $p^k - th$ power map:

$$C_2(1) \times \Omega^2 X \to C_2(p^k) \times (\Omega^2 X)^{p^k} \overset{\rho}{\to} \Omega^2 X$$

is the induced map and this is the $p^k - th$ power map.

Next, consider the map

$$C_2(j) \times X^j \to C_2(jp^k) \times X^{jp^k}$$

induced by $\hat{\theta}$ on little cubes and the diagonal map for X^k. These pass to quotients to give a self-map, θ, of May's construction, $C_2 X$ [20]. There is a map $\alpha : C_2 X \to \Omega^2 \Sigma^2 X$ which is a homotopy equivalence if X is of the homotopy type of a connected CW complex. Notice the diagram

$$
\begin{array}{ccc}
C_2 X & \overset{\alpha}{\longrightarrow} & \Omega^2 \Sigma^2 X \\
\theta \downarrow & & \downarrow \Omega^2 [p^k] \\
C_2 X & \overset{\alpha}{\longrightarrow} & \Omega^2 \Sigma^2 X
\end{array}
$$

homotopy commutes and so 4.1 follows.

To prove 4.2, observe that there is a chain equivalence

$$\overline{B}(P_k) \otimes_{\Sigma_k} (H_*(X, \mathbb{Z}/p))^{\otimes k} \overset{1 \otimes i^k}{\to} B(P_k) \otimes_{\Sigma_k} (C_* X)^{\otimes k}$$

where $i : H_*(X; \mathbb{Z}/p) \to C_* X$ is a chain equivalence and where $H_*(X; \mathbb{Z}/p)$ has trivial differential. Composing with ρ_*, we obtain a map

$$\sum_{k \geq 0} \overline{B}(P_k) \otimes_{\Sigma_k} H_*(X; \mathbb{Z}/p)^{\otimes k} \to C_*(\Omega^2 \Sigma^2 X).$$

which passes to quotients to give

$$\alpha_* : \Omega^2 \Sigma^2 (H_*(X; \mathbb{Z}/p)) \to C_*(\Omega^2 \Sigma^2 X).$$

That α_* induces a homology isomorphism follows directly from [11–III; Corollary 3.4] and its proof: $\Omega^2 \Sigma^2 (H_*(X; \mathbb{Z}/p))$ and $C_*(\Omega^2 \Sigma^2 X)$ are naturally filtered and the homology of the associated graded is given in each case by $H_*(B(P_k) \otimes_{\Sigma_k} \overline{H}_*(X; \mathbb{Z}/p)^{\otimes k})$. The map α_* evidently gives an isomorphism on the homology of the associated graded and so 4.2 follows.

8. PROOFS FOR SECTION 5. The mod-p homology of BB_∞ and $\Omega^2 S^3$ as Hopf algebra is given by

a.
$$\bigotimes_{i \geq 1} P[x_{2^i - 1}] \text{ if } p = 2$$

b.
$$\left(\bigotimes_{i \geq 0} \Lambda[x_{2p^i - 1}] \right) \otimes \left(\bigotimes_{j \geq 1} P[x_{2p^j - 2}] \right) \text{ if } p > 2$$

where x_k is of degree k. The Steenrod operations in homology are specified by

a'.
$$Sq_*^1 x_{2^i - 1} = (x_{2^{i-1} - 1})^2 \text{ if } i > 1,$$
$$Sq_*^{2^r} x_{2^i - 1} = 0 \text{ if } r > 0, \text{ and}$$

b'.
$$P_*^1 x_{2p^j - 2} = -(x_{2p^{j-1} - 2})^p \text{ if } j > 1,$$
$$\beta x_{2p^j - 1} = x_{2p^j - 2}$$
$$P_*^{p^r} x_{2p^j - 2} = 0 \text{ if } r > 0, \text{ and}$$
$$P_*^{p^r} x_{2p^j - 1} = 0.$$

Using this information, we prove 5.1 in case $p = 2$; the details for $p > 2$ are similar.

Notice that the module of primitives $PH_* BB_\infty$ has a basis given by $(x_{2^i - 1})^{2^k}$ and so there is at most one primitive in any fixed degree. Furthermore, $Sq_*^{2^{i+k-1}} \cdots Sq_*^{2^{k+1}} Sq_*^{2^k} (x_{2^i - 1})^{2^k} = (x_1)^{2^{i+k-1}}$. Thus, if $f_*(x_1^{2^{i+k-1}}) \neq 0$, we have $f_*(x_{2^i - 1})^{2^k} \neq 0$ by naturality of the Steenrod operations. 5.1 follows because f_* is monic when restricted to $PH_* BB_\infty$.

To prove 5.2, observe that the map $\rho : B_\infty \to 0$ satisfies $(B\rho_*)(x_1) \neq 0$, $(B\rho)_*$ is multiplicative, and the one-dimensional generator of $H_1 BO$ has infinite height in the Pontrjagin ring. Thus $(B\rho)_*$ is a monomorphism by 5.1. Since ρ factors as a composite

$$B_\infty \xrightarrow{\pi} \Sigma_\infty \to \Phi_\infty \to GL(\mathbb{Z}) \to GL(\mathbb{R}) \simeq 0 \, ,$$

it follows that $H_* B_\infty$ is isomorphic to a sub-Hopf algebra of $H_* G$ and so 5.2 follows from [22].

To prove 5.3 first observe that the Burau representation commutes with Whitney sums and thus induces a loop map $f : \Omega^2 S^2 \to \mathbb{Z} \times BGL(\mathbb{Z}[t^{\pm 1}])^+$. Next consider the composite

$$B_\infty \xrightarrow{b} GL(\mathbb{Z}[t^{\pm 1}]) \xrightarrow{\text{determinant}} \mathbb{Z}[t^{\pm 1}]^*$$

and observe that determinant $(b(\sigma_i)) = \pm t$. Hence, the map $\Omega^2 S^2 \to \mathbb{Z} \times BGL(\mathbb{Z}[t^{\pm 1}])^+$ cannot factor through $\mathbb{Z} \times BGL(\mathbb{Z})^+$ because it can't on π_1. If f were a double loop map, then it would be induced by a map $S^o \to \mathbb{Z} \times BGL(\mathbb{Z}[t^{\pm 1}])^+$ which factors through through $\mathbb{Z} \times BGL(\mathbb{Z})^+$. Since the standard inclusion $\mathbb{Z} \to \mathbb{Z}[t^{\pm 1}]$ induces a map of ∞-loop spaces $\mathbb{Z} \times BGL(\mathbb{Z})^+ \to \mathbb{Z} \times BGL(\mathbb{Z}[t^{\pm 1}])^+$, our map f factors through $\mathbb{Z} \times BGL(\mathbb{Z})^+$ were it a double loop map and 5.3 follows.

To prove 5.4, consider the composite

$$Z_k \xrightarrow{i} B_k \xrightarrow{\pi} \Sigma_k \, .$$

Z_k is generated by $(\sigma_1, \ldots, \sigma_{k-1})^k$ and thus this generator projects to $(1, k, k-1, \ldots, 3, 2)^k$ in Σ_k and is trivial. Thus π factors through G_k:

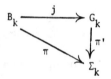

Since π_* is monic in mod-2 homology, so is j_*. Consider the Lyndon spectral sequence for the short exact sequence

$$1 \to Z_k \to B_k \to G_k \to 1 \, .$$

Since Z_k is central, the local coefficient system is trivial and $E^2_{p,q} = H_p G_k \otimes H_q Z_k$ (with $\mathbb{Z}/p$-coefficients for p prime). Since j_* is monic (with $\mathbb{Z}/2$-coefficients), $d^2 : E^2_{*,0} \to E^2_{*-2,1}$ is a split epimorphism and thus it follows that $H_{q+2}(G_k; \mathbb{Z}/2) \cong H_{q+2}(B_k; \mathbb{Z}/2) \oplus H_q(G_k; \mathbb{Z}/2)$.

If p is prime to $k(k-1)$, we compare the Lyndon spectral sequences for

$$\begin{array}{ccccc}
Z_k & \longrightarrow & B_k & \longrightarrow & G_k \\
\Big\downarrow{\scriptstyle \cong} & & \Big\downarrow{\scriptstyle h} & & \Big\downarrow{\scriptstyle \lambda} \\
\mathbb{Z} & \xrightarrow{k(k-1)} & \mathbb{Z} & \longrightarrow & \mathbb{Z}/k(k-1)
\end{array} \, .$$

In this case, $H_*(Z_k, \mathbb{Z}/p) \to H_*(B_k, \mathbb{Z}/p)$ is a split monomorphism and the fibre is totally non-homologous to zero. Thus, the homology spectral sequence collapses and $H_*(B_k; \mathbb{Z}/p) \cong H_*(G_k; \mathbb{Z}/p) \otimes H_*(Z_k; \mathbb{Z}/p)$.

We now consider B_3 and G_3. Consider the equivalence class of σ_1^6 in G_3. If this were trivial, then σ_1^6 must equal $(\sigma_1\sigma_2))^3$ by a comparison of the abelianization homomorphism. But

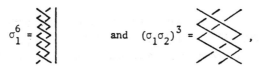

$$\sigma_1^6 = \qquad \text{and} \qquad (\sigma_1\sigma_2)^3 = \qquad ,$$

thus the equivalence class of σ_1^{6n} in G_3 is not that of the identity if $n > 0$.

A comparison of Lyndon spectral sequences together with the observation that $h : B_3 \to \mathbb{Z}$ is a homology isomorphism (with trivial local coefficients) gives the rest of the proposition.

9. This note scratches the surface. Discussion of the important and beautiful applications of Brown-Gitler spectra have been deliberately omitted. The interested reader should see papers of Brown and Peterson, G. Carlsson, R. Cohen, M. Mahowald, and H. Miller.

REFERENCES

[1] E. Artin, *Theory of braids*, Ann. of Math., **48**, (1947), 101–126.

[2] E. Artin, *Braids and permutations*, Ann.of Math., **48**, (1947), 643–649.

[3] J. S. Birman, *Braids, Links, and Mapping Class Groups*, Ann. of Math. Studies **82**, Princeton Univ. Press, 1974.

[4] J. M. Boardman and R. M. Vogt, *Homotopy everything H-spaces*, B.A.M.S., **74**, (1968), 1117–1122.

[5] E. H. Brown and F. P. Peterson, *The Brown-Gitler spectrum, $\Omega^2 S^3$, and $\eta_j \epsilon \pi_{2^j}(S)$*, to appear in Uspekhi Math. Nauk.

[6] S. Bullett, *Braid orientations and Stiefel-Whitney classes*, Quart. J. Math. Oxford, **32**, (1981), 267–285.

[7] S. Bullett, *Permutations and braids in cobordism theories*, Proc. Lond. Math. Soc., **38**, (1979), 517–531.

[8] W. L. Chow, *On the algebraic braid group*, Ann. of Math., **49**, (1948), 654–658.

[9] F. R. Cohen, *Braid orientations and bundles with flat connections*, Inv. Math., **46**, (1978), 99–110.

[10] F. R. Cohen, *Little cubes and the classifying space for n-sphere fibrations*, A.M.S. Proceedings of Symposia in Pure Math., **32**, (1978), 245–248.

[11] F. R. Cohen, T. J. Lada and J. P. May, *The homology of iterated loop spaces*, Springer-Verlag L.N.M. v. 533, Springer-Verlag, 1976.

[12] F. R. Cohen, M. E.Mahowald and R. J. Milgram, *The stable decomposition of the double loop space of a sphere*, A.M.S. Proceedings of Symposia in Pure Math., **32**, (1978), 225–228.

[13] E. Fadell and L. Neuwirth, *Configuration spaces*, Math. Scand., **10**, (1962), 111–119.

[14] R. Fox and L. Neuwirth, *The braid groups*, Math. Scand., **10**, (1962), 119–126.

[15] P. J. Hilton, *On the homotopy groups of the union of spheres*, J. London Math. Soc., **30**, (1955), 154–172.

[16] R. K. Lashof, *Poincaré duality and cobordism*, T.A.M.S., **109**, (1963), 257–277.

[17] I. Madsen and R. J. Milgram, *On spherical fibre bundles and their PL reductions*, New Developments in Topology, London Math. Soc. Lecture Notes **11**, Cambridge Univ. Press, 1974.

[18] W. Magnus, A. Karrass and D. Solitar, *Combinatorial Group Theory*, Interscience Publishers, 1966.

[19] M. E. Mahowald, *Ring spectra which are Thom spectra*, Duke Math. J., **46**, (1979), 549–559.

[20] J. P. May, *The geometry of iterated loop spaces*, Springer-Verlag L.N.M. v. 271, Springer-Verlag, 1972.

[21] J. Milnor, *On the construction FK*, Algebraic Topology, A Student's Guide, London Math. Soc. Lecture Notes **4**, Cambridge Univ. Press, 1972, 119–136.

[22] J. Milnor and J. C.Moore, *On the structure of Hopf algebras*, Ann. of Math., **81**, (1965), 211–264.

[23] F. P. Peterson and H.Toda, *On the structure of $H^*(BSF; \mathbb{Z}/p)$*, J.Math. Kyoto Univ., **7**, (1967), 113–121.

[24] R. Strong, *Notes on Cobordism Theory*, Mathematical Notes, Princeton University Press, 1968.

[25] R. Thom, *Quelques propriétiés globales des variétés différentiables*, Comm. Math. Helv., **28**, (1954), 17–86.

[26] J. B. Wagoner, *Delooping classifying spaces in algebraic K-theory*, Topology, **11**, (1972), 417–423.

[27] G. W. Whitehead, *Homotopy Theory*, Springer-Verlag Graduate Texts in Math. v. 61, Springer-Verlag, 1978.

[28] G. B. Segal, *Configuration spaces and iterated loop spaces*, Invent. Math., **21** (1973), 213–221.

UNIVERSITY OF KENTUCKY, LEXINGTON, KY 40506

Contemporary Mathematics
Volume 44, 1985

MORE COMPACTA OF INFINITE COHOMOLOGICAL DIMENSION

Leonard R. Rubin

Abstract

Three large classes S, $\overline{S}$, S_1 of subspaces of the Hilbert cube are stipulated. Each $\overline{X}$ in $\overline{S}$ is the closure of an $X \in S$ and is contained in an $X_1 \in S_1$. All spaces in S_1 are compact and are shown to have infinite cohomological dimension.

1. Introduction. The research in this paper can be considered a continuation of that begun in [Ru3]. In that paper three classes of spaces lying in the Hilbert cube, say S, S_1, S_2 were presented so that for each $X_2 \in S_2$ there were $X \in S$, $X_1 \in S_1$ with certain properties. First, X_1 and X_2 were always compacta, whereas X was strongly infinite dimensional and totally disconnected (hence not compact); second, $X \subset \overline{X} = X_1 \subset X_2$; finally, we proved in [Ru3] that the cohomological dimension (c-dim) of X_2 is ∞ for each such X_2.

What distinguishes this result from previous ones [W1] is that apparently X_2 is not (nor does it appear to contain) an intersection of separators or continuumwise separators as required in the theorems of [W1]. Spaces X_1 in the class S_1 are, a fortiori, not in this latter category either. Worse still, spaces X in S are not even compact.

To better deal with the theory to be developed in this paper, it will be preferable to broaden the class of spaces S. The larger class S will contain spaces that are compact as well as certain noncompact ones; it will include all X in the previous class S. The class $\overline{S}$ will consist of all $\overline{X}$, $X \in S$. In turn, there will be $X_1 \in S_1$ with $\overline{X} \subset X_1$, and we will prove that c-dim $X_1 = \infty$ for each $X_1 \in S_1$. The results will generalize those of [Ru3] and Theorem 3.1 of [W1].

AMS Subject Classification: 55M10, 54F45.

Key Words and Phrases: cohomological dimension, dimension intersection property, Eilenberg-MacLane space, essential family, strongly infinite dimensional.

Although some spaces in S or $\overline{S}$ will lie in S_1, there will be many that do not; for these the cohomological dimension has not yet been determined.

For the connection between the theory of cohomological dimension and that of cell-like mappings and dimension raising, one may consult [Ru3] and [W2]. The main point to make here is that it is unknown whether a separable metric space of infinite dimension must have infinite cohomological dimension. If there is a cell-like dimension raising map of a compactum, then there exists an infinite dimensional compactum of finite cohomological dimension, and conversely.

2. Preliminaries. The Hilbert cube Q is $\Pi\{I_k|k = 1, 2, \cdots\}$ where $I_k = [-1, 1]$. Throughout this paper spaces are separable and metrizable and hence all can be embedded topologically in Q. Let $\Pi_k : Q \to I_k$ be the coordinate projection, $A_k = \Pi_k^{-1}(-1)$, $B_k = \Pi_k^{-1}(1)$. For $0 \le t = t_k < 1$, let $A_k^{t_k} = A_k^t = \Pi_k^{-1}([-1, -1+t])$ and $B_k^{t_k} = B_k^t = \Pi_k^{-1}([1-t, 1])$.

The Eilenberg-MacLane space $K_n = K(\mathbb{Z}, n)$ is described in [W2]. We treat K_n as a CW-complex such that $S^n \subset K_n$ and,

$$\Pi_k(K_n, *) \approx \begin{cases} \Pi_k(S^n, *) & k \le n \\ 0 & k \ge n+1 \end{cases}$$

2.1. Definition. A map $f : X \to I^{n+1}$ is *stable* if $f|f^{-1}(S^n) : f^{-1}(S^n) \to S^n$ does not extend to a map of X to S^n. It is called *cohomologically stable* if it does not extend to a map of X to K_n.

2.2. Note. The existence of a stable map is equivalent to $\dim X \ge n+1$, while that of a cohomologically stable map implies $c\text{-}\dim X \ge n + 1$ (c-dim means cohomological dimension).

2.3. Notation. If Γ is a set of natural numbers, then by Q_Γ we mean the set of all $(x_1, x_2, \ldots, x_i, \ldots) \in Q$ such that $x_i = 0$ if $i \notin \Gamma$. Thus if Γ is finite, then Q_Γ is a copy of I^n for some n. If $Y \subset Q$, then Y_Γ denotes $Y \cap Q_\Gamma$.

2.4. Definition. A collection $\{(A_k', B_k')|k \in \Gamma\}$ of disjoint pairs of closed subsets of a space X is called an *essential family* for X provided that if S_k is a closed set separating A_k' and B_k' in X for each $k \in \Gamma$, then $\cap\{S_k|k \in \Gamma\} \ne \emptyset$. (This implies $\dim X \ge$ card Γ.) Any subset of an essential family is an essential family. A set such as S_k is often called a *separator* of (A_k', B_k').

The following proposition is similar to 5.5 of [R-S-W]; the first part does not require compactness. See also 2.5 of [Ru3].

2.5. Proposition. Let $\{(A_k', B_k')|k \in \Gamma\}$ be an essential family for a space X and let $J \subset \Gamma$. If for each $j \in J$, S_j is a separator of (A_j', B_j'), and $X^* = \cap\{S_j|j \in J\}$, then $\{(A_k' \cap X^*, B_k' \cap X^*)|k \in \Gamma - J\}$ is an essential family for X^*. If in addition X is compact, then for each $k \in \Gamma - J$, X^* contains a continuum meeting A_k' and B_k'.

2.6. Definition. A space is called *strongly infinite dimensional* if it has an infinite essential family. It is *hereditarily strongly infinite dimensional* if in addition each subspace is either 0-dimensional or strongly infinite dimensional. Such spaces are constructed in [Ru1, Ru2].

It will be convenient in a later part of the paper to have the following Lemmas available.

2.7. LEMMA. *Let X be a space of dimension $n+1$, let $\{(A'_k, B'_k)|1 \leq k \leq n+1\}$ be an essential family for X, and let $A = \cup\{A'_k \cup B'_k|1 \leq k \leq n+1\}$. Suppose $f : A \to S^n$ is a map such that $A'_k = f^{-1}(A_k)$, $B'_k = f^{-1}(B_k)$, $1 \leq k \leq n+1$. (Here $S^n \subset I^{n+1} \subset Q$, $A_k = A_k \cap S^n$, $B_k = B_k \cap S^n$). Then there does not exist an extension of f to a map of X to K_n.*

PROOF. The map f extends to a map of X to I^{n+1}. By 4.4 of [Ru3], f is stable. Suppose $F : X \to K_n$ is a map such that $F = f$ on $f^{-1}(S^n) = A$. Then F can be homotoped, rel A, to a map of X into the $(n+1)$-skeleton of K_n, which is precisely S^n. This contradicts the stability of f.

2.8. LEMMA. *Suppose $W \subset Q$ and S is a separator in W of $(A^t_k \cap W, B^t_k \cap W)$ where $0 < t_k < 1$. Then there exists a separator Z of (A_k, B_k) in Q such that $Z \cap W \subset S$.*

PROOF. See the proof of 6.2 of [Ru1].

Next we have Lemma 4.3 of [Ru3].

2.9. LEMMA. *Let $K \subset \tilde{X}$ be such that $\dim K \leq m < \infty$. Suppose $\{(A'_k, B'_k)|1 \leq i \leq n\}$ is a collection of disjoint pairs of closed subsets of $\tilde{X}$ and that for each i, S_i is a separator of (A'_i, B'_i) in $\tilde{X}$. Let U be a neighborhood of $S = \cap\{S_i|1 \leq i \leq n\}$. Then for each i there exists S^*_i so that S^*_i is a separator of (A'_i, B'_i) in $\tilde{X}$, $S^* = \cap\{S^*_i|1 \leq i \leq n\} \subset U$, and $\dim(S^* \cap K) \leq m - n$.*

3. Calculating Cohomological Dimension. In this section we will prove that c-$\dim X_1 = \infty$ for each X_1 in a certain class of spaces. In the next section there will be many examples.

3.1. Definition. Let Γ_1, Γ_2 be a disjoint pair of sets of natural numbers. Then a space $Y \subset Q$ satisfies $DIP(\Gamma_1, \Gamma_2)$ (dimension intersection property) if for each $\Gamma \subset \Gamma_2$, the following two conditions prevail.

3.1.1 $\dim(Y_{\Gamma_1 \cup \Gamma}) \leq \operatorname{card} \Gamma$, and

3.1.2 for any collection $\{t_k|k \in \Gamma, 0 < t_k < 1\}$, the set $\{(A^t_k \cap Y_{\Gamma_1 \cup \Gamma}, B^t_k \cap Y_{\Gamma_1 \cup \Gamma})|k \in \Gamma\}$ is an essential family for Y.

For the remainder of this section, assume Γ_1, Γ_2 and Y satisfying $DIP(\Gamma_1, \Gamma_2)$ are fixed and $\Gamma_1 \cup \Gamma_2$ is the set of all natural numbers. Also fix some collection $\{t_k|k \in \Gamma_2, 0 < t_k < 1\}$.

Let $\Gamma_0 \subset \Gamma_2$ and for each $k \in \Gamma_0$ let Z_k be a separator of (A^t_k, B^t_k) in Q. Let $\mathcal{F}$ be the collection of all sets which are finite intersections of Z_k's with the set Y; by $\overline{\mathcal{F}}$ we will mean the set of closures of elements of $\mathcal{F}$. One may consider three spaces: $X = \cap\mathcal{F}$; $\overline{X}$; $X_1 = \cap\overline{\mathcal{F}}$. Varying Y, Z_k, these spaces $X, \overline{X}, X_1$ determine the three classes $\mathcal{S}, \overline{\mathcal{S}}, \mathcal{S}_1$ respectively mentioned in the Introduction. We need to require $\Gamma_2 - \Gamma_0$ to be infinite.

3.2. THEOREM. *If Y satisfies $DIP(\Gamma_1, \Gamma_2)$ and the compactum X_1 is chosen as above, then c-dim $X_1 \geq card\,(\Gamma_2 - \Gamma_0)$.*

COROLLARY. *If $\Gamma_2 - \Gamma_0$ is infinite, then c-dim $X_1 = \infty$.*

PROOF. The Corollary is immediate. Let $\Gamma_3 \subset \Gamma_2 - \Gamma_0$ be a set whose cardinality is $n + 1$. We shall find a closed subset A of X_1 and a map f of A to S^n that cannot be extended to a map of X_1 to K_n. Identify Q_{Γ_3} with I^{n+1} and S^n with ∂I^{n+1}. Choose C to be a closed collar neighborhood of S^n in I^{n+1}. Define A to be $X_1 \cap \theta^{-1}(C)$ where $\theta : Q \to I^{n+1}$ is coordinate projection. The map f on A is given as θ followed by a map ρ of C to S^n. The collar C and map ρ are chosen so that ρ restricted to S^n is as close to the identity as desired and so that for $k \in \Gamma_3$, $\rho^{-1}(A_k) = A_k^t \cap I^{n+1}$, $\rho^{-1}(B_k) = B_k^t \cap I^{n+1}$ for some choice of $0 < t < 1$. Suppose f extends to a map $F : X_1 \to K_n$. Using the fact that K_n is an ANE, assume F is defined on a neighborhood V of X_1 in Q.

Since the elements of $\mathcal{F}$ are compact, some finite intersection of elements of $\mathcal{F}$ is contained in V. This implies that for some finite subset Γ of Γ_0, $\cap\{Z_k | k \in \Gamma\} \cap Y \subset V$. Let $\Lambda = \Gamma_1 \cup \Gamma \cup \Gamma_3$, let $U = V_\Lambda$, and let $S_k = Z_k \cap Y_\Lambda$, $k \in \Gamma$. By 3.1.1, dim $Y_\Lambda \leq card\,(\Gamma \cup \Gamma_3)$. Since S_k is a separator of $(A_k^t \cap Y_\Lambda, B_k^t \cap Y_\Lambda)$ in Y_Λ, and $\cap\{S_k | k \in \Gamma\} \subset U$, use 2.9 to select separators S_k^* in Y_Λ so that $S^* \subset U$ and dim $S^* \leq card\,\Gamma_3 = n + 1$. Employing 3.1.2 and 2.5, we see that $\{(A_k^t \cap S^*, B_k^t \cap S^*) | k \in \Gamma_3\}$ is an essential family for S^*, so dim $S^* = n + 1$.

The choice of the collar C and map ρ show that $A_k^t \cap S^* \subset F^{-1}(A_k) \cap S^* = A_k'$ and $B_k^t \cap S^* \subset F^{-1}(B_k) \cap S^* = B_k'$. Thus $\{(A_k', B_k') | k \in \Gamma_3\}$ is an essential family for S^*. Restricting F to S^* and applying Lemma 2.7, we arrive at a contradiction. This concludes the proof that c-dim $X_1 \geq card\,(\Gamma_2 - \Gamma_0)$.

4. Applications. Theorem 3.3 can be used to detect infinite cohomological dimension for many examples. Before presenting several, we give a slightly different statement of 3.2 of [Ru3] (which was derived from 3.1 of [Ru2]).

4.1. PROPOSITION. *Let Γ_0 be an infinite, proper subset of the set of natural numbers, and suppose $\{t_k | k \in \Gamma_0\}$ is a set of real numbers such that $0 \leq t_k < 1$. Then there exists a set $\{Z_k | k \in \Gamma_0\}$ of closed subsets of Q satisfying,*

4.1.1. Z_k *continuum-wise separates* (A_k^t, B_k^t),
4.1.2. $Z_k \cap A_k = \varnothing$ *and* $Z_k \cap B_k = \varnothing$, *and*
4.1.3. $Z = \cap\{Z_k | k \in \Gamma_0\}$ *is hereditarily strongly infinite dimensional.*

4.2. Note. 4.1.1 means that each continuum meeting both A_k^t and B_k^t intersects Z_k. Combining 4.1.1 and 4.1.2, it is easy to see that Z_k is a separator of (A_k, B_k).

4.3. Examples. In each of the following, we shall name a space Y that satisfies $DIP(\Gamma_1, \Gamma_2)$. After $\Gamma_0 \subset \Gamma_2$ is named, we may employ $\{Z_k | k \in \Gamma_0\}$ as in 4.1 to obtain the spaces $X, \overline{X}, X_1$ as prescribed in Section 3. Thus the examples can be hereditarily strongly infinite dimensional if desired, provided Γ_0 is an infinite, proper subset of Γ_2.

4.3.1. Let $Y = Q$, the Hilbert cube, $\Gamma_1 = \varnothing$, and $\Gamma_2 =$ all natural numbers. Easily Y satisfies $DIP(\Gamma_1, \Gamma_2)$. Choose Γ_0 to be any subset of Γ_2. The result 3.1 of [W1] is subsumed by this example combined with the proof of 3.3.

4.3.2. Let $\Gamma_1 = \{1\}$, $\Gamma_2 = \{2, 3, \ldots\}$. Let T be a 0-dimensional subspace of I_1 whose cardinality is that of the real line. Let $\mathcal{C}$ denote the set of continua in Q which intersect both A_1 and B_1. With the Hausdorff metric, it is known that $\mathcal{C}$ is a compactum, hence card $(\mathcal{C}) \leq$ card (T). Let $\alpha : T \to \mathcal{C}$ be a surjection. For each $t \in T$, choose $y_t \in \alpha(t) \cap \Pi_1^{-1}(t)$, and let $Y = \{y_t | t \in T\}$. One easily sees that Y is a totally disconnected space; 3.1.1 is satisfied because T is 0-dimensional. Further Y is strongly infinite dimensional and $\{(A_k^t \cap Y, B_k^t \cap Y) | k \in \Gamma_2\}$ is an essential family for Y whenever $0 < t_k < 1$ for each $k \in \Gamma_2$. To see why this is so, let S_k, $k \in \Gamma_2$, be a separator in Y of $(A_k^t \cap Y, B_k^t \cap Y)$. By Lemma 2.8, there exist sets $\tilde{S}_k$ closed in Q, separating (A_k, B_k) in Q, and such that $\tilde{S}_k \cap Y \subset S_k$. By 2.5, $\cap \{\tilde{S}_k | k \in \Gamma_2\}$ contains a continuum meeting A_1 and B_1, so $\varnothing \neq Y \cap (\cap \{\tilde{S}_k | k \in \Gamma_2\}) = \cap \{(Y \cap \tilde{S}_k | k \in \Gamma_2\} \subset \cap \{S_k | k \in \Gamma_1\}$. Thus by definition, $\{(A_k^t \cap Y, B_k^t \cap Y) | k \in \Gamma_2\}$ is an essential family for Y. A slight variation of this proof shows that 3.1.2 is satisfied.

4.3.3. Modify 4.3.2 by setting $\Gamma_1 = \{1, 2\}$, $\Gamma_2 = \{3, 4, \ldots\}$ and choosing $T \subset I_1 \times I_2$. Choose the set $\mathcal{C}$ to be the collection of all continua C for which $\{(A_k \cap C, B_k \cap C) | k \in \Gamma_1\}$ is an essential family. A modified version of 2.5 can be proved and used in this case to prove that $\{(A_k^t \cap Y, B_k^t \cap Y) | k \in \Gamma_2\}$ is an essential family for Y. We leave it to the reader to supply additional varieties of this example.

4.3.4. Modify 4.3.2 by setting $\mathcal{C}$ to be the collection of all G_δ-sets C in Q for which $\{(A_1^t \cap C, B_1^t \cap C)\}$ is an essential family. Here, choose any $0 < t < 1$, but then require that the set T be contained in the interval $[-1 + t, 1 - t]$. The cardinality of this set C is still that of the continuum. One may similarly modify 4.3.3 using G_δ-sets.

4.4. Comment. All these examples have a characteristic which has not been exploited. In 4.3.2, if $J_1 = I_1$ and for $k > 1$, $J_k = [a_k, b_k] \subset I_k$, then $Y \cap \Pi\{J_k | k = 1, 2, \ldots\}$ enjoys the same properties in $\Pi\{J_k | k = 1, 2, \ldots\}$ as Y did in Q. If you project Y into $Q_{\Gamma_1 \cup \Gamma}$ where $\Gamma \subset \Gamma_2$, then the resulting space has dimension $\leq$ card Γ.

REFERENCES

[Ru1] Leonard R. Rubin, *Hereditarily strongly infinite dimensional spaces*, Mich. Math. J. **27** (1980), 65–73.

[Ru2] Leonard R. Rubin, *Noncompact hereditarily strongly infinite dimensional spaces*, Proc. Amer. Math. Soc. **79** (1980), 153–154.

[Ru3] Leonard R. Rubin, *Totally disconnected spaces and infinite cohomological dimension*, Topology Proceedings **7**, (1982), 157–166.

[R-S-W]   Leonard R. Rubin, R. M. Schori, and John J. Walsh, *New dimension theory techniques for constructing infinite-dimensional examples*, General Topology and its Appls. **10** (1979), 83–102.

[W1]   John J. Walsh, *A class of spaces with infinite cohomological dimension*, Mich. Math. J. **27** (1980), 215–222.

[W2]   John J. Walsh, *Dimension, cohomological dimension, and cell-like mappings*, Shape Theory and Geometric Topology, Proceedings, Dubrovnik 1981, Springer-Verlag Lecture Notes in Mathematics, Berlin, Heidelberg, New York, 1981, 105–118.

UNIVERSITY OF OKLAHOMA

Contemporary Mathematics
Volume **44**, 1985

ENDOMORPHISMS IN THE HOMOTOPY CATEGORY

A. Zabrodsky

Introduction and Summary

The idea that self maps of a CW complex store a vast amount of information about the structure of the complex is spread throughout the literature. The fact that many self maps exist "in nature" is of course very useful.

In these notes we concentrate in three aspects regarding self maps (or endomorphisms) of CW complex (almost invariably simply connected and of finite type):

In Chapters 1 and 2 we bring some simple observations regarding the information one can obtain from $im[X,X] = End\ X \to End\ H_*(X,G)$ about the structure of $[X,X]$. In particular, we demonstrate the usefulness and stability of the characteristic roots of $H^*(f,Z_p)(Z_p = Z/pZ), f : X \to X$.

In Chapter 2, some simple cases of $ker[Aut\ X \to Aut(H_*(X,Z))]$ are studied. The strongest result we know along these lines, is given in [1].

Chapter 3 and 4 are devoted to a lifting theorem in the homotopy category of endomorphisms. In its slightly simplified version it states,

3.1. THEOREM. *Given a commutative diagram*

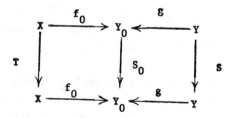

Suppose

$$\prod_{n=0}^{\infty} \exp H^n(X, \pi_n(fiberg)) = m_0 < \infty$$

and one of the following two assumptions hold:

(3.1.b)′ g is a principal fibration.

(3.1.b)″ Y, Y_0 are H-spaces, $g-$ an $H-$ map . (No assumptions about S_0, S).

Then, if f_0 lifts to $f' : X \to Y$ then there exists a lifting $f : X \to Y$ of f_0, and an integer t so that $f \circ T^{m_0^t} \sim S^{m_0^t} \circ f$.

In Chapter 4, some applications of the lifting theorem are given:

4.1. PROPOSITION. *Given a commutative diagram*

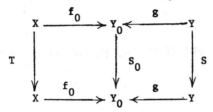

Suppose for every n $H^n(X, \pi_n(fiberg))$ is finite and for every prime p the characteristic polynomial of $H^{n+1}(T, Z_p)$ is relatively prime to those of $\pi_n(\hat{S}) \otimes Z_p$ and $[\pi_{n+1}(\hat{S})/ \text{ torsion }] \otimes Z_p$ (where $\hat{S}$:fiber $g \to$ fiber g is the map induced by S_o, S). Then f_0 lifts to $f : X \to Y$.

One application of this theorem is given by:

4.4. Let ξ be an n dimensional complex vector bundle on CP^n. The obstructions for ξ to have k sections (beyond the rational Chern classes $c_{n-r}(\xi), r < k$) are mod p obstructions only for p satisfying

$$(p-1)[p(p-1)-1] = (p-1)^3 + (p-1)(p-2) \leq k .$$

(Thus for < 10 sections, there are only mod 2 obstructions, for < 76 sections there are only mod 2 and mod 3 obstructions). Some consequences are derived for vector bundles on lens spaces as well.

There are some applications for H-spaces (4.5).

We adopt the following conventions and notations. Spaces are normally assumed to be pointed simply connected CW complexes of finite type. By commutative diagrams of maps we mean commutative up to homotopy. Homotopies $f_0 \sim f_1 : X \to Y$ are assumed to be maps F from X to the space PY of free paths in Y so that $\mathcal{E}_i F = f_i$ $i = 0, 1$ ($\mathcal{E}_t : PY \to Y$ the evaluation at t map). Finally Z_p is always denotes the integers mod $p : Z_p = Z/pZ$. We denote by $\mathbb{P}$-the set of primes.

1. The characteristic polynomial of a cohomology endomorphism.

Let X be a CW complex, $f : X \to X$ an endomorphism. In this section, we list some simple properties of the characteristic polynomials of $H^n(f, R)$, R - a field. Some slightly more involved properties will be dealt with in the next section.

1.1. Given a field R and a map $f : X \to X$. We denote by $_f\dot{P}_n$ the characteristic polynomial of $H^n(f, R)$. Put

$$_fP_n(x) = \prod_{m \leq n} {}_f\dot{P}_m(x).$$

Let $\otimes H^*(f,R) : \otimes H^*(X,R) \to \otimes H^*(X,R)$ be the morphism induced by f on the free associative graded algebra generated by $H^*(X,R)$. Denote by ${}_f\hat{P}_n(x)$ the characteristic polynomial of $\otimes H^*(f,R)$ in dim n and put ${}_f\dot{P}_n(x) = \prod_{m \leq n} {}_f\hat{P}_m(x)$.

Let $CR_n(f,R)$ be the set of characteristic roots of ${}_f P_n(x)$, $CR(f,R) = \bigcup_n CR_n(f,R)$. Similarly, $\hat{CR}_n(f,R)$ is the set of characteristic roots of ${}_f\hat{P}_m(x)$, $\hat{CR}(f,R) = \bigcup_n \hat{CR}_n(f,R)$. Finally, let $\overline{CR}_n(f,R), \overline{CR}(f,R)$ be the multiplicative closures of $CR_n(f,R)$ and $CR(f,R)$ respectively. One can easily see that $\hat{CR}_n(f,R) \subset \overline{CR}_n(f,R), \hat{CR}(f,R) = \overline{CR}(f,R)$.

Although the characteristic values lie in general in some extension field of R and $CR(f,R)$ may be contained in a non finite extension, most of the properties we use could be traced back to some properties of the characteristic polynomial, thus involving only a finite extension of R.

1.2. PROPOSITION. *Consider a commutative diagram*

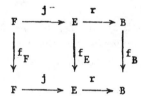

where $F \xrightarrow{j} E \xrightarrow{r} B$ *is a fibration. Then*

(a) ${}_{f_E}\dot{P}_n(x)|_{{}_{f_F \times f_B}\dot{P}_n(x)}$

(b) *If* B *is simply connected then* ${}_{\Omega f_B}\dot{P}_n(x)\mid {}_{f_B}\hat{P}_{2n}(x)$

(c) *If* B *is simply connected then* ${}_{f_F} P_n(x) \mid_{\Omega f_B \times f_E} P_n(x)$.

PROOF. The proof uses simple spectral sequence arguments: If A, ∂ is a differential vector space, $\varphi : A \to A$ a differential linear transformation, then the characteristic polynomial of $H_n(\varphi)$ divides that of φ and if $A = \cup F^n A$ is a filtered vector space, $E_0 A$ its associated graded vector space and $\varphi : A \to A$ is a linear transformation of filtered vector spaces then $E_0 \varphi$ and φ have the same characteristic polynomial.

(a) $f = (f_F, f_E, f_B)$ induces a linear transformation of spectral sequences of the Serre spectral sequences:

$$\mathcal{E}_r(f) : \mathcal{E}_r(F,E,B) \to \mathcal{E}_r(F,E,B)$$

$$\mathcal{E}_2^n(f) = H^n(f_F \times f_B, R) : H^n(F \times B, R) \to H^n(F \times B, R).$$

Hence ${}_{f_F \times f_B}\dot{P}_n(x)$ is divisible by the characteristic polynomial of $\mathcal{E}_\infty^n(f) = \mathcal{E}_0 H^n(f_E, R)$ etc.

(b) is proved similarly—using the Eilenberg-Moore spectral sequence:

$$\mathrm{Tor}^{H^*(B,R)}(R,R) \Rightarrow H^*(\Omega B, R) \text{ and } \mathcal{E}_1(f_B) : \overline{\mathrm{Bar}}(H^*(B)) \to \overline{\mathrm{Bar}} H^*(B)$$

$$\mathcal{E}_1^{-k,n} \sum_{\Sigma n_i = n} \overline{H}^{n_1}(B,R) \otimes \ldots \overline{H}^{n_k}(B,R) = \mathcal{E}_1^n(B) \to \mathcal{E}_1^n(B)$$

The characteristic polynomial of $\mathcal{E}_1^n(f_B)$ divides $f_B \hat{P}_{2n}(x)$ and proceed with the spectral sequence argument.

(c) is a direct consequence of (a).

1.3. COROLLARY. *If a multiplicatively closed subset A of an extension field of R contains any two of the sets*

$$CR(f_F, R), \ CR(f_E, R), \ CR(f_B, R)$$

it contains the third.

1.3. has an obvious finite version relating

$$CR_{n_1}(f_F, R), \ CR_{n_2}(f_E, R), \ CR_{n_3}(f_B, R).$$

1.4. PROPOSITION. *Let X be simply connected. Given $f : X \to X$ and let $Ht_m(f) : Ht_m(X) \to Ht_m(X)$ be its Postnikov homotopy approximation in dim $\leq m$.*

Then:

(a) $\overline{CR}(Ht_m(f), R) = \overline{CR}_m(Ht_m(f), R) = \overline{CR}_m(f, R)$

(b) The characteristic roots of $\pi_*(f, R)$ (or equivalently of $\pi_*(f) \otimes R$) and those of $H^*(f, R)$ have the same multiplicative closures.

(c) $\pi_*(f) \otimes R$ and $\pi_*(\Sigma f) \otimes R$ have the same multiplicative closure of characteristic roots.

PROOF. As $H^i(Ht_m(X), R) = H^i(X, R)$ for $i \leq m$ to prove (a) suffices to prove

$$\overline{CR}_m(Ht_m(f), R) = \overline{CR}(Ht_m(f), R).$$

This is quite obvious for $X = K(\pi, m)$. Consider inductively

$$
\begin{array}{ccccc}
K(\pi_m(X), m) & \to & Ht_m(X) & \to & Ht_{m-1}(X) \\
\downarrow & & \downarrow & & \downarrow \\
K(\pi_m(f), m) & & Ht_m(f) & & Ht_{m-1}(f)
\end{array}
$$

$$K(\pi_m(X), m) \ \to \ Ht_m(X) \ \to \ Ht_{m-1}(X)$$

by 1.2.(a) $\overline{CR}(Ht_m(f), R)$ is contained in the multiplicative closure of

$$CR_m(K(\pi_m(f), m), R) \bigcup CR_{m-1}(Ht_{m-1}(f), R)$$

By the Serre exact sequence the first set is contained in

$$CR_m(Ht_m(f), R) \bigcup CR_{m-1}(Ht_{m-1}(f), R)$$

and (a) follows from the equality

$$CR_{m-1}(Ht_{m-1}(f), R) = CR_{m-1}(Ht_m(F), R).$$

(b) follows from the same inductive observation.

(c) follows from (b) and from the simple observation:

$$\overline{CR}(f, R) = \overline{CR}(\Omega \Sigma f, R).$$

1.5. Examples: (1.5.1.) In [4-2], the notion of a power space was introduced. A Z_p power space is a pair X, f with $CR(f, Z_p) \subset Z_p$. 1.2, 1.3 and 1.4 imply that the property of being a power space is preserved by fibrations, looping, suspensions, etc.

1.5.2. Let $f : X \to X$ satisfy $H^*(f, R) = 1$. Then $\pi_*(f) \otimes R$ is not necessarily the identity morphism but 1 is its only characteristic root.

2. Endomorphisms and their cohomology.

In this section we shall study properties of $End(X)$ which could be detected by its images in $End\, H_*(X, Z)$ and $End\, \pi_*(X)$. Some further properties of the characteristic polynomials of $H^*(f, R), f : X \to X$, are derived.

We give here a simple proof that for a simply connected H_0 space X (with either $H_*(X, Z)$ or $\pi_*(X)$ finite dimensional)

$$\ker H_* : \operatorname{Aut} X \to \operatorname{Aut} H_* X$$

is a finite nilpotent group. (Compare with [3-1] corollary 9.10).

Moreover this proof yields an estimate of the degree of nilpotency.

It is worthwhile to mention here that the most general theorem along these lines is the following

2.0. THEOREM ([1]). *Let X be a nilpotent finite complex. If a subgroup $G \subset$ Aut X acts nilpotently on $H_*(X, Z)$ then G is nilpotent.*

2.1. Let $\Phi \subset End\,(X)$ be a subset. We shall use the obvious notations $H_*(\Phi, Z) = \{H_*(f, Z)|f \in \Phi\}$ etc. In an arbitrary pointed category C a subset $\Phi \subset End_C(A) = Hom_C(A, A)$ is said to be k-nilpotent if $\psi^k = \{*\}(\psi^k = \{\alpha_1\, o\, \alpha_2\, o...\alpha_k|\alpha_j \in \psi\})$. We say that ψ is nilpotent if it is k nilpotent for some k.

2.2. LEMMA. *Given an arbitrary space X. Let $\Phi \subset End(X)$ have the property that for every $n \leq n_0\, H_n(\Phi, Z)$ is k_n-nilpotent. Then for every simply connected space Y with $\pi_t(Y) = 0$ for $t \neq m_1, m_2, \ldots m_{s_0}, m_i \leq n_0, \Phi^* \subset End_{Set}([X, Y])$ is $k = 2^{s_0} \prod_{s=1}^{s_o} k_{m_s}$ nilpotent. (i.e.: if $f_1, f_2, \ldots f_k \in \Phi$, $g : X \to Y$ then $g \circ f_1 \circ f_2 \circ f_3 \ldots \circ f_k \sim *$). Moreover:*

(a) If $\pi_1(\Phi)$ is nilpotent the restriction on Y being simply connected may be removed (and the evaluation of the nilpotency of Φ^* will have to be changed in an obvious way).

(b) With an obvious change in the estimates of the nilpotency of Φ^* the simple connectivity of Y can be replaced by the solvability of $\pi_1(Y)$.

(c) The finiteness property on $\pi_n(Y)$ could be replaced by the finite dimensionality of X.

PROOF. First note that by the universal coefficient theorem—for any abelian group $G, H_n(\Phi, Z)\, k_n$ — nilpotent implies $H^n(\Phi, G)$ is $2k_n$ nilpotent. The proof is by induction on s_o: Let $\hat{Y} \xrightarrow{j} Y \xrightarrow{r} K(\pi_{m_1}(Y), m_1) = K_{m_1}$ be a connective fibering. For every $f_1, f_2, \ldots f_{2km_1} \in \Phi, r : X \to Y$ as $H^*(\Phi^{2k_{m_1}}, \pi_{m_1}(Y))[rog] =$

0 the composition $X \xrightarrow{f_1} X \xrightarrow{f_2} X \to \ldots \xrightarrow{f_{2k_{m_1}}} X \xrightarrow{g} Y \xrightarrow{r} K_{m_1}$ is null homotopic: hence, $\prod_{i \le 2k_{m_1}} f_i$ lifts to $\hat{g} : X \to \hat{Y}$ and by induction $\hat{g} \circ f_{i_1} \circ f_{i_2} \circ \ldots \circ f_{i_{\hat{k}}} \sim *$ for $\hat{k} = 2^{s_0-1} \prod_{i=2}^{s_o} k_{m_i}$. If $\pi_1(\Phi)$ is k_0 nilpotent then every k_0-product of elements of Φ lifts to the universal covering space $\tilde{Y}$ of Y and one can proceed as above with $\tilde{Y}$ replacing Y and $k = 2^{s_o-1} \prod_{s=2}^{s_o} k_{m_s}$. If $\pi_1(Y)$ is ℓ-solvable the proof is as above with the first ℓ-stages being liftings $X \to Y_i, \pi_1(Y_i) = [\pi_1 Y_{i-1}, \pi_1 Y_{i-1}]$. (c) is obvious.

2.3. COROLLARY. *Let X be a finite complex with $\pi_1(X)$ solvable. Then $f \in End(X)$ is nilpotent ($f^k \sim *$) if and only if $H_*(f, Z)$ is. One can replace the solvability of $\pi_1(X)$ by the nilpotency of $\pi_1(f)$.*

2.4. LEMMA. *Let $\Phi \subset End(X), X - \ell$ connected, $\ell \ge 1$. If $H_n(\Phi, Z_p)$ is nilpotent for $n \le n_0$ then $\pi_n(\Phi, Z_p)$ is nilpotent for $n \le n_0$ (with a different degree of nilpotency).*

PROOF. By induction on $n_0 - \ell$. For $n_0 = \ell + 1$ one has essentially $X = K(\Pi, n_0)$ (as all information on $H_{\le n_0}$ and $\pi_{\le n_0}(\ ,Z_p)$ is contained in Ht_{n_0}). In this case $H_{n_0}(X, Z_p) = \Pi \otimes Z_p = \pi_{n_0-1}(X, Z_p)$. $Tor(H_{n_0}(X), Z_p) \approx Tor(\Pi, Z_p) \approx \pi_{n_0}(X, Z_p)$ and 2.4 obviously holds.

Suppose 2.4 holds for $n_0 - \ell = m - 1$. Consider $\hat{\ell} = n_0 - m$ connected space X and maps induced on its $\hat{\ell} + 1$ connective fibering $j : \hat{X} \to X$:

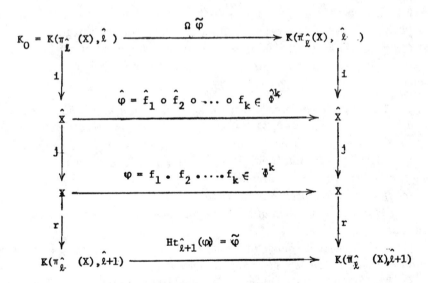

Suppose $\oplus_{n \le n_0} H_n(\Phi, Z_p)$ is k-nilpotent. In particular $H_{\hat{\ell}+1}(\varphi, Z_p) = 0$ and consequently $H_{\hat{\ell}+1}(\check{\varphi}, Z_p) = H_*(\check{\varphi}, Z_p) = 0$, as well as $\phi_*(\check{\varphi}, Z_p) = 0$. Using the Serre spectral sequence one obtains:

$$E^2(\Omega\tilde{\varphi}, \hat{\varphi}, \varphi) : E^2(i,j) = H_*(X, Z_p) \otimes H_*(K_0, Z_p) \to E^2(i,j).$$

$E^2(\Omega\tilde\varphi,\hat\varphi,\varphi) = H_*(\varphi,Z_p)\otimes H_*(\Omega\tilde\varphi,Z_p) = 0$ in total $dim \le n_0$. Hence, $H_*(\hat\Phi^k,Z_p)$ ($\hat\Phi$ the connective fiberings of Φ) lowers filtrations in $dim \le n_0$ and $H_*(\hat\Phi^{n_0 k},Z_p) = 0$. One can apply the induction hypothesis to $\hat\Phi$ to conclude that $\phi_n(\hat\Phi,Z_p)$ is nilpotent for $n \le n_0$. This, together with the nilpotency of $\pi_*(Ht_{\hat\ell+1}(\Phi),Z_p)$ implies the nilpotency of $\pi_*(\Phi,Z_p), n \le n_0$.

2.5. LEMMA. *Let X,Y be simply connected spaces. Given a $\mathbb{P}-p$ equivalence $g : X \to Y$ with cofiber of dimension $n_o < \infty$. If $\Phi \subset End(X)$ induces nilpotent families $H_n(\Phi,Z_p), n \le n_0$, then for some $k, \Phi^k \subset im\ g^*$.*

I.e.: for every k-sequence $f_1, f_2, \ldots, f_k \in \Phi$ there exists $h : Y \to X$ so that

$$f_1\ o\ f_2\ o \ldots o \ldots o\ f_k \sim h\ o\ g\ .$$

PROOF. g can be decomposed into a sequence of principal cofibrations:

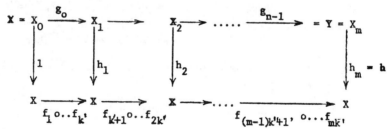

$M_i = M(Z_p,n_i)$—a Moore space, $M_i \xrightarrow{r_i} X_i \xrightarrow{g_i} X_{i+1}$ a cofibration, $n_i \le n_0$. By 2.4 $\pi_{\le n_0}(\Phi,Z_p)$ is k' nilpotent for some k' and for every sequence $f_1, f_2, \ldots, f_{mk'} \in \Phi$. One can complete a diagram

Put $k = mk'$.

2.6. COROLLARY. *Let M be a connected finite complex, $\Phi \subset End(\Sigma M)$. $H_*(\Phi,Z_p)$ is nilpotent if and only if for every $r > 0$ there exists $k = k(r)$ so that $(\Phi)^k \subset p^r[End(\Sigma M)]$, i.e: if $f_i \in \Phi$ $(f_1\ o \ldots o\ f_k) \sim p^r \cdot h$ for some $h = h(f_1, f_2, \ldots f_k) \in End(\Sigma M)$.*

PROOF. Obviously $(\Phi)^k \subset p[End(\Sigma M)]$ implies $[H_*(\Phi,Z_p)]^k = 0$ and $H_*(\Phi,Z_p)$ is nilpotent. Conversely, if $H_*(\Phi,Z_p)$ is nilpotent, apply 2.5 for $X = \Sigma M = Y, g = p^r \cdot 1$ to obtain $(\Phi)^k \subset (p^r \cdot 1)^* End(\Sigma M) = p^r[End(\Sigma M)]$.

2.7. PROPOSITION. *Let A be any functor from the homotopy category of finite connected complexes into the category of finite dimensional Z_p vector spaces. Suppose A is $\vee$-additive, i.e.*

$$A(X \vee Y) = A(X) \oplus A(Y), A(f \vee g) = A(f) \oplus A(g).$$

If $f \in End(X)$ then the characteristic roots of $A(\Sigma f)$ are contained in $CR(f,Z_p)$.

PROOF. Let $_fP(x)$ be the characteristic polynomial of $H_*(f, Z_p)$. Then $H_*(_fP(\Sigma f), Z_p) =_f P[H_*(f, Z_p)] = 0$ and $\{_fP(\Sigma f)\} = \Phi \subset End(\Sigma X)$ satisfies $H_*(\Phi, Z_p) = 0$. By 2.6 there exists k so that $[_fP(\Sigma f)]^k = ph, h \in End(\Sigma X)$. $0 = pA(h) = A(ph) = A[_fP(\Sigma f)] = [_fP(A(\Sigma f)]^k$. Hence, the minimal polynomial of $A(\Sigma f)$ divides $[_fP(x)]^k$ and 2.7 follows.

2.8. COROLLARY. *Let X be a connected CW complex. If for some functor A (as in 2.7) $A(\Sigma X) = Z_p$ then for any map $f : X \to X$ $CR(f, Z_p) \cap Z_p \neq \varnothing$.*

2.7 has a characteristic 0 version:

Let $f : \Sigma X \to Y$ satisfy $H_*(f, Q) = 0$. If $\pi_n(Y) = 0$ for $n > n_0$ then for some $\lambda \neq 0$ $\lambda f \sim *$. This implies a Q-version for 2.7.

Given a homomorphism $\alpha : A \to B$ of abelian groups, α is said to be $0(mod\, p)(\alpha \equiv 0(mod(p))$, if $im\alpha \otimes Z_p = 0$. This is equivalent to the existence of a p-isomorphism $\gamma : A' \to A$ with $\alpha \circ \gamma = 0$ (e.g: Take $A' = ker\,\alpha$).

If $\alpha_1, \alpha_2 : A \to B$ we say that $\alpha_1 \equiv \alpha_2(mod\, p)$ if $\alpha_1 - \alpha_2 \equiv 0\,mod\,p$. If G is a finite abelian group, we say that $\alpha_1 \equiv \alpha_2(mod\,G)$ if $\alpha_1 \equiv \alpha_2(mod\,p)$ for every $p|$ order G.

2.9. PROPOSITION. *Let $r : X \to X_0$ be a rational equivalence, $\pi_n(fiber\,g) = 0$ for $n > n_0$. Let M be a CW complex and $f, g : M \to M$ commuting endomorphisms with $H_*(f, Z) \equiv H_*(g, Z)\,mod\,\pi_*(\,fiber\,\psi)$.*

If $k : M \to X$ satisfies $r \circ k \circ f \sim r \circ k \circ g$ then for some $m, k \circ f^m \sim k \circ g^m$. Consequently, for an H_0 space X with $\pi_n(X) = 0$ for $n > n_0$, $H_*(f, Z) = H_*(g, Z)$, $f, g : M \to M$ commute imply $(f^*)^m = (g^*)^m \in End\,[M, X]$, i.e., for every $k : M \to X$ $k\,g^m \sim k\,f^m, m$—a product of primes dividing order $\pi_*(\,fiber\,g)$.

PROOF. The second part follows from the first by taking $r : X \xrightarrow{\sim 0} K(\pi_*(X)/\text{tor-sion})$. To prove the first part put

$$m = \{ \prod_{n \leq n_0} \exp[\pi_n(\,fiber\,r)]\}^2$$

m is the desired integer:

Decompose r into s principal filtrations:

$$X = X_s \xrightarrow{r_{s,s-1}} X_{s-1} \xrightarrow{r_{s-1,s-2}} X_{s-2} \to \ldots \to X_n K(G_n, n) \xrightarrow{r_{n,n-1}} X_{n-1} \to \ldots X_0$$

$G_n = \pi_n(\,fiber\,r)$. Put $k_n = r_{r+1,n} \circ r_{n,n-1}, \circ \ldots \circ r_{s,s-1} \circ k, m_n = \{\prod_{t \leq n} [exp\,G_t]^2\} \cdot [exp(G_{n+1})]$. Suppose inductively that $k_n \circ f^{m_n} \sim k_n \circ g^{m_n}$. One obtains:

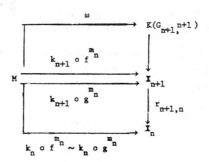

$k_{n+1}f^{mn} \sim \omega * (k_{n+l}o \ g^{mn})$ where $(*)$ denotes the actions of $[M, K(G_{n+1}, n+1)]$ on $[M, X_{n+1}]$.

By induction we shall prove:

(2.9.1) $k_{n+1}(f^{mn})^t \sim [t \cdot (\omega \ o \ f^{mn(t-1)})] * k_{n+1} \ o(g^{mn})^t.$

Indeed:

$$k_{n+1}(f^{mn})^t = k_{n+1}(f^{mn})^{t-1} \ o \ f^{mn} \sim \{[(t-1)(\omega \ o \ f^{mn(t-2)})] * k_{n+1}(g^{mn})^{t-1}\} o f^{mn}$$

$$\sim [(t-1)\omega \ o \ f^{mn(t-1)}] * [(k_{n+1} \ o \ g^{mn(t-1)})o \ f^{mn}]$$

$$\sim [(t-1)\omega \ o \ f^{mn(t-1)}] * [k_{n+1} \circ f^{mn} \circ g^{mn(t-1)}]$$

$$\sim [(t-1)\omega \ o \ f^{mn(t-1)}] * [(\omega * k_{n+1}g^{mn})o \ g^{mn(t-1)}]$$

$$\sim [(t-1)\omega \ o \ f^{mn(t-1)}] * [\omega \ o \ g^{mn(t-1)} * k_{n+1}g^{mn}t]$$

$$\sim [(t-1)\omega \ o \ f^{mn(t-1)} + \omega \ o \ g^{mn(t-1)}] * k_{n+1}g^{mn}t$$

2.9.1. will follow if we prove that $\omega \ o \ g^{mn(t-1)} \sim \omega \ o \ f^{mn(t-1)}$. This follows from the following observation:

(2.9.2) Given a finite group G, $exp \ G = q$ and let $\varphi_1, \varphi_2 : M \to M$. Suppose $H^*(\varphi_1, G)$ and $H^*(\varphi_2, G)$ commute. Then $H_*(\varphi_1, Z) \equiv H_*(\varphi_2, Z)$ (mod G) implies $H^*(\varphi_1^q, G) = H^*(\varphi_2^q, G)$.

PROOF OF (2.9.2). For abelian groups A, B and a homomorphism $\alpha : A \to B$ $\alpha \equiv O(G)$ implies $\text{Ext}(\alpha, G) = 0 = \text{Hom}(\alpha, G)$.

Put $\overline{\varphi}_i = \text{Ext}(H_{k-1}(\varphi_i, Z), G), \overline{\varphi}_i = \text{Hom}(H_k(\varphi_i, Z), G)$ and consider

Hence $[H^k(\varphi_1, G) - H^k(\varphi_2, G)]^2 = 0$

$H^k(\varphi_1^q, G) = [H^k(\varphi_2, G) + H^k(\varphi_1, G) - H^k(\varphi_2, G)]^q = H^k(\varphi_2^q, G) + q \ H^k(\varphi_2^{q-1}, G)$
$[H^k(\varphi_1, G) - H^k(\varphi_2, G)] + \hat{\varphi}[H^k(\varphi_1, G) - H^k(\varphi_2 G)]^2 = H^k(\varphi_2^q, G).$

As $\exp G_{n+1}|m_n, H^{n+1}(g^{mn(t-1)}, G_{n+1}) = H^{n+1}(f^{mn(t-1)}, G_{n+1}), \omega \ o \ f^{mn(t-1)} \sim \omega \ o \ g^{mn(t-1)}$ and 2.9.1. follows.

2.9.1 for $t = \exp G_{n+1}$ implies $k_{n+1}(f^{mn})^{\exp G_{n+1}} \sim k_{n+1}(g^{mn})^{\exp G_{n+1}}$ and $k_{n+1}f^{m_{n+1}} \sim k_{n+1}g^{m_{n+1}}$. For $n = s$ one gets $k \ o \ f^m \sim k \ o \ g^m$.

2.10 COROLLARY. *Let X be a simply connected H_0 space, $\pi_n(X) = 0$ for $n > n_0$. If $f, g \in \text{End}(X)$ commute, $H^*(f, Z) = H^*(g, Z)$, then $f^m \sim g^m$. Consequently, $f \in \text{Aut}(X)$ is of finite order if and only if $H_*(f, Z)$ is. Moreover, if $r : X \to K(\pi_*(X) \ / \ torsion)$ is a $\mathbb{P} - \mathbb{P}_1$ equivalence $H^*(f, Z) = H^*(g, Z)(\text{mod } Z_{P_1})$ -order $f \ / \ order \ H_*(f, Z)$ is divisible only by primes in $\mathbb{P}_1$. One has the following analogue of 2.10.*

2.11. PROPOSITION. *Let X be a finite connected complex. If $f, g \in \mathrm{End}(X)$ satisfy:*

(a) $\Sigma f, \Sigma g$ commute.

(b) $H_*(\Sigma f, Z) = H_*(\Sigma g, Z)$.

Then for some m $\Sigma f^m \sim \Sigma g^m$.

PROOF. The proof is similar to that of 2.9 provided one can find a suitable rational model for ΣX:

Let $a_i^{(2n)} \in H^{2n}(X, Z)$ and $b_j^{(2-n-1)} \in H^{2n-1}(X, Z); 1 \leq i \leq r_{wn}, 1 \leq j \leq r_{2n-1}$, represent bases for $H^{2n}(X, Z)$ /torsion and $H^{2n-1}(X, Z)$ /torsion respectively. Let $f_i : \Sigma X \to K(Z, 2n+1)^{(N)}$ be a map into a high skeleton of $K(Z, 2n+1)$ representing (the image of) $a_i^{(2n)} \in H^{2n+1}(\Sigma X, Z)$. Follow f_i by a rational equivalence to obtain

$$\alpha_i^{(2n)} : \Sigma X \to S^{2n+1}$$

Let $\beta_i^{(2n-1)} : \Sigma X \to S^{2n}$ be obtained by suspending $X \xrightarrow{b_i^{(2n-1)}} K(Z, 2n-1)^{(N)} \overset{\sim}{\to}^O S^{2n-1}$. $\alpha_i^{(2n)}, \beta_i^{(2n-1)}$ induce a rational equivalence $\psi : \Sigma X \to \vee_n(\vee_{r_n} S^n) = \tilde{S}$. If $H^*(f, Z) = H^*(g, Z)$ then $\psi \Sigma f \sim \psi \Sigma g$. Now proceed as in 2.9 by decomposing

$$\psi : X \to \tilde{S}.$$

The fact that 2.10 and 2.11 do not hold in general (even not for suspensions and non suspension maps) could be illustrated as follows:

2.12. Example: Let $X = S^{2n} \vee S^{4n-1}, g = 1$, and let $f : X \to X$ be given by $\pi_{2n}(f) = 1$

$$\pi_{4n-1}(f)/\text{torsion} = \begin{pmatrix} 1 & 1 \\ 0 & 1 \end{pmatrix}$$

Then $\pi_{4n-1}(f^m)/\text{torsion} = \begin{pmatrix} 1 & m \\ 0 & 1 \end{pmatrix} \neq 1$ and $f^m \neq 1$ though $H^*(f, Z) = 1$. Moreover, $(\hat{f})^m = Ht_{4n-2}(f^m) \neq 1$, hence, by 2.10 $H_*(\hat{f}, Z) \neq 1$. This illustrates the well known fact that there is no functor Φ from abelian groups to abelian groups with $\Phi H_*(\_, Z) = H_*(\Omega\_, Z)$. (Such a functor exists if one limits oneself to the category of suspensions and suspension maps).

So far we have studied algebraic properties of $End(X)$ that can be detected by its images in $End(H_*(X, Z))$ and $End(\pi_*(X))$. We are about to study the complementary case.

Let $E_0(X) = ker(H_* : Aut(X) \to Aut H_*(X))$, $E_0(X)$ could be referred to as the algebraically non distinguishable maps.

Here we study only the case where X is an H_0 space. The more general result could be found in [1].

2.13. PROPOSITION. *Let X be a nilpotent H_0 space, X-finite dimensional (or alternatively $\pi_n(X) = 0$ $n > n_0$). Then $E_0(X) = \ker[H_* : Aut(X) \to Aut(H_*(X, Z))]$ is a finite nilpotent group, moreover:*

Let $t = \prod_{n \leq n_0} \exp(\ker \sigma_n \oplus im\ \sigma_n)$, $\sigma_n : \pi_n(X) \to PH_n(X, Z)$ then the order of $E_0(X)$ divides a power of t.

PROOF. Let $r : X \xrightarrow{\sim 0} X_0 = K(\pi_*(X)/ \text{ torsion }$ induce an isomorphism of $QH^*(\ ,Z)/$ torsion . Then $t = \prod_{n \leq n_0} \exp \pi_n(\text{ fiber } r)$. If $f \in E_0(X)$, then $r \circ f \sim r$. Apply 2.9 for $M = X, g = 1, k = 1$ to deduce that $f^{t^2} = 1$. If one shows that $E_0(X)$ is finite then order $E_0(X)$ divides a power of t as all its elements have order dividing t^2. The finiteness of $E_0(X)$ follows from the following observation: If $\psi : Y \to Y_0$ is a map, define $E(Y, \psi) = \{f \in \text{Aut}(Y) | \psi \circ f \sim \psi\}$. Let $Y = \to Y_m \to Y_{m-1} \to \ldots \to Y_k \to Y_{k-1} \to \ldots \to Y_0$ be a Postnikov approximation of ψ. Let ψ_k denote the composition $Y_k \to Y_0$.

One has homomorphisms $\alpha_k : E(Y_k, \psi_k) \to E(Y_{k-1}, \psi_{k-1}))$, Order $\ker \alpha_k \leq$ order $H^k(Y_k, \pi_k(\text{ fiber } \psi))$, $\ker \alpha_1 = E(Y_1, \psi_1)$, hence, order $E(Y, \psi) \leq \prod_k$ order $H^k(Y_k, \pi_k(\text{fiber } \psi))$ and if $\pi_*(\text{fiber } \psi))$ is finite $E(Y, \psi)$ is finite. For every $p | t$ let

$$X \xrightarrow{\psi'(p)} X(p) \xrightarrow{\psi''(p)} X_0$$

be a decomposition of ψ, $\psi'(p)$ and $\psi''(p)$ are $\mod p$ and $\mod(P - p)$ equivalences respectively. As X is the pull back of $\psi''(p)$, $p | t$ one can easily see that if

$$_pE(X(p), \psi''(p)) = \{f \in \text{Aut } X(p) | \psi''(p)f \sim \psi''(p), H_*(f) \equiv 1 \mod Z_p\}$$

then one has an isomorphism

$$E_0(X) \to \prod_{p|t} {}_pE(X(p), \psi''(p)).$$

By 2.9 order $_pE(X(p), \psi''(p))$ is a power of p and $E(X)$ is a direct product of finite p-groups.

2.14. Remark: For non H_0 spaces $E_0(X)$ or even

$$\hat{E}_0(X) = \ker[\text{Aut}(X) \to \text{Aut}(H_*(X) \times \text{Aut}(\pi_*(X))]$$

are not necessarily finite: Let $k_\lambda : S^{2n} \times S^{2n-1} \to S^{2n} \times S^{2n-1}$ be the composition $S^{2n} \times S^{2n-1} \to S^{2n} \times S^{2n-1} \vee S^{4n-1} \xrightarrow{1 \vee \lambda [c_{2n}, c_{2n}]} S^{2n} \times S^{2n-1} \vee S^{2n-1} \to S^{2n} \times S^{2n-1}$ then $k_\lambda \circ k_\nu = k_{\lambda+\nu}, k_\lambda \in \hat{E}_0(S^{2n} \times S^{2n-1})$ and it can be shown that if $\lambda \neq 0$ $k_\lambda \not\sim 1$. Hence k_1 is of infinite order. (For $n = 1$ for example $\hat{E}_0(S^1 \times S^2) = Z$ and $\text{Aut}(S^1 \times S^2) \approx (Z_2 * Z_2) \oplus Z_2$).

3. Lifting obstructions and the lifting theorem.

Consider the lifting problem in the category $\text{End}(H)$, H-the homotopy category:

Given a commutative diagram:

(3.1.1)

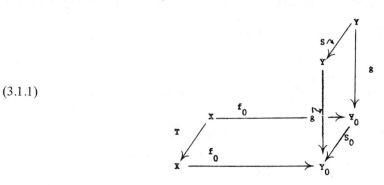

Can one find a mapping $f : X \to Y$ with $S \circ f \sim f \circ T$? Or more precisely:

(3.1.2.). Given (3.1.1.) and homotopies $F_0 : S_0 \circ f_0 \sim f_0 \circ T$ and $G : S_0 \circ g \sim g \circ S$ are there a mapping $f : X \to Y$ and homotopies $r : g \circ f \sim f_0$ and $F : S \circ f \sim f \circ T$ so that $-PS_0 \circ r + G \circ f + Pg \circ F + r \circ T \sim F_0$ corel E_0, E_∞? (i.e.: the two are homotopic as homotopies $S_0 \circ f_0 \sim f_0 \circ T$).

The lifting problem in $\operatorname{End} H$ contains the lifting problem in H (the existence of $f : X \to Y, g \circ f \sim f_0$) and thus, it is natural to study the lifting problem in $\operatorname{End}(H)$ modulo that in H. This relative problem is given as follows:

If in (3.1.1.) there exists $f : X \to Y$ with $g \circ f \sim f_0$ does there exist a lifting $\hat{f} : X \to Y$ with $S \circ \hat{f} \sim \hat{f} \circ T$?

The answer in general may be negative as can be seen by the following simple example:

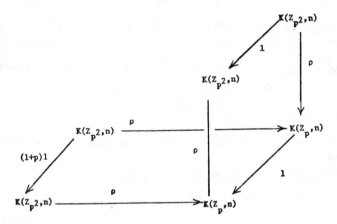

There is no $\hat{f} : K(Z_{p^2}, n) \to K(Z_{p^2}, n)$ so that $\rho \circ \hat{f} \sim \rho$ and $\hat{f}[(1+p)1] \sim \hat{f}$. But if one raises all endomorphisms to the p-th power (and as $(1 + p)^p = 1 (\operatorname{mod} p^2)$) such a lifting obviously exists: $\hat{f} = 1$.

However, if one replaces $K(Z_{p^2}, n)$ by $K(Z, n)$ in the above diagram one can see no remedy to the lifting problem. This observation holds in a more general situation as stated in the following lifting theorem the proof of which is the main goal of this section:

3.1. THE LIFTING THEOREM. *Given the diagram (3.1.1) with the (3.1.2) homotopies: $F_0 : S_0 \circ f_0 \sim f_0 \circ T, G : S_0 \circ g \sim g \circ S$.*

Suppose:

(3.1.a) $H^n(X, \pi_n(\text{fiber } g))$ *are finite and*

$\infty > m_0 = \prod_{n=0}^{\infty} \exp H^n(X, \pi_n(\text{fiber } g))$

(3.1.b) *One of the following two assumptions holds:*

Either

(3.1.b)′ *g is a principal fibration.*

or

(3.1.b)″ *Y, Y_0 are H-spaces and g is an H-map.*

If there exists a lifting $f' : X \to Y$ of f_0 then there exists an integer m (divisible only by primes dividing m_0, i.e. $m | m_0^t$ for some t), a lifting $f : X \to Y$ and homotopies $r : f_0 \circ g \sim f$, $F : S^m \circ f \sim f_0 \circ T^m$ covering the homotopy $S_0^n \circ f \sim f_0 \circ T^m$ induced by F_0 and G.

i.e:

$$F_0^m = \sum_{k=0}^{m-1} PS_0^k \circ F_0 \circ T^{m-k-1}, G^m = \sum_{k=0}^{m-1} PS_0^k \circ G_0 \circ S^{m-k-1}$$

$$F_0^m : S_0^m \circ f_0 \sim f_0 \circ T^m, G^m : S_0^m \circ g \sim g \circ S^m$$

then

$$-P\,S_0 \circ r + G^m \circ f + Pg \circ F + r \circ T^m \sim F_0^m \text{ corel } E_0, E_\infty$$

(Note that (3.1.b)' does not require S, S_0 to be a map of principal fibrations and that (3.1.b)'' does not restrict S, S_0 to be H-maps).

(3.1.3) Remark: Actually one can slightly generalize the (3.1.b) conditions and replace it by the assumption that g is w-principal (See [4-1] p.53). This covers both (3.1.b)' and (3.1.b)'' but the proof will have to include an analysis of the properties of Postnikov decomposition of w-principal fibrations (properties that are known to hold for the two special cases of (3.1.b)).

The proof of 3.1. will occupy the rest of this section.

Let C be a category. Suppose $\text{Hom}_C(A, B)$ is an algebraic looop. Define $W : \text{End}_C(B) \times \text{End}_C(A) \to \text{End}_{\text{Set}} [\text{Hom}_C(A, B)]$ by $W_{S,T}\varphi = S \circ \varphi - \varphi \circ T, S, T \in \text{End}_C(A) \times \text{End}_C(B), \varphi \in \text{Hom}_C(A, B), (\ ) - (\ )$ denotes the difference in the algebraic loop.

Alternatively:

If $T^*, S_* : \text{Hom}_C(A, B) \to \text{Hom}_C(A, B)$ are the right and left compositions with T and S respectively then T^*, S_* commute and $W_{S,T} = S_* - T^*$. If $\text{Hom}_C(A, B)$ is an abelian group and S_*, T^* are homomorphisms then one obviously has the following binomial formula and its consequences:

3.2. $W_{S,T}^n \varphi = \sum_{k=0}^n \binom{n}{k}(-1)^k\, S^{n-k}\varphi T^k$

3.3. $W_{S,T}^{p^r} = W_{S^{p^r}, T^{p^r}} \pmod{p}$

3.4. Definition: An element $g \in \text{End}_C(A)$ is said to be p-periodic iff for some integer $r > 0$ $g^{p^r} = g$.

3.5. If $\text{End}_C(A)$ is finite then for any $g \in \text{End}_C(A)$ there exists $t \geq 0$ so that g^{p^t} is p-periodic.

3.6. LEMMA. *Suppose $\text{Hom}_C(A, B)$ is a Z_p-vector space and S_*, T^* are linear transformations. Define $\phi = \text{Per}(S, T)_{s,r} = \{\varphi : A \to B | S_\varphi^{p^s} = S\varphi, \varphi \circ T^{p^r} = \varphi \circ T\}$ then $W_{S,T}(\phi) \subset \phi$ and $W_{S,T}|\phi$ is p-periodic. In particular, if S_*, T^* are p-periodic so is $W_{S,T}$.*

PROOF. As S_*, T^* commute with $W_{S,T}\ \phi$ is obviously $W_{S,T}$ invariant. $S^{p^{sr}} \circ \phi = S \circ \phi$ and $\phi \circ T^{p^{sr}} = \phi \circ T$ imply $W_{S^{p^{sr}}, T^{p^{sr}}}|\phi = W_{S,T}|\phi$ and apply 3.3.

By 2.10, one has

3.7. LEMMA. *If* $\pi_n(X) = 0$, $n > N(X)$ $X \overset{\sim P-p}{\to} K(\pi_*(X)/\text{torsion})$ *then* $f : X \to X$ *is p-periodic if and only if* $H^*(f, Z)$ *is.*

3.8. The lifting obstruction:
Consider the following commutative diagram

(3.8.0)

with homotopies $F_0 : S_0 \circ f_0 \sim f_0 \circ T$, $F_1 : \hat{S} \circ h \sim h \circ S_0$, $\ell : * \sim h \circ f_0$. ℓ induces a lifting $f : X \to Y = \text{fiber } h$, $S_0, \hat{S}, F_1$ induce a map $S : Y \to Y$.

Consider the obstruction for the existence of a homotopy $F : S \circ f \sim f \circ T$ covering F_0 (i.e: If $r : Y \to Y_0$ is the natural projection then obviously $r \circ S = S_0 \circ r$. The homotopy F is then required to satisfy

$$Pr \circ F \sim F_0 \text{ corel } E_0, E_\infty).$$

This obstruction is precisely the map
$$\alpha(\ell, F_0, F_1) : X \to \Omega B$$
$$\alpha(\ell, F_0, F_1) = L\hat{S} \circ \ell + F_1 \circ f + Ph \circ F_0 - \ell \circ T.$$

Assuming that B is an H-space with multiplication μ_B then one can easily prove the following:

(3.8.1) If $\gamma : X \to \Omega B$ then
$$\alpha(\gamma + \ell, F_0, F_1) = W_{\Omega\hat{S},T}\gamma + \alpha(\ell, F_0, F_1)$$

(3.8.2) Given $\xi : Y_0 \to \Omega B$ let $\xi \bullet F_1 : \hat{S} \circ h \sim h \circ S_0$ be given by $\xi \bullet F_1 = P\mu_B \circ (\xi \times F_1) \circ \Delta_{Y_0}$ then:
$$\alpha(\ell, F_0, \xi \bullet F_1) = \xi \circ f_0 + \alpha(\ell, F_0, F_1).$$

(3.8.3) Given a composition of two (3.8.0) type diagram

(3.8.0)[1])

Define
$$F_0' * F_0 = PS_0' \circ F_0 + F_0' \circ T, F_1' * F_1 = P\hat{S} \circ F_1 + F_1' \circ S_0$$
$$F_0' * F_0 : S_0' \circ S_0 \circ f_0 \sim f_0 \circ T' \circ T, F_1' * F_1 : \hat{S} \circ \hat{S} \circ h \sim h \circ S_0' \circ S_0$$
then:

$$\alpha(\ell, F_0' * F_0, F_1' * F_1) = \Omega\hat{S}' \circ \alpha(\ell, F_0, F_1) + \alpha(\ell, F_0', F_1') \circ T$$

In particular, if $F_0^n = \underbrace{F_0 * F_0 * \ldots * F_0}_{n}, F_1^n = \underbrace{F_1 * F_1 * \ldots * F_1}_{n}$

then

$$\alpha(\ell, F_0^n, F_1^n) = \sum_{k=0}^{n-1} \Omega\hat{S}^k \circ \alpha(\ell, F_0, F_1) \circ T^{n-k-1}$$

3.8.4. Example: The obstruction α is a special case of a "ladder" Toda brackets discussed briefly in [4-5] section 2:

Given a commutative diagram

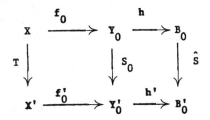

and homotopies

$$\ell : * \sim h \circ f_0, \ell' : * \sim h' \circ f_0'$$

$$F_0 : S_0 \circ f_0 \sim f_0' \circ T \quad F_1 : \hat{S} \circ h \sim h' \circ S_0$$

Define

$$\alpha(\ell, \ell', F_0, F_1) : X \to \Omega B_0'$$

by

$$\alpha(\ell, \ell', F_0, F_1) = L\hat{S} \circ \ell + F_1 \circ f_0 + Ph' \circ F_0 - \ell' \circ T$$

3.8.4.0. The composition of 3.8.3 induces a monoid structure Γ on the set of 3.8.0 diagrams. If one lets the homotopy $\ell : * \sim h \circ f_0$ vary 3.8.1, 3.8.3 actually show that the obstructions $\{\alpha(-, F_0, F_1)\}$ represent an element in $H^1(\Gamma, [X, \Omega B_0])$.

3.8.4.1. *Given a commutative diagram*

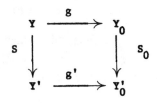

with fiber $g = K(G,n)$, fiber $g' = K(G',n)$ then S, S_0 are maps of principal fibrations. (Compare with [4-1] example 2.4.5).

PROOF OF 3.8.4.1. g is induced by the following map $h : Y_0 \to K(G, n+1)$:

$$Y_0 \xrightarrow{j} C_g \xrightarrow{\tilde{h}} Ht_{n+1}(C_g) \simeq K(G, n+1)$$

C_g – the cone on g.

And similarly for g'. Consider the following ladder:

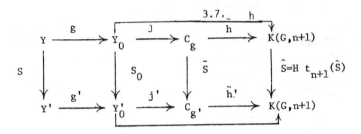

Let $F_0 : S_0 \circ g \sim g' \circ S$, let $F_1' = k \circ \tilde{S} \circ j : \tilde{S} \circ j \sim j' \circ S_0$ be the constant homotopy and let $F_1'' : \hat{S} \circ \tilde{h} \sim \tilde{h}' \circ \tilde{S}$.

Denote by $\ell : * \sim h \circ g$ and $\ell' : * \sim h' \circ g'$ the homotopies induced by the natural homotopies $\tilde{\ell} : * \sim j \circ g$, $\tilde{\ell}' : * \sim j' \circ g'$. $\ell = L\tilde{h} \circ \tilde{\ell}$, $\ell' = L\tilde{h}' \circ \tilde{\ell}'$.

ℓ, ℓ' induces a map $\tilde{g} : Y \to Y_1 = \text{fiber}\, h$ ($\tilde{g}' : Y' \to Y_1' = \text{fiber}\, h'$) $\tilde{g}, \tilde{g}'$ are homotopy equivalences. Let $S_1 : Y_1 \to Y_1'$ be induced by $\tilde{S}, \hat{S}$ and $F_1'' * F_1' :$ $\hat{S} \circ h \sim h' \circ S_0$. Our assertion is that $S_1 \circ g \sim \tilde{g} \circ S$. To see that one can check directly that $* \sim \alpha(\tilde{\ell}, \ell', F_0, F_0') : Y \to \Omega C_{g'}$, and consequently $\alpha(\tilde{\ell}, \tilde{\ell}', F_0, F_1'' * F_1') \sim \Omega\tilde{h}' \alpha(\tilde{\ell}, \tilde{\ell}', F_0, F_1') \sim *$ which completes the proof of 3.8.4.1.

Our $\alpha(\ell, F_0, F_1)$ is the special case of $\alpha(\ell, \ell', F_0, F_1)$ where $X = X', Y_0 = Y_0', B_0 = B_0', f_0 = f_0', h = h'$ and $\ell = \ell'$. Hence, if

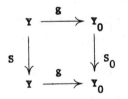

is commutative, fiber $g = K(G,n)$ then g is principal and S, S_0 is a map of principal fibrations.

3.9. *A special case of the lifting theorem: If in 3.1.1 fiber $g = K(G,n)$, G-finite, then 3.1 holds.*

PROOF. By 3.8.4 S, S_0 are maps of principal fibrations induced by $h : Y_0 \to K(G, n+1)$ and one has a commutative diagram

$$
\begin{array}{ccccc}
X & \xrightarrow{f_0} & Y_0 & \xrightarrow{h} & K(G,n+1) \\
\downarrow{\scriptstyle T} & {\scriptstyle F_0} & \downarrow{\scriptstyle S_0}\ {\scriptstyle F_1} & & \downarrow{\scriptstyle \hat{S}} \\
X & \xrightarrow{f_0} & Y_0 & \xrightarrow{h} & K(G,n+1)
\end{array}
$$

As f_0 lifts to $f' : X \to Y$ there exists $\ell : * \sim h \circ f_0$. Our obstruction here is the 3.8 $\alpha(\ell, F_0, F_1)$.

The proof is by induction on $\exp G$. Let $p \mid \exp G$. One may assume that $H^n(T, Z_p)$ and $\pi_{n+1}(\hat{S}) \otimes Z_p$ are p-periodic, otherwise replace T, S_0 and $\hat{S}$ by $T^{p^r}, S_0^{p^r}, \hat{S}^{p^r}$ for suitably chosen r (3.5).

Consider

$$
\begin{array}{ccccccc}
X & \xrightarrow{f_0} Y_0 & \xrightarrow{h} & K(G,n+1) & \xrightarrow{\rho} & K(G \otimes Z_p, n+1) \\
\downarrow{\scriptstyle T} & \downarrow{\scriptstyle F_0} & \downarrow{\scriptstyle S_0}\ {\scriptstyle F_1}\ {\scriptstyle \hat{S}} & & {\scriptstyle F_1'} & & \downarrow{\scriptstyle \hat{S}_1} \\
X & \xrightarrow{f_0} Y_0 & \xrightarrow{h} & K(G,n+1) & \xrightarrow{\rho} & K(G \otimes Z_p, n+1)
\end{array}
$$

Then

$$\alpha(\mathcal{L}\rho \circ \ell, F_0, F_1' * F_1) = \Omega\rho \circ \alpha(\ell, F_0, F_1)$$

As $\Omega\hat{S}_1$ and $H^n(T, Z_p)$ are p-periodic by 3.6

$$W^{p^j}_{\Omega\hat{S}_1, T}\alpha(\mathcal{L}\rho \circ \ell, F_0, F_1' * F_1) = W_{\Omega\hat{S}_1, T}$$

for some $j > 0$.

Replace ℓ by $\tilde{\ell} = -W^{p^j-2}_{\Omega\hat{S}, T}\alpha(\ell, F_0, F_1) + \ell$

by 3.8.1

$$\alpha(\mathcal{L}\rho \circ \tilde{\ell}, F_0, F_1' * F_1) = \alpha[(-W^{p^j-2}_{\Omega\hat{S}_1, T}\alpha(\mathcal{L}\rho \circ \ell, F_0, F_1' * F_1) + \mathcal{L}\rho \circ \ell), F_0, F_1' * F_1)]$$
$$= -W^{p^j-1}_{\Omega\hat{S}_1, T}\alpha(\mathcal{L}\rho \circ \ell, F_0, F_1' * F_1) + \alpha(\mathcal{L}\rho \circ \ell, F_0, F_1' * F_1) =$$
$$= (1 - W^{p^j-1}_{\Omega\hat{S}_1, T})\,\alpha(\mathcal{L}\rho \circ \ell, F_0, F_1' * F_1) \in \ker W_{\Omega\hat{S}_1, T}$$

Hence,

$$\Omega\hat{S}_1 \circ \alpha(\mathcal{L}\rho \circ \tilde{\ell}, F_0, F_1' * F_1) = \alpha(\mathcal{L}\rho \circ \tilde{\ell}, F_0, F_1' * F_1) \circ T$$

and by 3.8.3.

$$\alpha(\mathcal{L}\rho \circ \tilde{\ell}, F_0^m, (F_1' * F_1)^m) = \sum_{k=0}^{m-1} \Omega\hat{S}_1^k \circ \alpha(\mathcal{L}\rho \circ \tilde{\ell}, F_0, F_1' * F_1)T^{m-k-1} = m\,\alpha(\mathcal{L}\rho \circ \tilde{\ell}, F_0, F_1' * F_1)T^{m-1}$$

and

$$\alpha(\mathcal{L}\rho \circ \tilde{\ell}, F_0^p, (F_1' * F_1)^p) = 0.$$

It follows that f_0 lifts to $f_1 : X \to Y_1 = $ fiber $\rho \circ h$ and for some t $f_1 \circ T^{p^t} \sim S_1^{p^t} \circ f_1, S_1 : Y_1 \to Y_1$ induced by S_0 and $\hat{S}_1$. Now fiber $f_1 = K(\tilde{G}, n)$ $\tilde{G} = \ker(G \to G \otimes Z_p), \exp \tilde{G} = \frac{1}{p} \exp G$ and one proceeds by induction.

3.10. THE PROOF OF 3.1. (For g a 0-equivalence.) One uses induction on $\prod_m \exp \pi_m(\text{fiber } g)$, or equivalently, on the primitive Postnikov decomposition of g. As in 3.8.4, one has

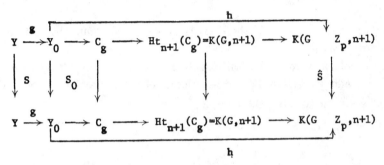

where $n - 1$ is the connectivity of fiber $g, G = \pi_n(\text{fiber } g)$ and $p | \exp G$. Then g, S, S_0 factor as

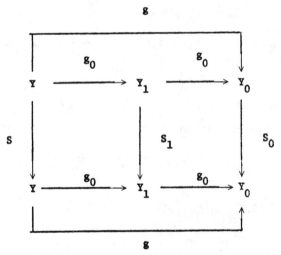

$Y_1 = $ fiber $h, g_0 : Y_1 \to Y_0, \prod_m \exp \pi_m(\text{fiber } g_1) = \frac{1}{p} \prod_m \exp \pi_m(\text{fiber } g)$. By 3.9 $f_0 : X \to Y_0$ lifts to $f_1; X \to Y_1$ with $f_1 \circ T^{p^t} = S_1^{p^t} \circ f_1$. To proceed by induction one has to show that f_1 can be lifted to $f' : X \to Y$: If f_0 is liftable to say $f' : X \to Y$, in the 3.9 process one may have to alter the lifting $g_1 \circ f' : X \to Y_1$ in order to obtain the desired $f_1 : X \to Y_1$ and it has to be checked that the change does not destroy the liftability property.

By the proof of 3.9 the alteration of the original lifting is of the form

$$ W_{\Omega \hat{S}_1, T}^{p^j - 2} \alpha(\tilde{\ell}_1, F_0, \hat{F}_1), \text{ where } \alpha(\tilde{\ell}_1, F_0, \hat{F}_1) $$

is the 3.8 obstruction of the diagram

$$
\begin{array}{ccccc}
X & \xrightarrow{\ f_0\ } & Y_0 & \xrightarrow{\ h\ } & K(G \otimes Z_p, n+1) \\
\downarrow{\scriptstyle T} & F_0 & \downarrow{\scriptstyle S_0\ \ \hat F_1} & & \downarrow{\scriptstyle \hat S_1} \\
X & \xrightarrow{\ f_0\ } & Y_0 & \xrightarrow{\ h\ } & K(G \otimes Z_p, n+1)
\end{array}
$$

with $\tilde{\ell}_1 : * \sim h \circ f_0$ induced by a lifting $f' : X \to Y$ of f_0 and the homotopy $* \sim h \circ g : h \circ f_0 \sim h \circ g \circ f' \sim *$.

An alternative and completely equivalent description of α is given as follows:

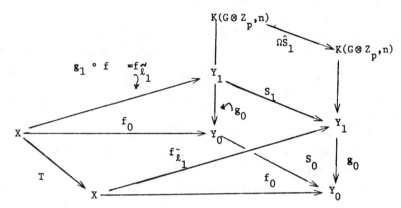

$$
g_0 \circ S_1 \circ f_{\tilde{\ell}_1} = S_0 \circ g_0 \circ f_{\tilde{\ell}_1} = S_0 \circ f_0 \sim f_0 \circ T = g_0 \circ f_{\tilde{\ell}_1} \circ T
$$

$$
g_0 \circ S_1 \circ f_{\tilde{\ell}_1} = S_0 \circ g_0 \circ \tilde f_{\ell_1} = S_0 \circ f_0 \sim f_0 \circ T = g_0 \circ f_{\tilde{\ell}_1} \circ T
$$

There exists a (unique up to homotopy) map $\omega = \alpha(\ell_1, F_0, F_1) : X \to K(G \otimes Z_p, n)$ so that

$$
S_1 \circ f_{\tilde{\ell}_1} \sim \omega * (f_{\tilde{\ell}_1} \circ T)
$$

via a homotopy covering F_0. (*–the action of $[X, K(G, n)]$ on $[X, Y_1]$). Now we use the 3.1.b assumptions and that both (3.1.b)$'$ and (3.1.b)$''$ are inherited by $g_1 : Y \to Y_1$ so one can continue the proof by induction once one proves that $f_1 : X \to Y_1$ with $f_1 \circ T^{p^s} \sim S_1^{p^s} \circ f_1$ can be lifted to $X \to Y$.

3.10.1. The (3.1.b)$'$ case: If $f' : X \to Y$ is a lifting of f_0 then f' defines $\ell_1 : * \sim h \circ f_0$ and a lifting $f_{\tilde{\ell}_1} = g_1 \circ f' : X \to Y_1$. As above, as g is principal with fiber, say V, one obtains

$$
\tilde\omega : X \to V
$$

so that

$$
S \circ f' \sim \tilde\omega * (f' \circ T)
$$

and the homotopy covers S_0. Now, the following diagram

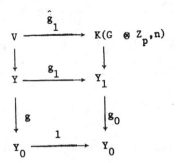

describes a map of principal fibrations. One must have $\hat{g}_1 \circ \tilde{\omega} = \omega = \alpha(\tilde{\ell}_1, F_0, \hat{F}_1)$.
If $S_V : V \to V$ is induced by S, S_0, as $\hat{g}_1$ is a loop map $\hat{g}_1 W_{S_V,T}^{p^j-2} \tilde{\omega} = W_{\Omega \hat{S}_1,T}^{p^j-2} \alpha(\tilde{\ell}, F_0, \hat{F}_1) = u$ and the alteration of $g_1 \circ f'$ by u can be obtained by
altering f' by $v = W_{S_V,T}^{p^j-2} \tilde{\omega} \in [X, V]$ hence, $f_1 = (-u) * (g_1 \circ f')$ lifts to $(-v) * f'$:
$X \to Y$.

3.10.2. The case of $(3.1.b)''$: The lifting $g_1 \circ f' : X \to Y_1$ is altered by
$W_{\omega \hat{S}_1,T}^{p^j-2} \omega : f_1 = -W_{\Omega \hat{S}_1,T}^{p^j-2} \omega * (g_1 \circ f')$. But one can easily see that in this
case one can write f_1 as

$$f_1 = -W_{S_1,T}^{p^j-2} D_{S_1 \circ g_1 \circ f', g_1 \circ f' \circ T} + g_1 \circ f'$$

(Sums and differences in the algebraic loop $[X, Y_1]$: $-h = D_{*,h}$. $W_{S_1,T}h = D_{S_1 \circ h, h \circ T}$). As g_1 is an H-map, $S_1 \circ g_1 \sim g_1 \circ S$ one has
$$f_1 = -W_{S_1,T}^{p^j-2} D_{g_1 \circ S \circ f', g_1 \circ f' \circ T} + g_1 \circ f' =$$
$$= -W_{S_1,T}^{p^j-2} g_1 D_{S \circ f', f' \circ T} + g_1 \circ f' =$$
$$= -g_1 W_{S,T}^{p^j-2} D_{S \circ f', f' \circ T} + g_1 \circ f' = g_1 [-W_{S,T}^{p^j-2} D_{S \circ f', f' \circ T} + f']$$
and f_1 is liftable.

3.10.3. Remark: The $(3.1.b)''$ case of 3.1 has an extension.
Suppose $h : A \to X, \hat{T} : A \to A, h \circ \hat{T} \sim T \circ h$. *If* $f_0 : X \to Y_0$ *lifts to* $\hat{f} : X \to Y$
so that $\hat{f} \circ h \circ \hat{T} \sim S \circ \hat{f} \circ h$ *then* $\hat{f}$ *lifts to* $f : X \to Y$, $f \circ T^m \sim S^m \circ f$ *and*
$f \circ h \sim \hat{f} \circ h$.

This could be seen as follows. By 3.10.2, the lifting f is obtained by

$$f = -W_{S,T}^q D_{S \circ \hat{f}, \hat{f} \circ T} + \hat{f} \text{ for some } q.$$

$f \circ h = -W_{S,T}^q D_{S \circ \hat{f}, \hat{f} \circ T} \circ h + \hat{f} \circ h$. But
$W_{S,T}^q D_{S \circ \hat{f}, \hat{f} \circ T} \circ h = W_{S,\hat{T}}^q D_{S \circ \hat{f} \circ h, \hat{f} \circ h \circ \hat{T}} = W_{S,\hat{T}}^q(*) = *$
3.11. The proof of 3.1 (g not a 0 equivalence): Here we have

(3.1.a) $$\prod_n \exp H^n(X, \pi_n(\text{fiber } g)) = m_0 < \infty.$$

One has to make the adjustment in the proof of 3.9, i.e: instead of $\exp G < \infty$
one has to assume $\exp H^n(X, G) < \infty$. If V is any finite group, Y_1 the fiber of
$Y_0 \to K(G, n+1) \to K(G \otimes V, n+1)$ then one has

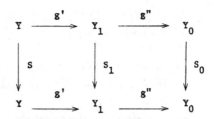

and by 3.9 f_0 lifts to $f_1 : X \to Y_1, S_1^m \circ f_1 \sim f_1 \circ T^m$. Moreover, by the 3.10.1 process f_1 still lifts to $X \to Y$. g' is principal induced by $Y_1 \to K(\ker(G \to G \otimes V), n+1)$. It can be argued that if $\rho_* : H^n(X, G) \to H^n(X, G \otimes V)$ then the obstruction $\alpha(\ell, F_0, F_1) \in H^n(X, G)$ satisfies $\rho_* \alpha(\ell, F_0, F_1) = 0$. Now choose $V = Z_m$ $m = \exp H^n(X, G)$, then

$$\rho_* : H^n(X, G) \approx H^n(X, G) \otimes V \varinjlim H^n(X, G \otimes V)$$

ρ_* injective implies $\alpha(\ell, F_0, F_1) = *$.

4. The lifting theorem: Applications.

The lifting theorem in $\mathrm{End}(H)$ can be used to solve lifting problems in H. The principal method is given by the following proposition:

4.1. PROPOSITION. *Given a commutative diagram and homotopies F_0, F_1 as in 3.1.1., 3.1.2:*

$$
\begin{array}{ccccc}
X & \xrightarrow{f_0} & Y_0 & \xrightarrow{g} & Y \\
\downarrow{\scriptstyle T} & {\scriptstyle F_0} & \downarrow{\scriptstyle S_0} \;\; {\scriptstyle F_1} & & \downarrow{\scriptstyle S} \\
X & \xrightarrow{f_0} & Y_0 & \xrightarrow{g} & Y
\end{array}
$$

Suppose for every n $H^n(X, \pi_n(\mathrm{fiber}\, g))$ is finite and for every prime p and integer n the characteristic polynomial of $H^{n+1}(T, Z_p)$ is relatively prime to the characteristic polynomials of $\pi_n(\hat{S}) \otimes Z_p$ and $\pi_{n+1}(\hat{S})/\mathrm{torsion} \otimes Z_p$ where $\hat{S} : \mathrm{fiber}\, g \to \mathrm{fiber}\, g$ is the map induced by S, S_0 and F_1. Then f_0 lifts to $f : X \to Y$.

PROOF. First we prove 4.1 for the case where g is a $\mathbb{P} - p$ equivalence and $\pi_n(\mathrm{fiber}\, g) = 0$ for $n > N$. In this case we decompose g, S, S_0 into its Postnikov decomposition

$$
\begin{array}{ccccccc}
Y = Y_m & \xrightarrow{g_m} & \cdots & \to & Y_1 & \xrightarrow{g_1} & Y_0 \\
{\scriptstyle S = S_m}\downarrow & & & & {\scriptstyle S_1}\downarrow & & {\scriptstyle S_0}\downarrow \\
Y = Y_m & \xrightarrow{g_m} & \cdots & \to & Y_1 & \xrightarrow{g_1} & Y_0
\end{array}
$$

$$
\begin{array}{ccccc}
Y_n & \xrightarrow{g_{n-1}} & Y_{n-1} & \xrightarrow{k_n} & K(G_n, n+1) \\
\downarrow{\scriptstyle S_n} & & \downarrow{\scriptstyle S_{n-1}} & & \downarrow{\scriptstyle \tilde{S}_n} \\
Y_n & \xrightarrow{\phantom{g_{n-1}}} & Y_{n-1} & \xrightarrow{k_n} & K(G_n, n+1)
\end{array}
$$

$G_n = \pi_n(\text{fiber } g), \tilde{S}_n = K(\pi_n(\hat{S}), n+1), G_n$-a p group. Suppose inductively that f_0 lifts to $f_{n-1} : X \to Y_{n-1}$ and that for some $r_{n-1}, f_{n-1} \circ T^{p^{r_{n-1}}} \sim S_{n-1}^{p^{r_{n-1}}} \circ f_{n-1}$.

If one shows that $k_n \circ f_{n-1} \sim *$ then f_{n-1} lifts to $f_n : X \to Y_n$ and by 3.9 f_n may be assumed to satisfy $f_n \circ T^{p^{r_n}} \sim S_n^{p^{r_n}} \circ f_n$ and one may proceed by induction.

Using a simple inductive argument and the exact sequence

$$H^{n+1}(X, G_n \otimes Z_{p^{r-1}}) \otimes Z_p \to H^{n+1}(X, G_n \otimes Z_{p^r}) \otimes Z_p \to H^{n+1}(X, G_n \otimes Z_p)$$

one can see that the characteristic roots of $H^{n+1}(T, G_n \otimes Z_p) = H^{n+1}(T, Z_p) \otimes 1_{G_n \otimes Z_p}$ contain those of $H^{n+1}(T, G_n) \otimes Z_p$ and the characteristic roots of $H^{n+1}(X, \pi_n(\hat{S}) \otimes Z_p) = 1_{H^{n+1}(X, Z_p)} \otimes \pi_n(\hat{S}) \otimes Z_p$ contain those of $H^{n+1}(X, \pi_n(\hat{S})) \otimes Z_p$. Hence $T^* \otimes Z_p = H^{n+1}(T, G_n) \otimes Z_p$ and $(\hat{S}_n)_* \otimes Z_p = H^{n+1}(X, \pi_n(\hat{S})) \otimes Z_p$ have relatively prime characteristic polynomials. This implies that $T^* \otimes Z_p - (\hat{S}_n)_* \otimes Z_p = [T^* - (\hat{S}_n)_*] \otimes Z_p$ is an isomorphism, so is $[(T^{p^{r_{n-1}}})^* - (S_n^{p^{r_{n-1}}})_*] \otimes Z_p$ and therefore $(T^{p^{r_{n-1}}})^* - (S_n^{p^{r_{n-1}}})_*$ is an isomorphism. Now $k \circ f_{n-1} \circ T^{p^{r_{n-1}}} \sim \tilde{S}_n^{p^{r_{n-1}}} \circ k_n \circ f_{n-1}$ implies $[k_n \circ f_{n-1}] \in \ker[(T^{p^{r_{n-1}}})^* - (S_n^{p^{r_{n-1}}})_*]$ and $k_n \circ f_{n-1} \sim *$.

Now suppose g is a rational equivalence and $\pi_n(\text{fiber } g) = 0$ for $n > N$. Then for every pg, S, S_0 factors as

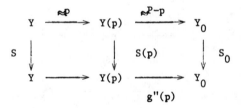

By the first part for every p f_0 lifts to $Y(p)$. As Y is the pull back of finitely many of the $g''(p)$ f_0 lifts to Y.

Now we shall remove the rational equivalence restriction but still assume $\pi_n(\text{fiber } g) = 0$ for $n > N$. In this case we shall show the existence of a factorization and a lifting $\hat{f} : X \to \hat{Y}$ of $f_0, \hat{S}_1 \circ \hat{f} \sim \hat{f} \circ T$ so that g_1 is a rational equivalence, and $\pi_m(\text{fiber } g_0)$ are free abelian groups. Once this is done the fibration

$$
\begin{array}{ccccc}
\text{fiber } g_1 & \to & \text{fiber } g & \xrightarrow{\underset{0}{\alpha}} & \text{fiber } g_0 \\
\downarrow \hat{S}_1 & & \downarrow \hat{S} & & \downarrow \hat{S}_0 \\
\text{fiber } g_1 & \to & \text{fiber } g & \xrightarrow[\tilde{\alpha}]{\underset{0}{\ }} & \text{fiber } g_0
\end{array}
$$

implies that $\pi_r(\hat{S})/\text{torsion}$ and $\pi_r(\hat{S}_0)$ have the same integral characteristic polynomials and the same holds for $\pi_r(\hat{S})/\text{torsion} \otimes Z_p$ and $\pi_r(\hat{S}_0) \otimes Z_p$. As

$$0 \to \pi_{m+1}(\text{fiber } g)/\text{torsion} \xrightarrow{\underset{\ }{0}\tilde{\ }} \pi_{m+1}(\text{fiber } f_0) \to \pi_m(\text{fiber } g_1) \to$$

$$\text{torsion } \pi_m(\text{fiber } g) \to 0$$

is exact, the characteristic polynomial of $\pi_m(\hat{S}_1) \otimes Z_p$ divides those of $^p\pi_m(\hat{S}) \otimes Z_p$ and $\pi_{m+1}(\hat{S})/\text{torsion} \otimes Z_p$ and is relatively prime to the characteristic polynomial of $H^{m+1}(T, Z_p)$ and by the rational equivalence case f lifts to $f : X \to Y$. The construction of $\hat{Y}$ is done by induction: Suppose one has:

$$
\begin{array}{ccccccccc}
Y & \xrightarrow{(g_1)_n} & \hat{Y}_n & \xrightarrow{(g_0)_n} & Y_0 & & \hat{Y}_n & \xleftarrow{f_n} & X \\
\downarrow S & & \downarrow \tilde{S}_{1,n} & & \downarrow S_0 \quad \tilde{S}_{1,n}\downarrow & & \hat{f}_n\downarrow & & \downarrow T \\
Y & \longrightarrow & \hat{Y}_n & \longrightarrow & Y_0 & & \hat{Y}_n & \xleftarrow{f_n} & X
\end{array}
$$

where the cone on $(g_1)_n$ is rationally $n-1$ connected, $\pi_*(\text{fiber}(g_0)_n)$ free and $\pi_k(\text{fiber}(g_0)_n) = 0$ for $k \geq n - 1$. Let $V = \text{cone on } (g_1)_n, \tilde{S}' : V \to V$ induced by $\tilde{S}_{1,n}$ and S. Let $G = H^n(V, Z)/\text{torsion}$. One can find maps $h_1 : V \to K(G, n), (H^n(h_1, Z) \otimes Q$ an isomorphism$)$ and $\tilde{S}'_0 : K(G, n) \to K(G, n)$ satisfying $\tilde{S}'_0 \circ h_1 \sim h_1 \circ \tilde{S}'$. (One can easily construct $\tilde{S}'_0$ with $[\tilde{S}'_0 \circ h_1 - h_1 \circ \tilde{S}']$ of finite order. Then replace h_1 by λh_1). One can easily see that $G \approx \text{Hom}(G, Z) \approx \pi_{n-1}(\text{fiber } g)/\text{torsion}$ and that $\tilde{S}'_0 = K(\pi_{n-1}(\tilde{S})/\text{torsion}, n)$.

As for every p $H^n(T, Z_p)$ and $\pi_n(\tilde{S}'_0) \otimes Z_p \approx \pi_{n-1}(\tilde{S})/\text{torsion} \otimes Z_p$ have relatively prime characteristic polynomials so do $H^n(T, Z)/\text{torsion}$ and $\pi_n(\tilde{S}'_0) = \pi_n(\tilde{S}'_0)/\text{torsion}$. Considering

$$
\begin{array}{ccccccc}
X & \xrightarrow{\hat{f}_n} & \hat{Y}_n & \longrightarrow & V & \xrightarrow{h_1} & K(G,n) \\
\downarrow T & & \downarrow \tilde{S}_{1,n} & & \downarrow & & \downarrow \tilde{S}'_0 \\
X & \xrightarrow{\hat{f}_n} & \hat{Y}_n & \longrightarrow & V & \xrightarrow{h_1} & K(G,n)
\end{array}
$$

$$\underset{h}{\longrightarrow}$$

$[h \circ \hat{f}_n] \otimes Q \in \ker[(\tilde{S}_0')_* \otimes Q - T^* \otimes Q)] = 0$, hence, $h \circ \hat{f}_n$ is of finite order and replacing h by $\lambda_1 h$ if necessary one may assume $h \circ \hat{f}_n \sim *$. If $\ell : * \sim h \circ \hat{f}_n$, $F_0 : \tilde{S}_{1,n} \circ \hat{f}_n \sim \hat{f}_n \circ T$, $F_1 : \tilde{S}_0 \circ h \sim h \circ \tilde{S}_{1,n}$ then $\alpha(\ell, F_0, F_1) \in H^{n-1}(X, G) = H^{n-1}(X, \pi_{n-1}(\text{fiber } g))$ which if finite. Again replacing h by $\lambda_2 h$ if necessary one may assume $\alpha(\ell, F_0, F_1) = 0$.

Put $\hat{Y}_{n+1} = \text{fiber } h$, one obtains commutative diagrams

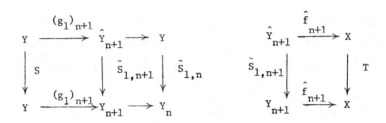

(the left hand side diagram is commutative by 3.8.4 type argument, as $h : \hat{Y}_n \to K(G, n)$ factors through the cone on $(g_1)_n$). This completes the inductive step and the case where $\pi_n(\text{fiber } g) = 0$ for $n > N$.

To remove this last restriction, note that as $H^n(X, \pi_n(\text{fiber } g))$ is finite and considering the Postnikov decomposition of g $Y \to \ldots \to Y_n \overset{g^n}{\to} Y_{n-1} \to \ldots \to Y_0$ Y - the homotopy inverse limit of $\{g_n\}$; the map $[X, Y_n] \to [X, Y_{n-1}]$ is finite to one. Consequently, $\lim_{\to}[X, Y_n] \to \bigcap_n \text{im}\{[X, Y_n] \to [X, Y_0]\}$ is surjective and $X \to Y_0$ lifts to $X \to Y$ if and only if for every n f_0 lifts to $f_n : X \to Y_n$.

4.2. Power spaces (See [4-2]): Suppose given a pair X, T where $T : X \to X$ and an odd prime p so that $H^*(X, Z_p) = \Lambda(x_1, x_2, \ldots, x_k)$, $|x_i| \le |x_{i+1}|$, $|x_i|$-odd, $k < p$. Let $\lambda \in Z$ be a primitive root of unity mod p. Given a commutative diagram

$$
\begin{array}{ccccccc}
V = \text{fiber } g & \longrightarrow & Y & \overset{g}{\underset{P-p}{\to}} & Y_0 & \overset{f_0}{\leftarrow} & X \\
\downarrow{\scriptstyle S_V} & & \downarrow{\scriptstyle S} & & \downarrow{\scriptstyle S_0} & & \downarrow{\scriptstyle T} \\
V = \text{fiber } g & \longrightarrow & Y & \overset{g}{\to} & Y_0 & \overset{f_0}{\leftarrow} & X
\end{array}
$$

Suppose λ is the only characteristic root of $QH^*(T, Z_p)$ and $\pi_*(S_V) \otimes Z_p$. If V is $|x_k| - 1$ connected then f_0 lifts to $f : X \to Y$ and for some $r \ge 1$ $f \circ T^{p^r} \sim S^{p^r} \circ f$. Indeed, in dim $> |x_k|$ the characteristic roots of $H^*(T, Z_p)$ are λ^i, $1 \le i \le k < p$, as $\lambda^i \neq \lambda \pmod{p}$ one can apply 4.1.

To apply this argument in order to prove the spherical decomposition of X (4.2.2) we need the following.

4.2.1. LEMMA. *Let X be an H_0 space, $H^*(X, Z)$ p-torsion free. Assume either X is finite dimensional or $\pi_n(X) = 0\ n > N$. If $T : X \to X$ is such that λ is the only characteristic root of $QH^*(T, Z_p)$ then X, T can be replaced by a mod p equivalent $\hat{X}, \hat{T}$, so that $QH^*(\hat{X}, Z)/$ torsion can be represented by $\tilde{\lambda}$ characteristic vectors of $H^*(\hat{T}, Z), \tilde{\lambda} \equiv \lambda$ (mod p). i.e. $\hat{X}, \hat{T}$ is given by the following diagrams*

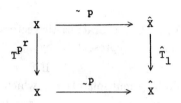

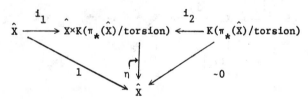

for some map h where $\eta : \hat{X} \times K(\pi_*(\hat{X})/$ torsion$) \to \hat{X}$ satisfies

(see [4-4] 1.7).

PROOF. One can replace T by T^{p^r} so that $QH^*(T^{p^r}, Z_p) = \lambda^{p^r} 1 = \lambda 1$. Then one can represent the basis of $QH^*(T^{p^r}, Z_p)$ by characteristic vectors in $H^*(X, Z_p)$. One has a diagram

Decompose the above diagram ([4-1]), 4.3)

$$
\begin{array}{ccccc}
X & \xrightarrow{\;\sim p\;} & X(p) = \hat{X} & \xrightarrow[\psi'']{\;\approx\;P\text{-}p\;} & K_0 \\
\downarrow {\scriptstyle T^{p^r}} & & \downarrow {\scriptstyle \hat{T}_1} & {\scriptstyle \approx P\text{-}p} & \downarrow {\scriptstyle T_1} \\
X & \xrightarrow{\;\sim p\;} & \hat{X} & \longrightarrow & K_0
\end{array}
$$

Let $x_1, x_2, \ldots, x_m \in H^*(\hat{X}, Z_p)$ be λ characteristic vectors generating $H^*(\hat{X}, Z_p)$ as an algebra. For any r one can represent x_i by $x_{i,r} \in H^*(X, Z_{p^r})$ and $x_{i,r}$ is a $\lambda^{p^{t^r}}$ characteristic vector of $H^*(\hat{T}_1^{p^{s^r}}, Z_{p^r})$. If $\tilde{x}_i$ $i = 1, \ldots, m$ are integral classes representing (a prime to p multiple of) $x_{i,r}$ which reduce to a basis for $QH^*(X, Z)/$ torsion then $H^*(\hat{T}_1^{p^{s^r}}, Z)\tilde{x}_i = \lambda^{p^{t^r}}\tilde{x}_i + p^r d$. If r is chosen properly and as p is the only torsion prime in $\pi_*(\text{fiber}\,\psi'')$ by [4-4] 1.8 replace $\hat{T}_1^{p^{s^r}}$ by $\hat{T}$ in the described way so that $\tilde{x}_i$ are $\tilde{\lambda} = \lambda^{p^{t^r}}$ characteristic vectors.

4.2.2. Example: (See [4-2].) Let X be a finite complex, $H^*(X, Z_p) = \Lambda(x_1, \ldots, x_m), |x_i| \le |x_{i+1}| = \text{odd}, m < p$. If $T : X \to X$ is such that λ is the only characteristic root of $QH^*(T, Z_p)$, λ–a primitive root of unity, then X can be decomposed mod p as follows:

$$
\begin{array}{ccccccccc}
X_1 \sim_p S & \xrightarrow{\;|x_1|\;} & X_2 & \xrightarrow{\;j_2\;} & X_3 & \to & \cdots & \to & X_m = X \\
& & \downarrow {\scriptstyle |x_2|} & & \downarrow {\scriptstyle h_3 \atop |x_3|} & & & & \downarrow {\scriptstyle |x_m|} \\
& & S & & S & & & & S
\end{array}
$$

where $X_i \xrightarrow{j_i} X_{i+1} \xrightarrow{h_{i+1}} S^{|x_{i+1}|}$ is a fibration, $H^*(X_i, Z_p) = \Lambda(x_{1,i}, \ldots, x_{i,i})$, $|x_{i,m}| = |x_i|, H^*(j_i, Z_p)\,x_{j,i+1} = x_{j,i}$ for $j \le i$.

PROOF. Replace X, T as in 4.2.1 so that one may assume that x_i can be represented by integral λ characteristic vectors $\tilde{x}_i$ then apply 4.2 for

$$
\begin{array}{ccccc}
Y = S & \xrightarrow{\;|x_m|\;} & Y_0 = K(Z, |x_m|) & \xleftarrow{\;\tilde{x}_m\;} & X \\
\downarrow {\scriptstyle S = \lambda 1} & & {\scriptstyle S_0 = \lambda 1}\; \downarrow & & \downarrow {\scriptstyle T} \\
Y = S & \xrightarrow{\;|x_m|\;} & Y_0 = K(Z, |x_m|) & \xleftarrow{\;\tilde{x}_m\;} & X
\end{array}
$$

to obtain

$$
\begin{array}{ccc}
X & \xrightarrow{\ h_m\ } & S^{|x_m|} \\
{\scriptstyle T^{p^r}}\downarrow & & \downarrow{\scriptstyle \lambda^{p^r}1} \\
X & \xrightarrow{\ h_m\ } & S^{|x_m|}
\end{array}
$$

X_{m-1}, T_{m-1} where $T_{m-1} : X_{m-1} = \text{fiber}\, h_m \to X_{m-1}$ satisfies the same properties as X, T with $H^*(h_m, Z)\tilde{x}_i = \tilde{x}_{i,m-1}$ etc. and one can proceed by induction.

4.3 Geometric dimension of vector bundles over CP^n and $L_n(p)$. Let $h_\nu : CP^n \to CP^n$ be the map satisfying $H^2(h_\nu, Z) = \nu 1, \nu \in Z$. Given a (complex) vector bundle ξ of (complex) dimension n on CP^n, if the (rational) Chern classes of ξ, $c_k(\xi)$, vanish for $m < k$, then for some integer ν $h^*\nu\xi$ has a geometric dimension $m : h^*\nu\xi = \xi_m \oplus C^{n-m}$.

Let $ch_m : B\,U(n) \to \prod_{k=1}^{n-m} K(Z, 2m+2k)$ represent the (integral) Chern classes of dim $> m$. As fiber $ch_m \approx_p BU(m)$ in dim $\leq n$ for all $p > n - m$, one can choose ν to be of the form $\nu = \prod_{p_i \leq n-m} p_i^{r_i}$. We shall show here that actually ν involves considerably less primes p.

4.4. PROPOSITION. *The obstructions for ξ to have a geometric dimension m are the rational Chern classes $c_k(\xi), k > m$, and mod p obstructions for p satisfying*

$$(4.4.1.1) \qquad (p-1)[p(p-1)-1] = (p-1)^3 + (p-1)(p-2) \leq n - m$$

(e.g. for the existence of $r, r < 10$, sections the obstructions beyond the rational Chern classes are only mod 2 obstructions. Mod 5 obstructions will enter only for $r \geq 76$ sections).

PROOF. Again, put $ch_m : BU \to \prod_{k>0} K(Z, 2m + 2k)$ for the realization of $> m$ dim integral Chern classes. Let $\tilde{B}(m) = \text{fiber}\, ch_m$. Then $BU(m) \underset{0}{\sim} \tilde{B}(m)$. Fix a prime p and decompose ϕ as

$$BU(m) \xrightarrow[\sim\, p]{\phi'} \tilde{B}(m, p) \xrightarrow[\sim\, \mathbb{P}-p]{\phi''} \tilde{B}(m).$$

If ξ satisfies $c_k(\xi) = 0$ for $k > m$, then $f_\xi : CP^n \to BU$ lifts to $\tilde{f} : CP^n \to \tilde{B}(m)$. To prove 4.4, one has to show that $\tilde{f}$ lifts to $\tilde{B}(m, p)$ for every p not satisfying (4.4.1.1).

By [2] Corollary 5.11, there exists a commutative diagram

$$
\begin{array}{ccc}
\mathrm{BU(m)} & \xrightarrow{\ \mathbf{1}\ } & \mathrm{BU} \\
\downarrow{\psi_{\lambda,\mathrm{m}}} & & \downarrow{\psi_\lambda} \\
\mathrm{BU(m)} & \xrightarrow{\ \mathbf{1}\ } & \mathrm{BU}
\end{array}
$$

provided $\lambda = \prod_{\tilde{p}_i > m} \tilde{p}_i{}^{r_i}$ and one can choose λ to be a primitive root of unity mod p. Recall that $\psi_\lambda, \psi_{\lambda,m}$ satisfy: $H^{2r}(\psi_\lambda, Z) = \lambda^r 1 = H^{2r}(\psi_{\lambda,m}, Z)$. By the commutativity of

$$
\begin{array}{ccc}
\mathrm{BU} & \xrightarrow{\ \mathrm{ch}_{\mathrm{m}}\ } & \prod\limits_{k=1}^{\infty} K(Z, 2m+2k) \\
\downarrow{\psi_\lambda} & & \downarrow{\prod\limits_k (\lambda^{m+k} 1)} \\
\mathrm{BU} & \xrightarrow{\ \mathrm{ch}_{\mathrm{m}}\ } & \prod\limits_{k=1}^{\infty} K(Z, 2m+2k)
\end{array}
$$

one obtains a map $\tilde{\psi}_\lambda(m) : \tilde{B}(m) \to \tilde{B}(m)$ and as fiber $[\tilde{B}(m) \to BU] = \prod_k K(Z, 2m + 2k - 1)$ and $H^{2m+2k-1}(BU(m), Z) = 0$ one obtains commutative diagrams

$$
\begin{array}{ccc}
\mathrm{BU(m)} & \xrightarrow{\ \phi\ } & \tilde{\mathrm{B}}(\mathrm{m}) \\
\downarrow{\psi_{\lambda,\mathrm{m}}} & & \downarrow{\tilde{\psi}_\lambda(\mathrm{m})} \\
\mathrm{BU(m)} & \xrightarrow{\ \phi\ } & \tilde{\mathrm{B}}(\mathrm{m})
\end{array}
$$

and

$$
\begin{array}{ccccc}
\mathrm{BU(m)} & \xrightarrow[\sim p]{\ \phi\,\prime\ } & \tilde{\mathrm{B}}(\mathrm{m,p}) & \xrightarrow[\sim P-p]{\ \phi''\ } & \tilde{\mathrm{B}}(\mathrm{m}) \\
\downarrow{\psi_{\lambda,\mathrm{m}}} & & \downarrow{\tilde{\psi}_\lambda(\mathrm{m,p})} & & \downarrow{\tilde{\psi}_\lambda(\mathrm{m})} \\
\mathrm{BU(m)} & \xrightarrow[\sim p]{\ \phi'\ } & \tilde{\mathrm{B}}(\mathrm{m,p}) & \xrightarrow[\sim P-p]{\ \phi''\ } & \tilde{\mathrm{B}}(\mathrm{m})
\end{array}
$$

Now, $f_\xi : CP^n \to BU$ satisfies $f_\xi \circ h_\lambda \sim \psi_\lambda \circ f_\xi$. Indeed, $H^{2k}(h_\lambda, Q) = \lambda^k 1 = H^{2k}(\psi_\lambda, Q)$, hence, $H^{2k}(f_\xi \circ h_\lambda, Q) = H^{2k}(\psi_\lambda \circ f_\xi, Q)$ and as $H^*(CP^n, Z)$ is torsion free the following is injective

$$
[CP^n, BU] \to \mathrm{Hom}[H^*(BU, Q), H^*(CP^n, Q)]
$$

As $H^{2m+2k-1}(CP^n, Z) = 0$ the lifting $\tilde{f} : CP^n \to \tilde{B}(m)$ of f_ξ satisfies a commutative diagram:

$$
\begin{array}{ccc}
CP^n & \xrightarrow{\ \tilde{f}\ } & \tilde{B}(m) \\
\Big\downarrow{h_\lambda} & & \Big\downarrow{\tilde{\psi}_\lambda(m)} \\
CP^n & \xrightarrow{\ \tilde{f}\ } & \tilde{B}(m)
\end{array}
$$

Let $V'' = $ fiber ϕ'' and $\hat{\psi}''V'' \to V''$ be induced by $\tilde{\psi}_\lambda(m)$ and $\tilde{\psi}_\lambda(m,p)$.
To prove 4.4, one can apply 4.1 to

$$
\begin{array}{ccccccc}
(\text{fiber } \phi'')=V'' & \longrightarrow & Y=\tilde{B}(m,p) & \xrightarrow{g=\phi''} & Y_0=\tilde{B}(m) & \xleftarrow{f_0=\tilde{f}} & X=CP^n \\
\hat{\psi}'' \Big\downarrow & & S=\tilde{\psi}_\lambda(m,p)\Big\downarrow & & S_0=\tilde{\psi}_\lambda(m)\Big\downarrow & & \Big\downarrow T=h_\lambda \\
V'' \longrightarrow & & Y=\tilde{B}(m,p) & \xrightarrow{g=\phi''} & Y_0=\tilde{B}(m) & \xleftarrow{f_0=\tilde{f}} & X=CP^n
\end{array}
$$

provided one shows that the characteristic polynomials of $\pi_{k-1}(\hat{\psi}'') \otimes Z_p$ and of $H^k(h_\lambda, Z_p)$ are relatively prime for $2m < k \le 2n$. Now, the characteristic polynomial of $H^{2k}(h_\lambda, Z_p)$ is $x - \lambda^k(H^{2k+1}(h_\lambda, Z_p) = 0)$. One has to study now the characteristic roots of $\pi_{2k-1}(\hat{\psi}'') \otimes Z_p, m < k \le n$.

As $V = $ fiber $\phi \approx_p V''$ one may study $\pi_{2k-1}(\hat{\psi}) \otimes Z_p, \hat{\psi} : V \to V$ induced by $\hat{\psi}_\lambda(m)$ and $\psi_{\lambda,m}$. Comparing

$$
\begin{array}{ccccccc}
& & V & \longrightarrow & BU(m) & \xrightarrow{\phi} & \tilde{B}(m) \\
& & \downarrow & & \| & & \downarrow \\
\text{fiber } i \ \ = & & U/U(m) & \longrightarrow & BU(m) & \xrightarrow{i} & BU \\
& & \downarrow & & & &
\end{array}
$$

one can easily see that $\pi_{2k-1}(V'') = {}^p\pi_{2k-1}(V) \approx {}^p\pi_{2k-1}(U/U(m))$. If $\bar{\psi} : U/U(m) \to U/U(m)$ is induced by $\psi_{\lambda,m}$ and ψ_λ it suffices to study ${}^p\pi_{2k-1}(\bar{\psi}) \otimes Z_p$. Now

$$H^*(U/U(m), Z) = \Lambda(x_{2m+1}, x_{2m+3}, \ldots)$$
$$H^*(\bar{\psi}, Z)x_{2k-1} = \lambda^k x_{2k-1}.$$

One can construct coherent system of maps

$$
\begin{array}{ccc}
BU(k-1) & \longrightarrow & BU(k) \\
\Big\downarrow{\psi_{\lambda,k-1}} & & \Big\downarrow{\psi_{\lambda,k}} \\
BU(k-1) & \longrightarrow & BU(k)
\end{array}
$$

thus inducing systems of fibrations

$$
\begin{array}{ccccccc}
S^{2m+1} & \longrightarrow & U/U(m) & \longrightarrow & U/U(m+1) & \to \cdots \to & U/U(n) \\
\downarrow{\scriptstyle \lambda^{m+1}\cdot 1} & & \downarrow{\scriptstyle \tilde{\psi}} & & \downarrow{\scriptstyle \tilde{\psi}} & & \\
S^{2m+1} & \longrightarrow & U/U(m) & \longrightarrow & U/U(m+1) & \to \cdots \to & U/U(n)
\end{array}
$$

Thus, the spheres of dimensions $2k-1, n \geq k > m\ k \equiv k_0(p-1)$ carry the λ^{k_0} eigenspace in $^P\pi_t(U/U(m)), 2m < t \leq 2n$. But $^P\pi_{2k-1+2s(p-1)}(S^{2k+1}) = 0$ for $s < p(p-1) - 1$ and the characteristic polynomials of $H^{2k}(h_\lambda, ^P\pi_{2k-1}(V''))$ and $H^{2k}(\mathbb{C}P^n, ^P\pi_{2k-1}(\hat{\psi}''))$ are relatively prime for

$$ m < k \leq n $$

In order to investigate complex vector bundles on lens spaces $L^n(p), (p > 2)$, one uses the maps $u = u_r : L^r(p) \to CP^{[\frac{r}{2}]}$ given by

$$ L^{2n}(p) \to L^{2n+1}(p) \overset{S^1}{\to} CP^n. $$

As these maps yield surjections on $H^{2k}(, Z)$ every $[\frac{r}{2}]$ dim complex vector bundle ξ on $L^r(p)$ is of the form $u^*(\hat{\xi})$ for some vector bundle $\hat{\xi}$ on $CP^{[\frac{r}{2}]}$. Using 4.4 one has the following:

4.4.1. PROPOSITION. *Let ξ be an $[\frac{n}{2}]$ dimensional (complex) vector bundle over $L^n(p)$. If $\xi = u^*(\hat{\xi}), c_k(\hat{\xi}) = 0, k \geq [\frac{n}{2}] - r$ and $r \leq (p-1)^3 + (p-1)(p-2)$, then ξ has a geometric dimension $[\frac{n}{2}] - r$.*

PROOF. The map $h_\nu : CP^{\bar{n}} \to CP^{\bar{n}}$ can be given by $h_\nu : S^{2\bar{n}+1} \to S^{2\bar{n}+1}$

$$ \tilde{h}_\nu(r_0\,e^{i\tau_0}, r_1 e^{i\tau_1}, \ldots, r_{\bar{n}} e^{i\tau_{\bar{n}}}) = (r_0 e^{i\nu\tau_0}, r_1 e^{i\nu\tau_1}, \ldots, r_{\bar{n}} e^{i\nu\tau_{\bar{n}}}) $$

Now

$g_\nu : S^1 \to S^1 (g_\nu z = z^\nu)$ is a homomorphism and $\tilde{h}_\nu$ is a g_ν map. This induces $h_\nu : S^{2\bar{n}+1}/S^1 \approx CP^{\bar{n}} \to S^{2\bar{n}+1}/S^1 \approx CP^{\bar{n}}$. All maps restrict naturally to the $Z_p = \{e^{2r\pi i/p}|0 \leq r < p\} \subset S^1$ action on $S^{2\bar{n}+1}$ to induce

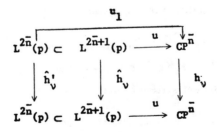

By 4.4. there exists an integer ν prime to p so that $h_\nu^*(\hat{\xi})$ has a geometric dimension $\bar{n} - r$ and so do $u^* h_\nu^*(\hat{\xi}) = \hat{h}_\nu^* \xi$ and $\hat{h}_\nu^{'*}(\xi|L^{2\bar{n}}(p))$. Now $H^1(\hat{h}_\nu, Z_p) =$

$\nu 1$ is an isomorphism and consequently $\hat{h}'_{\nu}$ is a homotopy equivalence, hence $\xi | L^{2\bar{n}}(p)$ has a geometric dimension $\bar{n} - r$.

To deal with $u^*(\hat{\xi}) = \xi$ one can choose s so that $\nu^s \equiv 1 \bmod p$. Hence $H^*(h'_{\nu^s}, Z) = 1$ and by 2.9, there exists t so that $h'_{\nu^t} = (h'_{\nu})^t \sim 1$. Consider the principal cofibration: $L^{2\bar{n}}(p) \to L^{2\bar{n}+1}(p) \to S^{2\bar{n}+1}$ with coaction $\hat{\eta} : L^{2\bar{n}+1}(p) \to L^{2\bar{n}+1}(p) \bigvee S^{2\bar{n}+1}$. $\hat{h}_{\nu^t} | L^{2\bar{n}}(p) \sim 1$ implies $\hat{h}_{\nu^t} \sim \mathcal{F}(1 \bigvee w)\hat{\eta}$ for some $w : S^{2\bar{n}+1} \to L^{2\bar{n}+1}(p)$. If $f_{\xi} : CP^n \to BU$ classifies $\hat{\xi}$, $f_{\hat{\xi}} \circ u$ classifies ξ and

$$f_{\hat{\xi}} \circ u \circ \hat{h}^t_{\nu} \sim f_{\hat{\xi}} \circ u \circ \mathcal{F} \circ (1 \bigvee w) \circ \hat{\eta} \sim \mathcal{F}[f_{\hat{\xi}} \circ u \bigvee f_{\hat{\xi}} \circ u \circ w] \circ \hat{\eta}.$$

As $\pi_{2\bar{n}+1}(BU) = 0 f_{\hat{\xi}} \circ u \circ w \sim *$ and

$$f_{\hat{\xi}} \circ u \circ \hat{h}^t_{\nu} \sim \mathcal{F}(f_{\hat{\xi}} \circ u \bigvee f_{\hat{\xi}} \circ u \circ w)\hat{\eta} \sim \mathcal{F}[f_{\hat{\xi}} \circ u \bigvee (*)]\hat{\eta} = f_{\hat{\xi}} \circ u$$

and $\xi = (h^t_{\nu})^* \xi$ has geometric dimension $\bar{n} - r$.

4.5. H-spaces "with identities": Let X, μ be an H-space, $\pi_m(X) = 0$ for $m > N(X)$. Suppose $H^*(\mu, Q)$ satisfies an identity in two variables, i.e: There exists $\phi_2 : X \times X \to X \times X$, $\phi_1 : X \to X$ so that $\phi_2(X \bigvee X) \subset X \bigvee X$, $\mathcal{F} \circ \phi_2 | X \bigvee X = \phi_1 \circ \mathcal{F}$ and $H^*(\phi_2, Q) \circ H^*(\mu, Q) = H^*(\mu, Q) \circ H^*(\phi_1, Q)$. Then there exists a multiplication $\mu_0 : X \times X \to X$, $H^*(\mu_0, Q) = H^*(\mu, Q)$ and $\phi_1^m \circ \mu_0 \sim \mu_0 \phi_2^m$ for some integer m depending on X.

PROOF. Let $h_0 : X \xrightarrow{\sim_0} K(\pi_*(X)/\text{torsion})$. Let $\phi_0 : K(\pi_*X/\text{torsion}) \to K(\pi_*X/\text{torsion})$ be induced by $\pi_*(\phi_1)/\text{torsion}$. Then $h_0 \circ \phi_1 - \phi_0 \circ h_0$ and $h_0 \circ \mu \circ \phi_2 - \phi_0 \circ h_0 \circ \mu$ have finite order, hence, replacing h_0 by $\nu 1 \circ h_0$ one may assume that the following diagram commutes:

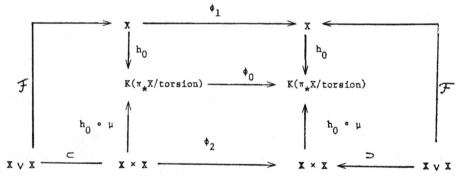

As μ is a lifting of $h_0 \circ \mu$ by 3.1 one can lift $h_0 \mu$ to $\mu_0 : X \times X \to X$ with $\phi_1^m \circ \mu_0 \sim \mu_0 \circ \phi_2^m$, $h_0 \mu \sim h_0 \mu_0$ and by 3.10.3 for $A = X \vee X$ μ_0 is an H-structure.

4.5.1. Example: If $H^*(\mu, Q)$ is cocommutative then mod odd primes one can replace μ by a homotopy commutative multiplication: $\phi_1 = 1, \phi_2 = T : X \times X \to X \times X$ and in this case one can choose m odd (as $p = 2$ is excluded). As $T^m = T$ one has $\mu_0 \circ T \sim \mu_0$.

4.5.2. For $p = 2$ one can apply the same method to obtain the following: If X, μ is an H-space $H^*(X, Q)$ coassociative and cocommutative, then mod $p \neq 3$

there exists $M : X \times X \times X \to X$ so that $M(x, y, *) = M(y, *, x) = M(*, x, y) = \mu(x, y)$, and $M(x, y, z) \sim M(z, x, y) \sim M(y, z, x)$.

5. Realizing cohomology subalgebras

Given a self map $T : X \to X$. We consider the following question: If $A \subset H^*(X, Z_p)$ is $H^*(T, Z_p)$ invariant, can A be realized. i.e.: is there a space Y with a self map $S : Y \to Y$ and a commutative diagram

$$
\begin{array}{ccc}
X & \xrightarrow{f} & Y \\
\downarrow T & & \downarrow S \\
X & \xrightarrow{f} & Y
\end{array}
$$

so that $H^*(f, Z_p)$ is injective with $im H^*(f, Z_p) = A$. There are ample examples where such a space does not exist. However, one has:

5.1. A REALIZATION THEOREM. *Let* $T : X \to X$ *be an endomorphism. Suppose there exist sequences of polynomials* $^n P_1(x), ^n P_2(x) \in Z[x]$, $^n P_i |^{n+1} P_i, {}^n P_1, {}^n P_2$ - *relatively prime so that*

(i) $^n P_1(H^n(T, Z)) \cdot {}^n P_2(H^n(T, Z)) = 0$

(ii) *For every n the multiplicative closure of the set of characteristic roots of* $^n P_1 \otimes Z_p$ *contains no characteristic root of* $^n P_2 \otimes Z_p$.

Then $A = \bigoplus_n im {}^n P_2(H^n(T, Z_p)) \subset H^*(X, Z_p)$ *is realizable, i.e.: There exist an endomorphism* $S : Y \to Y$ *and a map* $f : X \to Y$ *so that* S *of* $\sim f \circ T, H^*(f, Z_p)$ *injective,* $H^*(Y, Z_p) \approx im H^*(f, Z_p) = A$.

5.1.1. **Remark**: There exist some earlier versions of 5.1: Wilkerson's retraction theorem (see [3-2] and 5.8 of this chapter) stating that under some finiteness conditions and restrictions on rational cohomology the "stable" image of an endomorphism $T : X \to X$ is realizable by a mod p retract: $Y \xrightarrow{i} X \xrightarrow{r} Y$, ($r \circ i$ a mod p equivalence)$im H^*(r, Z_p) = \bigcap_{n=1}^{\infty} im H^*(T^n, Z_p)$.

Another version, due to Cooke and Smith, considers suspension spaces. It states that if $T : \Sigma X \to \Sigma X$ then ΣX decomposes mod p as $\Sigma X \approx_p \vee Y_i$ corresponding to the decomposition of the characteristic polynomial of $H^*(T, Z_p)$ into powers of prime factors.

The proof of 5.1 and its applications will follow from the following:

5.2. PROPOSITION. *Given the commutative diagram*

(5.2.1)

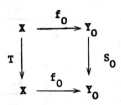

Suppose the multiplicative closure of the characteristic roots of $H^*(S_0, Z_p)$ contains no characteristic root of

$$\mathrm{Coker}\, H^*(f_0, Z_p) \xrightarrow{\hat{T}^*} \mathrm{Coker}\, H^*(f_0, Z_p)$$

and of

$$[\mathrm{Coker}\, H^*(f_0, Z)]/\,\mathrm{torsion} \otimes Z_p \xrightarrow{\hat{T}^*_0}$$
$$[\mathrm{Coker}\, H^*(f_0, Z)]/\,\mathrm{torsion} \otimes Z_p$$

$(\hat{T}^*, \hat{T}^*_0$ induced by T and $S_0)$.

Then (5.2.1) factors as

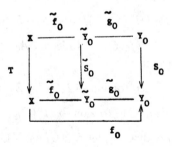

$H^*(f, Z_p)$ injective, $H^*(g, Z_p)$ surjective.

The first step improving 5.2 is by reducing it to the case where $H^*(f_0, Q)$ is injective.

5.3. LEMMA. With the notations and hypotheses of 5.2 there exists a commutative diagram

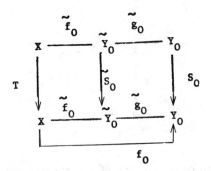

so that $im\, H^*(\tilde{f}_0, Z_p) = im\, H^*(f_0, Z_p), H^*(S_0, Z_p)$ and $H^*(\tilde{S}_0, Z_p)$ have the same multiplicative closures of characteristic roots (and consequently, the left hand square of 5.3.1 satisfies the hypothesis of 5.2) and in addition $H^*(\tilde{f}_0, Q)$ is injective.

PROOF. By the hypothesis of 5.2, the characteristic polynomials of $[H^n(S_0, Z)/$ torsion$] \otimes Z_p$ and of

$$\hat{T}_0^{n-1} : [\operatorname{Coker} H^{n-1}(f_0, Z)]/\operatorname{torsion} \otimes Z_p \to$$
$$[\operatorname{Coker} H^{n-1}(f_0, Z)]/\operatorname{torsion} \otimes Z_p$$

are relatively prime and therefore, so are the characteristic polynomials of $H^n(S_0, Z)/$ torsion and of $[\operatorname{Coker} H^{n-1}(f_0, Z)]/$ torsion $\to [\operatorname{Coker} H^{n-1}(f_0, Z)]/$ torsion.

Let $\hat{S}_0 : C_{f_0} \to C_{f_0}$ be the map induced on the mapping cone of f_0 by T and S_0. Let ${}_Q^n P_{S_0}(x)$ be the characteristic polynomial of $H^n(S_0, Z)/$ torsion. Then $H^n(C_{f_0}, Z)/$ torsion $\to H^n(Y_0, Z)/$ torsion induces a monomorphism

$$A_0 = \ker{}_Q^n P_{S_0}(H^n(\hat{S}_0, Z)/\operatorname{torsion} \to H^n(Y_0, Z)/\operatorname{torsion}$$

realizing rationally $\ker H^n(f_0, Q)$. Indeed, $A_0 \subset H^n(C_{f_0}, Z)/$ torsion is $H^n(\hat{S}_0, Z)/$ torsion invariant and

$$im(H^{n-1}(X, Z)/\operatorname{torsion} \to H^n(C_{f_0}, Z)/\operatorname{torsion} \cap A_0 =$$
$$= im\{[\operatorname{Coker} H^{n-1}(f_0, Z)]/\operatorname{torsion} \to H^n(C_{f_0}, Z)/\operatorname{torsion}\} \cap A_0 = 0.$$

Let $r : C_{f_0} \to K(A_0, N)$ realize A_0, let $\hat{S}' : K(A_0, N) \to K(A_0, N)$ be induced by $H^n(\hat{S}_0, Z)$. Then $\hat{S}' \circ r$ and $r \circ \hat{S}_0$ may fail to be homotopic, however, $\hat{S}' \circ r - r \circ \hat{S}_0$ has a finite order and replacing r by its non zero multiple, one may assume

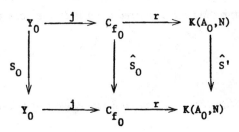

is commutative. $H^n(r \circ j, Q)$ is injective with image $\ker H^n(f_0, Q)$. As A_0 realizes $\ker_Q^n P_{S_0}(H^n(\hat{S}_0, Z)/\text{torsion})$, $_Q^n P_{S_0}(H^n(\hat{S}', Z)) = 0$ and $_Q^n P_{S_0}(\hat{S}') \sim *$. Consequently, the characteristic polynomial of $H^n(\hat{S}', Z_p)$ and thus of $Q\, H^m(\hat{S}', Z_p)$ for all m, divides that of $H^n(S_0, Z_p)$ and the multiplicative closure of characteristic roots of $H^*(\hat{S}', Z_p)$ is contained in that of $H^*(S_0, Z_p)$. Put $\hat{Y} = \text{fiber}\, r \circ j$ then one has a commutative diagram

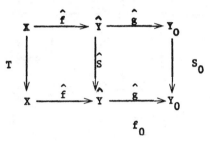

$$f_0$$

(See 3.8.4.1, and note that one may assume that the homotopy $\hat{S} \circ \hat{f} \sim \hat{f} \circ T$ covers that of $S_0 \circ f_0 \sim f_0 \circ T$). $H^n(\hat{Y}, Q) \approx H^n(Y_0, Q)/\ker H^n(f_0, Q)$ hence $H^n(\hat{f}, Q)$ is injective.

By 3.1, the multiplicative closure of characteristic roots of $H^*(\hat{S}, Z_p)$ is contained in the union of those of $H^*(\hat{S}', Z_p)$ and of $H^*(S_0, Z_p)$ and consequently it is contained in the multiplicative closure of characteristic roots of $H^*(S_0, Z_p)$. Consequently,

$$
\begin{array}{ccc}
X & \xrightarrow{\hat{f}} & \hat{Y} \\
\downarrow{\scriptstyle T} & & \downarrow{\scriptstyle \hat{S}} \\
X & \xrightarrow{\hat{f}} & \hat{Y}
\end{array}
$$

satisfies the hypothesis of 5.2. Obviously

$$\text{im}\, H^*(\hat{f}, Z_p) \supset \text{im}\, H^*(f_0, Z_p).$$

for every m the following is commutative

$$
\begin{array}{ccccc}
H^m(\hat{Y}, Z_p) & \xrightarrow{H^m(\hat{f}, Z_p)} & H^m(X, Z_p) & \xrightarrow{\rho} & \text{Coker } H^m(f_0, Z_p) \\
\downarrow{\scriptstyle H^m(\hat{S}, Z_p)} & & \downarrow{\scriptstyle H^m(T, Z_p)} & & \downarrow{\scriptstyle \hat{T}(m)} \\
H^m(\hat{Y}, Z_p) & \xrightarrow{H^m(\hat{f}, Z_p)} & H^m(X, Z_p) & \xrightarrow{\rho} & \text{Coker } H^m(f_0, Z_p)
\end{array}
$$

As $H^m(\hat{S}, Z_p)$ and $\hat{T}^{(m)}$ have relatively prime characteristic polynomials $\rho \circ H^m(\hat{f}, Z_p) = 0$ and $\operatorname{im} H^m(f, Z_p) \subset \operatorname{im} H^*(f_0, Z_p)$.

One proves 5.3 by annihilating inductively $\ker H^n(f_n, Q)$ as described above, obtaining the following commutative diagram:

(5.3.2)

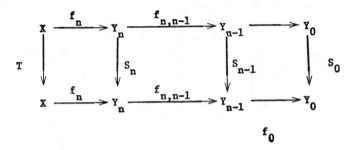

$\operatorname{im} H^*(f_n, Z_p) = \operatorname{im} H^*(f_0, Z_p)$, $H^*(S_n, Z_p)$ and $H^*(S_0, Z_p)$ have the same multiplicative closure of characteristic roots and $H^k(f_n, Q)$ is injective for $k < n$.

Now passing to a limit one has

$$
\begin{array}{ccc}
X & \xrightarrow{\tilde{f_0} = \varprojlim f_n} & \tilde{Y}_0 = \varprojlim Y_n \\
{\scriptstyle T}\downarrow & & \downarrow {\scriptstyle \tilde{S}_0 = \varprojlim S_n} \\
X & \xrightarrow{\tilde{f_0}} & \tilde{Y}_0
\end{array}
$$

The idea of the proof of 5.2 is fairly obvious. One tries to annihilate inductively $\ker H^{m_n}(f_n, Z_p)$ in a (5.3.2) type diagram where $H^k(f_n, Z_p)$ is injective for $k < m_n$ (and $S_n \circ f_n \sim f_n \circ T$ covers $S_{n-1} \circ f_{n-1} \sim f_{n-1} \circ T$, $g_{n,n-1} \circ f_n \sim f_{n-1}$).

There are three difficulties in such an argument: The first is to show that the elimination of $\ker H^{m_n}(f_n, Z_p)$ will indeed yield a commutative diagram

$$
\begin{array}{ccccc}
X & \xrightarrow{f_{n+1}} & Y_{n+1} & \xrightarrow{g_{n+1,n}} & Y_n \\
{\scriptstyle T}\downarrow & & \downarrow {\scriptstyle S_{n+1}} & & \downarrow {\scriptstyle S_n} \\
X & \xrightarrow{f_{n+1}} & Y_{n+1} & \xrightarrow{g_{n+1,n}} & Y_n
\end{array}
$$

$[(g_{n+1,n} : Y_{n+1} \to Y_n) = \operatorname{fiber}(Y_n \to K(\ker H^{m_n}(f_n, Z_p), m_n)].$
The second is to show that $\operatorname{im} H^*(f_{n+1}, Z_p) = H^*(f_n, Z_p)$.

Finally one has to show that $\lim_{n\to\infty} m_n = \infty$. The latter is the most technical and will be dealt with first:

5.4. LEMMA. *(1) Given a space M. For every integer n there exists an integer t so that for every r*

$$\mathrm{im}[H^n(M, Z_p r + t) \to H^n(M, Z_p r)] = \mathrm{im}[H^n(M, Z) \to H^n(M, Z_p r)]$$

(2) For any map $f : M \to L$ and integer n there exists s so that for every r

$$\mathrm{im}[\ker H^n(f, Z_p r + s) \to \ker H^n(f, Z_p r)] =$$

$$\mathrm{im}[\ker H^n(f, Z) \to \ker H^n(f, Z_p r)]$$

PROOF. Let $p^t = \exp {}^p H^{n+1}(M, Z)$. As

$$\mathrm{im}(\delta_r : H^n(M, Z_p r) \to H^{n+1}(M, Z)) \subset {}^p H^{n+1}(M, Z)$$

(1) follows from the commutative diagram

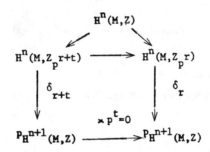

(2) Let $\rho_{k,m} : H^n(, Z_p k) \to H^n(, Z_p m)$ be the reduction $1 \le m \le k \le \infty (Z = Z_{p^\infty})$. Put $s = 2\hat{s} = 2\max(s_0, s_1)$ where $p^{s_0} = \exp^p H^{n+1}(L, Z) p^{s_1} = \exp^p[\mathrm{Coker}\, H^n(f, Z)]$.

Let $\mathcal{U} \in \mathrm{im}\ker H^n(f, Z_p r + s) \xrightarrow{\rho_{r+s,r}} \ker H^n(f, Z_p r)$
say $\mathcal{U} = \rho_{r+s,r}\mathcal{U}_1, \mathcal{U}_1 \in \ker H^n(f, Z_p r + s)$.

Put $\tilde{\mathcal{U}} = \rho_{r+s,r+\hat{s}}\mathcal{U}_1$. As $s - \hat{s} = \hat{s} \le s_0$ by part (1) $\tilde{\mathcal{U}} = \rho_{\infty,r+\hat{s}}\mathcal{U}_0, \mathcal{U}_0 \in H^n(L, Z)$. $0 = \rho_{r+s,r+\hat{s}}H^n(f, Z_p r + s)\mathcal{U}_1 = H^n(f, Z_p r + \hat{s})\tilde{\mathcal{U}} = \rho_{\infty,r+\hat{s}}H^n(f, Z)\mathcal{U}_0$. $H^n(f, Z)\mathcal{U}_0 \in \ker \rho_{\infty,r+\hat{s}}$ and $H^n(f, Z)\mathcal{U}_0 = p^{r+\hat{s}}\omega_0$. Let $\hat{\omega}_0$ be the image of ω_0 in $\mathrm{Coker}\, H^n(f, Z)$. $p^{r+\hat{s}}\hat{\omega}_0 = 0$ implies $\hat{\omega}_0 \in {}^p[\mathrm{Coker}\, H^n(f, Z)]$ and as $\hat{s} \le s_1 p^{\hat{s}}\hat{\omega}_0 = 0$: $p^{\hat{s}}\omega_0 = H^n(f, Z)\tilde{\mathcal{U}}_0$. $\mathcal{U}_0 - p^r\tilde{\mathcal{U}}_0 \in \ker H^n(f, Z)$ and $\rho_{\infty,r}(\mathcal{U}_0 - p^r\tilde{\mathcal{U}}_0) = \rho_{\infty,r}\mathcal{U}_0 = \mathcal{U}$.

5.5. COROLLARY. *If $f : M \to L$ is as in 5.4 then $H^n(f, Z)$ injective implies $\ker H^n(f, Z_p s) \to \ker H^n(f, Z_p)$ is zero.*

5.6. PROPOSITION. *Let $f : M \to L$. Suppose $H^n(f, Z)$ is injective. Consider a $\ker H^n(f, Z_p)$ annihilating process:*

264   A. ZABRODSKY

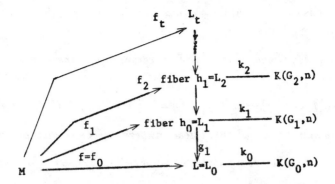

$H^n(k_i, Z_p)$ injective, $G_i = G_i \otimes Z_p \approx \ker H^n(f_i, Z_p) = \operatorname{im} H^n(k_i, Z_p)$. *Then for some $t, t < s$ (s of 5.4(2)) the tower stabilizes, i.e.: $H^n(f_t, Z_p)$ is injective.*

PROOF. Let $F_r = \operatorname{fiber}(L_r \to L_0)$. We shall show inductively that $F_r = K(V_r, n-1)$ for some p-group V_r, the natural map $F_r \to F_{r-1}$ yields an exact sequence

$$O = \to G_{r-1} \xrightarrow{i_{r-1}} V_r \xrightarrow{\tau_r} V_{r-1} \to 0$$

$p^r V_r = 0$ and $\operatorname{im} i_{r-1} \subset p^{r-1}V^r$.

The first observations follow by induction from

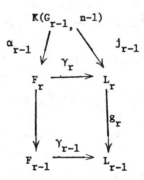

$F_1 \approx K(G_0, n-1), V_0 = 0, V_1 = G_0$. The same holds for the observation $p^r V_r = 0$. By exactness of

$$H^n(L_{r-1} Z_p) \xrightarrow{H^n(g_r, Z_p)} H^n(L_r, Z_p) \xrightarrow{H^n(j_{r-1}, Z_p)} H^n K(G_{r-1}, n-1)$$

if $A_r = \ker H^n(k_r \circ j_{r-1}, Z_p)$ one has a commutative diagram:

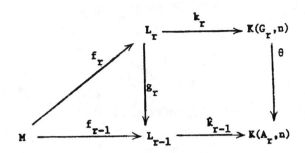

$H^n(\tau, Z_p) : A_r \to G_r$ injective. $* \sim \tau o\, k_r \, o \, f_r \sim \hat{k}_r o \, f_{r-1}$, $\operatorname{im} H^n(\hat{k}_{r-1}, Z_p) \subset$ $\ker H^n(f_{r-1}, Z_p) = \operatorname{im} H^n(k_{r-1}, Z_p) = \ker H^n(g_r, Z_p)$. It follows that $* \sim \hat{k}_{r-1} o\, g_r \sim$ $\tau \, o \, k_r$ and $A_r = \operatorname{im} H^n(\tau, Z_p) \subset \ker H^n(k_r, Z_p) = 0$: Consequently $H^n(k_r \, o \, j_{r-1}, Z_p)$ is injective. One has a commutative diagram:

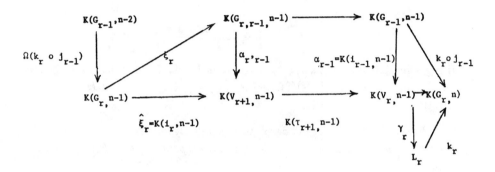

which in turn yields a diagram

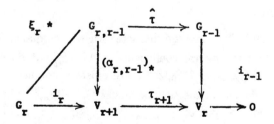

$H^n(k_r o\, j_{r-1}, Z_p)$ is injective and so is $H^{n-1}(\Omega k_r o\, j_{r-1}, Z_p)$, hence, $H^{n-1}(\xi_r, Z_p) = 0 = H_{n-1}(\xi_r, Z_p) = H_{n-1}(\xi_r, Z) \otimes Z_p = \xi_{r*} \otimes Z_p$ and $\operatorname{im} \xi_{r*} \subset p\, G_{r,r-1}$. Suppose by induction $\operatorname{im} i_{r-1 \subset p}{}^{r} - 1_{V_r}$. For any $v \in G_{r,r-1}(\alpha_{r,r-1})_* v = p^{r-1}\hat{v} + i_{r*}\omega_0$, $\omega_0 \in G_r, p^{r-1}\tau_{r+1}\hat{v} = i_{r-1} o \hat{\tau} v$.

Let $\mathcal{U} \in G_r$. $\xi_{r*}\mathcal{U} = p\tilde{\mathcal{U}}, \tilde{\mathcal{U}} \in G_{r,r-1}, i_r \mathcal{U} = (\alpha_{r,r-1*})\xi_{r*}\mathcal{U} = (\alpha_{r,r-1})_* p\tilde{\mathcal{U}} = p(\alpha_{r,r-1})_*\tilde{\mathcal{U}} = p[p^{r-1}\hat{\mathcal{U}} + i_{r*}\hat{\omega}_0] = p^r\hat{\mathcal{U}} + i_{r*}p\hat{\omega}_0 = p^r\hat{\mathcal{U}}.$

Hence, im $i_r \subset p^r V_{r+1}$. $T_r \otimes Z_p$ is an isomorphism, $V_r \otimes Z_p \approx G_0$. Fiber $(L_r \to L_0) = K(V_r, n-1)$ implies that $L_r \to L_0$ is principal and one has

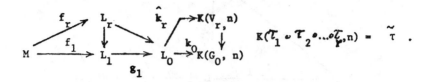

Now if $\sigma : V_r \to Z_p r$ is a projection on a direct summand $K(\sigma, n) o \hat{k}_r : L_0 \to K(Z_p r, n)$ represents an element in $\ker H^n(f_0, Z_p r)$ and $K(\sigma \otimes Z_p, n) o k_r : L_0 \to K(Z_p, n)$ represents its image in $\ker H^n(f_0, Z_p)$:

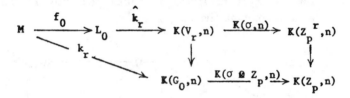

If $r \geq s$ by 5.5, $K(\sigma \otimes Z_p, n) o k_r \sim *$ and as $H^n(k_r, Z_p)$ is injective $\sigma \otimes Z_p = 0 = \sigma$ and V_s has no $Z_p s$ direct factors $p^{s-1} V_s = 0$, $G_{s-1} = 0$ and $H^n(k_{s-1}, Z_p)$ is injective.

5.7. PROPOSITION. *With the 5.2 notations and hypothesis*

(1) for every $n \geq 0$ there exist relatively prime polynomials $^n P_1(x), ^n P_2(x) \in Z[x]$ so that

$$^n P_1[H^m(S_0, G)] = 0, ^n P_1 \bullet^n P_3[H^m(T, G)] = 0 \text{ for } G = Z, Z_p r, m \leq n.$$

(2) If $H^m(f_0, F)$ are injective $F = Z_p, Q, m < n$ then $H^m(f_0, G)$ are mod p injective $G = Z, Z_p r, m < n$. If in addition, $H^n(f_0, Q)$ is injective then $H^n(f_0, Z)$ is mod p injective.

PROOF. (1) Let $^n P_1^{(0)}(x)$ be the characteristic polynomial of $\bigoplus_{m \leq n} H^m(S_0, Q)$. ($^n P_1^{(0)}$ obviously has integral coefficients). As $^n P_1^{(0)}$ divides the characteristic polynomial of $\bigoplus_{m \leq n} H^m(S_0, Z_p)$ there exists a polynomial $^n P_1^{(0)}(x) \in Z[x]$ with leading coefficient prime to p so that $^n P_1^{(p)} \bullet^n R_1^{(0)} =^n \hat{P}_1$ represents the characteristic polynomial of $\bigoplus_{m \leq n} H^m(S_0, Z_p)$ and im$^n \hat{P}_1[H^m(S_0, Z)] \subset$ $^p H(Y_0, Z)$. As $(^n \hat{P}_1)^t$ annihilates $H^m(S_0, Z_p t) m \leq n$ and for $t = \log_p \exp^p H^m(Y_0, Z)$ $H^{m-1}(Y_0, Z_p t) \to^p H^m(Y_0, Z)$ is injective $(^n \hat{P}_1)^{t+1}$ annihilates $H^m(S_0, Z), m \leq n$, and $(^n \hat{P}_1)^{2(t+1)}$ annihilates $H^m(S_0, G), G = Z, Z_p r$ for all r.

Let $^n P_2^{(0)}(x)$ be the characteristic polynomial of $\bigoplus_{m \leq n} \hat{T}_Q^{(m)}$, $\hat{T}_Q^{(m)}$: Coker $H^m(f_0, Q) \to$ Coker $H^m(f_0, Q)$ induced by T, S_0. By hypothesis $^n P_2^{(0)}(x)$ is prime to $^n \hat{P}_1(x)$. Let $^n P_2^{(p)}(x) \in Z[x]$ represent the characteristic polynomial of $\bigoplus_{m\leq n}\hat{T}_{Z_p}^{(m)}, \hat{T}_{Z_p}^{(m)} : \operatorname{Coker} H^m(f_0, Z_p) \to \operatorname{Coker} H^m(f_0, Z_p)$. Then using the arguments leading to $({}^nP_1)^{t+1}$ one has $s_1, s_2 \geq t+1$ so that $[{}^nP_2^{(0)}\,{}^nP_2^{(p)}]^{s_1} \cdot ({}^n\hat{P}_1)^{s_2}$ annihilates $H^m(T, G)$, $m \leq n$, $G = Z, Z_{p^r}$.

Put ${}^nP_2 = [{}^nP_2^{(0)}\bullet^n P_2^{(p)}]^{s_1}$, ${}^nP_1 = ({}^n\hat{P}_1)^{s_2}$.

(2) As ${}^nP_1, {}^nP_2$ are relatively prime for any endomorphism of an abelian group $\varphi : G \to G$ satisfying $({}^nP_1\bullet^n P_2)(\varphi) = 0$ one has $G \approx \ker{}^n P_1\oplus\ker{}^n P_2 = \operatorname{im}{}^n P_2 \oplus \operatorname{im}{}^n P_1$ and if

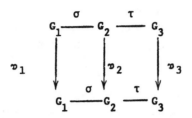

has exact rows, ${}^nP_1 \bullet^n P_2(\varphi_i) = 0$, then σ, τ induce exact sequences

$$\ker{}^n P_i(\varphi_1) \to \ker{}^n P_i(\varphi_2) \to \ker{}^n P_i(\varphi_3).$$

Obviously, $H^m(f_0, G)[H^m(Y_0, G)] \subset \ker{}^n P_1(H^m(T, G))m \leq n$, $G = Z, Z_{p}r$, $\operatorname{im} H^m(f_0, Z_p) = \ker{}^n P_1[H^m(T, Z_p]$. Now assume inductively $H^m(f_0, Z_pr)$ is injective for $m < n, r < r_0$. Using the 4-lemma for

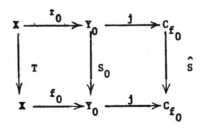

one concludes that $H^m(Y_0, Z_{p_0^r}) \to \ker{}^n P_1[H^m(T, Z_{p^{r_0}})]$ is injective. $H^m(f_0, Q)$ injective implies that $\ker H^m(f_0, Z)$ is finite. There exists r so that

$$\operatorname{torsion}[H^m(Y_0, Z)] \to H^m(Y_0, Z_pr)$$

and $\operatorname{torsion}[H^m(X, Z)] \to H^m(X, Z_pr)$ are mod p injections. Consequently $H^m(f_0, Z)$ is a mod p injective for $m < n$. Now assume $H^n(f_0, Q)$ is injective and let where C_{f_0} is the mapping cone of $f_0, \hat{S}$ induced by T and S_0. $H_n(j, Q) = 0$, $H^n(C_{f_0}, Q) \approx \operatorname{Coker} H^{n-1}(f_0, Q)$ and $^n P_1(H^n(\hat{S}, Q))$ is an isomorphism. It follows that $\ker^n P_1(H^n(\hat{S}, Z)) \subset \operatorname{torsion} H^n(C_{f_0}, Z)$.

Let $x \in \ker H^n(f_0, Z)$, $x = H^n(j, Z)y$ and one may assume $y \in \ker^n P_1(H^n(\hat{S}, Z))$ and for some q prime to p $qy \in^p H^n(\hat{S}, Z)$.

As for some r $\delta : H^{n-1}(C_{f_0}, Z_p r) \to^p H^n(C_{f_0}, Z)$ is injective, $H^{n-1}(f_0, Z_p r)$ injective implies $H^{n-1}(j, Z_p r) = 0$ and consequently $^p H^n(j, Z) = 0$ $0 = H^n(j, Z)qy = qx$.

PROOF OF 5.2. First one uses 5.3 so that one may assume that $H^*(f_0, Q)$ is injective. Now one uses a simple annihilation procedure for $\ker H^*(f_0, Z_p)$: Suppose

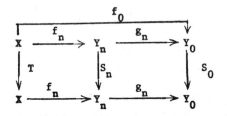

is commutative with $S_{no} f_n \sim f_{no} T$ covering $S_0 f_0 \sim f_{oo} T$. Further, suppose the 5.2 hypothesis hold for f_n, Y_n, S_n replacing f_0, Y_0, S_0 and $\operatorname{im} H^*(f_n, Z_p) = \operatorname{im} H^*(f_0, Z_p), H^m(f_n, Z_p)$ injective for $m < m_n$. Consider

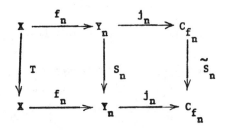

C_{f_n} – the cone on $f_n, \tilde{S}_n$ induced by T, S_n. For every $m \le m_0$ $H^m(j_n, Z_p)$: $\ker^{m_0} P_1(H^m(\tilde{S}_n, Z_p)) \to H^m(Y_n, Z_p)$ is injective and its image is $\ker H^m(f_n, Z_p)$:

$$0 \to \operatorname{Coker} H^{m-1}(f_n, Z_p) \to H^m(C_{f_n}, Z_p) \xrightarrow{H^m(j_n, Z_p)} H^m(Y_n, Z_p) \xrightarrow{H^m(f_n, Z_p)} H^m(X, Z_p) \to$$

$$0 \to \ker {}^{m_0}P_1 = 0 \to \ker {}^{m_0}P_1(H^m(\tilde{S}_n, Z_p)) \to H^m(Y_n, Z_p) \to \ker {}^{m_0}P_1 \ H^m(T, Z_p) \to 0$$

Put $G_{m_n} = \ker^{\hat{m}} P_1(H^{m_n}(\tilde{S}_n, Z_p))$ $(\hat{m} \ge m_n)$ one has

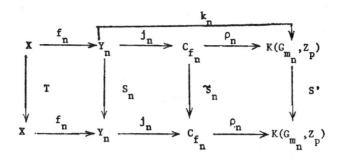

and as in 3.8.4.1 f_n lifts to $f_{n+1} : X \to Y_{n+1} = \text{fiber } k_n$ to obtain

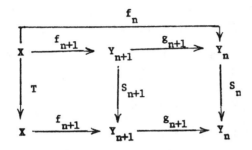

S_{n+1} induced by $S', S_n, S_{n+1} o\, f_{n+1} \sim f_{n+1} T$ covers $S_n o\, f_n \sim f_n o\, T$.

As in the proof of 5.3, the multiplicative closure of characteristic roots of $H^*(S', Z_p)$ is contained in that of $H^*(S_n, Z_p)$ by 1.3. The latter contains the multiplicative closure of characteristic roots of $H^*(S_{n+1}, Z_p)$. The obvious surjections $[\text{Coker } H^m(f_n, Z)]/\text{torsion} \to [\text{Coker } H^m(f_{n+1}, Z)]/\text{torsion}$ Coker $H^m(f_n, Z_p) \to \text{Coker } H^m(f_{n+1}, Z_p)$ imply the validity of the 5.2 hypothesis for $f_{n+1}, Y_{n+1}, S_{n+1}$ replacing f_n, Y_n, S_n. As in the proof of 5.3 using the following

$$
\begin{array}{ccccc}
H^m(Y_{n+1}, Z_p) & \xrightarrow{H^m(f_{n+1}, Z_p)} & H^m(X, Z_p) & \xrightarrow{\rho_m} & \text{Coker } H^m(f_n, Z_p) \\
\downarrow{\scriptstyle H^m(S_{n+1}, Z_p)} & & \downarrow{\scriptstyle H^m(T, Z_p)} & & \downarrow{\scriptstyle \hat{T}_n^{(m)}} \\
H^m(Y_{n+1}, Z_p) & \longrightarrow & H^m(X, Z_p) & \longrightarrow & \text{Coker } H^m(f_n, Z_p)
\end{array}
$$

as the characteristic polynomials of $H^m(S_{n+1}, Z_p)$ and of $\hat{T}^{(m)}$ are relatively prime $\rho_m H^m(f_{n+1}, Z_p) = 0$ and $\operatorname{im} H^*(f_{n+1}, Z_p) \subset \operatorname{im} H^*(f_n, Z_p)$ (the other inclusion is obvious).

Now, by 5.7, $H^{m_n}(f_n, Z)$ is injective so one can apply 5.6 to conclude that $\lim_{n \to \infty} m_n = \infty$. Put $Y = \lim_{\leftarrow} Y_n$, $f = \text{Lim}_{\leftarrow} f_n$.

PROOF OF 5.1. One may assume $^nP_1 \bullet {}^nP_2[H^m(X,G)] = 0$ for $m \le n$, $G = Z, Z_pr$, $r \le r_n = \max_{m \le n} \log_p \exp {}^pH^m(X,Z)$.

Suppose by induction one has a commutative diagram

$$
\begin{array}{ccc}
X & \xrightarrow{\ \ f_{0,n-1}\ \ } & \mathrm{I\!I}K(A_m,m) = Y_{0,n-1} \\
& & m < n \\
T \downarrow & & \downarrow S_{0,n-1} \\
X & \xrightarrow[m<n]{\ \ f_{0,n-1}\ \ } & \mathrm{I\!I}K(A_m,m) = Y_{0,n-1}
\end{array}
$$

So that $[^{n-1}P_1]^{n-1}[H^m(S_{0,n-1},G)] = 0$

$\operatorname{im} H^m(f_{0,n-1},G) = \ker^{n-1} P_1[H^m(T,G)]$ for $m < n$, $G = Z, Z_pr, r \le r_{n-1}$.

Now, $H^{n-1}(X, Z_pr_n) \xrightarrow{\delta} {}^pH^n(X,Z)$ is surjective and consequently so is $^pH^n(Y_{0,n-1},Z) \to \ker^n P_1[^pH^n(T,Z)]$.

Let $A_{n,0} = \ker^n P_1[H^n(T,Z)/\text{torsion}]$ and choose a splitting $A_{n,0} \to \ker^n P_1[H^n(T,Z)]$. Realize this splitting by

$$\tilde{f}_{0,n,0}: X \to K(Z_{n,0},n)$$

If $\tilde{S}_{0,n,0}: K(A_{n,0},n) \to K(A_{n,0},n)$ is induced by $H^n(T,Z)/\text{torsion}$, then $\alpha_n = \tilde{S}_{0,n,0} \circ \tilde{f}_{0,n,0} - \tilde{f}_{0,n,0} - o\,T$ has a finite order and replacing $\tilde{f}_{0,n,0}$ by some prime to p multiple of it one may assume that α_n has order power of p. One can obtain a map $\sigma_n: \prod_{m<n} K(A_m,m) \to K(A_{n,0},n)$ so that the following is commutative:

$$
\begin{array}{ccc}
X & \xrightarrow{\ \hat{f}_{0,n} = (f_{0,n-1} \times \tilde{f}_{0,n,0}) \circ \Delta\ } & \mathrm{I\!I}\,K(A_m,m) \times k(A_{n,0},n) \\
& & m < n \\
T \downarrow & & \downarrow \tilde{S}_{n,0} \\
X & \xrightarrow[m<n]{\hat{f}_{0,n}} & \mathrm{I\!I}K(A_m,m) \times K(A_{n,0},n)
\end{array}
$$

$p_1 \circ \tilde{S}_{n,0} = S_{n-1,0} \circ p_1$, $_2 \circ \tilde{S}_{n,0} = \sigma_n \circ p_1 + \tilde{S}_{0,n,0} p_2$. One can easily argue that $(^{n-1}P_1)^{n-1n}P_1$ (consequently $(^nP_1)^n$) annihilates $\tilde{S}_{n,0}$, $\operatorname{im} H^n(\hat{f}_{0,n},Z) = \ker^n P_1[H^n(T,Z)]$. Now one can complete $A_n = A_{n,0} \oplus A_{n,p}$, $X \to K(A_{n,p}\,n)$ realizing $\ker^n P_1[H^n(T,Z_pr)]$, $r \le r_{n+1} = \log_p \exp^p H^{n+1}(X,Z)$,

$$S_{n,0}: \prod_{m \le n} K(A_m,m) \to \prod_{m \le n} K(A_m,m),$$

$$f_{0,n}: X \to \prod_{m \le n} K(A_m,m), \qquad {}^nP_1(H^m(S_{n,0},G)) = 0$$

$$\operatorname{im} H^m(f_{0,n},G) = \ker^n P_1[H^m(T,G)]$$

$m \leq n$, $G = Z, Z_p$ r $r \leq r_{n+1}$. Put $Y_0 = \lim_{\leftarrow} Y_{0,n}$, $S_0 = \lim_{\leftarrow} S_{0,n}$, $f_0 = \lim_{\leftarrow} f_{0,n}$ to obtain

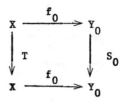

$\operatorname{im} H^m(f_0, Z_p) = \ker^m P_1(H^m(T, Z_p)) = A_m$, $^n P_2(\operatorname{Coker}[H^n(f_0, Z)/\operatorname{torsion}]) = 0$, hence, by the properties of $^n P_1, ^n P_2$ the hypothesis of 5.2 are satisfied and one has

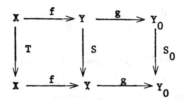

with f, Y, S realizing A.

5.8. Examples:

5.8.1.. Sullivan's sphere groups ([2]):

Let $X = K(Z, 2), T = \lambda 1, \lambda-$ a primitive root of unity mod p. If $k|p-1$ put $^n P_1(x) = \prod_{2ki \leq n}(x - \lambda^{ki})$, $^n P_2(x) = \prod_{\substack{j \not\equiv 0(k) \\ 2j \leq n}}(x - \lambda^j)$. $A = \bigoplus_i H^{2ki}(X, Z_p) \approx$ $Z_p[\omega^k]$ is realizable by a space Y and one has a commutative diagram

$$
\begin{array}{ccc}
K(Z,2) & \longrightarrow & Y \\
\downarrow{\scriptstyle \lambda 1} & & \downarrow{\scriptstyle \psi_\lambda} \\
K(Z,2) & \longrightarrow & Y
\end{array}
$$

$H^{2m}(\psi_\lambda, Z_p) = \lambda^m 1$.

5.8.2.. One can repeat the argument of 5.8.1 with $BSU(n)$ replacing $K(Z, 2)$, $\lambda 1$ replaced by $\psi_\lambda (\lambda = \Pi\, p_i^{r_i}$ $p_i > n$, λ-primitive root of unity mod p)–see [2]. Again let $k|p-1$ then one obtains a space $Y, \hat{\psi}_\lambda : Y \to Y$,

$$
H^m(Y, Z_p) = \begin{cases} H^m(BSU(n), Z_p) & m \equiv 0\,(2k) \\ 0 & \text{otherwise}, \end{cases}
$$

$H^{2t}(\hat{\psi}_\lambda, Z_p) = \lambda^t 1$.

5.8.3.. "Thin Lens Spaces": Consider the lens space $L^{2n+1}(p^r) = S^{2n+1}/Z_p r$. Let $h_\lambda : L^{2n+1}(p^r) \to L^{2n+1}(p^r)$ be induced by

$$g_{\lambda n+1} : S^{2n+1} \to S^{2n+1}, S^{2n+1} = \{z_0, \ldots, z_n \in C^{n+1} | \Sigma|z_i|^2 = 1\}$$

$$g_{\lambda n+1}(z_0, \ldots, z_n) = \frac{z_0^\lambda}{|z_0|^{\lambda-1}}, \ldots, \frac{z_n^\lambda}{|z_n|^{\lambda-1}}.$$

$H^1(h_\lambda, Z_p) = \lambda 1$. Let λ be a primitive root of unity mod p, $k|(n+1, p-1)$. Consider the diagram

$$
\begin{array}{ccc}
X = L^{2n+1}(p^r) & \xrightarrow{f_0} & [\prod_{ki< n} K(Z_p r, 2ki-1)] \times K(Z,2n+1) = Y_0 \\
T = h_\lambda \downarrow & & [\prod (\lambda^{ki}1)] \times \lambda^{n+1}1 = S_0 \quad \downarrow \\
X = L^{2n+1}(p^r) & \xrightarrow{f_0} & [\prod_{ki< n} K(Z_p r, 2ki-1)] \times K(Z,2n+1) = Y_0
\end{array}
$$

$f_0 = [(\prod_{ki<n} \times g_i) \times g_0]\Delta^{n+1/k}$ where $g_i : L^{2n+1}(p^r) \to K(Z_{p^r}, 2ki - 1)$ $g_0 : L^{2n+1}(p^r) \to K(Z, 2n + 1)$ satisfy $H^{2k-1}(g_i, Z_{p^r}), H^{2n+1}(g_0, Z))$– isomorphisms. The multiplicative closure of the characteristic roots of $H^*(S_0, Z_p)$ is the set $\{\lambda^{ik} | 1 \leq i \leq p-1\}$. The characteristic roots of Coker $H^*(f_0, Z_p) \to$ Coker $H^*(f_0, Z_p)$ are λ^m, $m \not\equiv 0(k)$, [Coker $H^*(f_0, Z)$]/ torsion $= 0$ and the hypothesis of 5.2 hold to obtain

$$
\begin{array}{ccccc}
L^{2n+1}(p^r) & \xrightarrow{f} & Y & \xrightarrow{g} & Y_0 \\
h_\lambda = T \downarrow & & s \downarrow & & \downarrow s_0 \\
L^{2n+1}(p^r) & \xrightarrow{f} & Y & \xrightarrow{g} & Y_0
\end{array}
$$

$H^*(f, Z_p)$ (and consequently $H^*(f, Z_p r)$) are injective:

$$H_m(Y, Z) = \begin{cases} Z_{p^r} & m+1 \equiv 0\,(2k), \quad m < 2n-1 \\ Z & m = 2n+1 \\ 0 & \text{otherwise .} \end{cases}$$

For $n = \infty$ one obtains for every $k|p-1$ a space Y with

$$H_m(Y, Z) = \begin{cases} Z_{p^r} & m+1 \equiv 0\,(2k) \\ 0 & \text{otherwise .} \end{cases}$$

One has a converse to 5.7.3:

5.7.3.1.. *Theorem (Peter Hoffman): If a space Y satisfies*

$$H_m(Y, Z) = \begin{cases} Z_{p^r} & m+1 \equiv 0(2k), m \leq r \\ 0 & \text{otherwise .} \end{cases}$$

then $k|p-1$.

5.8.4.. *Let $T : X \to X$. Suppose for every m $\operatorname{Coker}[H^m(T, Z)/\text{torsion}$ is p-torsion free. Then $\bigcap_r \operatorname{im}[H^*(T^r, Z_p)]$ is realizable.*

PROOF. For every m there exists an integer k_m so that $\operatorname{rank} H^m(T^{k_m}, Q) = \operatorname{rank} H^m(T^{k_m+t}, Q)$ for every $t \geq 0$. Let

$$A_m = \operatorname{im}\{[H^m(T, Z)/\text{torsion}]^{k_m}\}$$

and let $f_m : X \to K(A_m, m)$ be a realization of a splitting $A_m \to \operatorname{im}\{H^m(T, Z))]^{k_m}\} \subset H^m(X, Z)$.

The fact that $\operatorname{Coker}[H^m(T, Z)/\text{torsion}]$ is p torsion free implies that

$$\hat{T}^{(m)} = H^m(T, Z)/\text{torsion}\,|A_m : A_m \to A_m$$

is a mod p isomorphism. One can replace f_m by its non zero multiple if necessary to obtain a commutative diagram

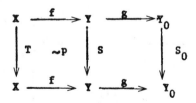

where $G_m = \bigcap_r \operatorname{im} H^m(T^r, Z_p)$, $X \to K(G_m, m)$ a realization of G_m, S_0 induced by $G_m \to G_m$ and $S_{m,0}$.

The above satisfies 5.2 as S_0 is a mod p isomorphism while T induces a nilpotent homomorphism on $\operatorname{Coker} H^m(f_0, Z_p)$ and on $[\operatorname{Coker} H^m(f_0, Z)]/\text{torsion}$. Thus applying 5.2 one obtains:

$$
\begin{array}{ccccc}
X & \xrightarrow{f} & Y & \xrightarrow{g} & Y_0 \\
\downarrow{\scriptstyle T} & {\scriptstyle \sim p} & \downarrow{\scriptstyle S} & & \downarrow{\scriptstyle S_0} \\
X & \xrightarrow{f} & Y & \xrightarrow{g} & Y_0
\end{array}
$$

$f : X \to Y$ is the desired realization.

5.8.5. Using the same line of arguments to the 5.8.4. hypothesis one can obtain a realization of $\bigcup_r \ker H^*(T^r, Z_p)$.

5.8.6. WILKERSON'S RETRACTION THEOREM ([3-2]).

Given $T : X \to X$ *with* $\operatorname{Coker}[H^m(f, Z)/\text{torsion}]$ *p-torsion free for every* m. *For every n there exists a space* Y_n, *maps* $f'_n : X \to Y_n$, $f''_n : Y_n \to X$, $f''_n o\, f'_n \sim$ T^{m_n} *so that for* $m \leq n$ $H^m(f'_n, Z_p)$ *is injective*, $H^m(f''_n, Z_p)$– *surjective*.

PROOF. We shall construct inductively a diagram

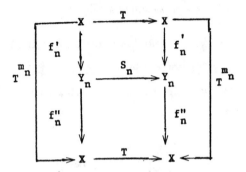

so that for $m \leq n$ $H^m(f'_n, F)$, $H^m(f''_n, F)$ are an injection and a surjection respectively for $F = Z_p, Q$, for every r and group of coefficients G S_n, T induce a nilpotent map on $\operatorname{Coker} H^r(f''_n, G)$. Moreover, f''_n factors as $f''_{n-1}o\, g_n$, $g_n : Y_n \to Y_{n-1}$, $g_n o\, S_n \sim S_{n-1}o\, g_n$ (and obviously $H^m(S_m, F)$ is an isomorphism for $m \leq n$, $F = Z_p, Q$).

Now, for every group G $\ker H^m(S^t_n, G) \subset \ker H^m(S^t_n o\, f'_n, G) = \ker H^m(f'_n o\, T^t, G)$. On the other hand, as

$$\operatorname{Coker} H^m(f''_n, G) \to \operatorname{Coker} H^m(f'_n, G)$$

is nilpotent, for every $x \in H^m(Y_n, G)$ there exists $t = t(G, m, n)$ so that $H^m(S^t_n, G)x \in \operatorname{im} H^m(f'_n, G)$. Let $x \in \bigcup_t \ker H^m(f'_n o\, T^t, G)$ and say $x \in \ker H^m(f'_n o\, T^{t_0}, G)t_0 \geq t(G, m, n)$, hence, $H^m(S^{t_0}_n, G)x = H^m(f'_n, G)y$ thus $y \in \ker[H^m(f'_n, G)o\, H^m(f''_n, G)] = \ker H^m(T^{m_n}, G)$, $x \in \ker H^m(S^{t_0+m_n}_n, G)$: It follows that

$$\bigcup_t \ker H^m(f'_n o\, T^t, G) = \bigcup_t \ker H^m(S^t_n, G).$$

For every m let $\alpha_m : Y_n \to K(A_m, m)$ realize $\bigcup_t \ker H^m(S^t_n, F)$, $F = Z_p, Q$, one can complete a commutative diagram

(5.8.6.1)

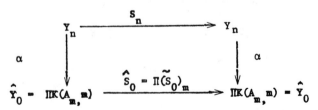

$(\tilde{S}_0)_m$ nilpotent, $S_n, \hat{S}_0$ induce isomorphisms

$$\operatorname{Coker} H^*(\alpha, F) \to \operatorname{Coker} H^*(\alpha, F), F = Z_p, Q .$$

For every m one has an injection

$$\operatorname{Coker} H^m(\alpha, F) \to H^m(X, F) \quad F = Z_p, Q:$$
$$H^m(Y_n, F) \to \operatorname{Coker} H^m(\alpha, F) \to H^m(X, F)$$
$$H^m(f'_n o\, T^{tm}, F)$$

and S_n, T induce an injection

$$\hat{S}^{(m)} : [\operatorname{Coker} H^m(\alpha, Z)]/\text{torsion} \to [\operatorname{Coker} H^m(\alpha, Z)]/\text{torsion}$$

with p-torsion free cokernel, hence $\hat{S}^{(m)}$ is a mod p isomorpohism. Thus (5.8.6.1) satisfies the hypothesis of 5.2 and one obtains

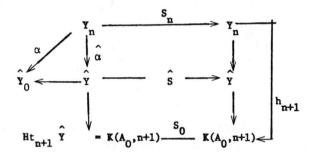

$H^m(h_{n+1}, G)$ is a mod p injection for $m = n+1, n+2$ $G = Z, Z_p r, Q$, $\operatorname{im} H^{n+1}(h_{n+1}, G) = \bigcup_t \ker H^{n+1}(S_n^t, G) = \bigcup_t \ker H^{n+1}(f'_n o\ T^t, G) = \ker H^{n+1}(f'_n o\, T^{t(G)}, G)$. Put $t_0 = t(A)$ to obtain (5.8.6.2)

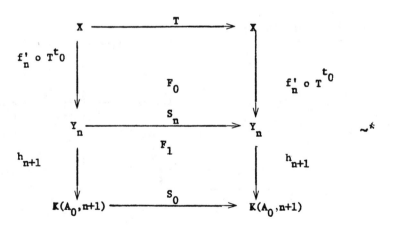

Now let $g_{n+1} : Y_{n+1} \to Y_n$ denote the (inclusion of) the fiber of h_{n+1}. Then one has a lifting of $f'_n o\ T^{ta} : X \xrightarrow{f_{n+1}} Y_{n+1} \xrightarrow{g_{n+1}} Y_n$ and as $H^{n+2}(h_{n+1}, Z_p)$ is injective so is $H^{n+1}(f'_{n+1}, Z_p)$. Now S_0 nilpotent implies that so is $\operatorname{Coker} H^m(g_{n+1}, G) \to \operatorname{Coker} H^m(g_{n+1}, G)$ (induced by S_0, S_n and by their induced $S_{n+1} : Y_{n+1} \to Y_{n+1}$). Indeed, the last observation follows from a simple

Serre spectral sequence argument: the maps ΩS_0, S_n, S_{n+1} induce a spectral sequence morphism $E_r^{s,t} \to E_r^{s,t}$ which is nilpotent on $\bigoplus_{t>0} E_r^{s,t}$ hence on $\bigoplus_{t>0} E_\infty^{s,t}$. The latter represents the associated graded module of $\text{Coker } H^*(g_{n+1}, G)$. As for each m, this filtration induces a finite filtration on $\text{Coker } H^m(g_{n+1}, G)$ our assertion follows.

It remains to show that one has a commutative diagram

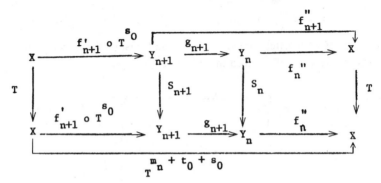

or equivalently if we consider (5.8.6.2) with $t_0 + s_0$ replacing t_0 (and possibly altering h_{n+1} without changing its mod p effects) one can choose $\hat{\ell} : * \sim h_{n+1} o f'_n o T^{t_0 + s_0}$, $\hat{F}_0 : f'_n o T^{t_0 + s_0 + 1} \sim S_n o f'_n o T^{t_0 + s_0}$, $\hat{F}_1 : h_{n+1} o S_n \sim S_0 o h_{n+1}$ with $\alpha(\hat{\ell}, F_0, F_1) = 0$ (see 3.8).

Now, $\text{im } H^n(T^{mn}, G) \subset \text{im } H^n(f'_n, G)$ with cokernel being torsion group of order prime to p and as $\text{Coker}[H^n(T, Z)/\text{torsion}]$ is p torsion free there exists s_0 so that $\text{im } H^n(T^{s_0}, G)/\text{im } H^n(T^{s_0+r}, G)$ is a torsion group of order prime to p too.

Thus, for some prime to p integers

$q_1, q_2 \qquad q_1 q_2 \text{ im } H^n(T^{s_0}, A_0) \subset q_2 \text{ im } H^n(T^{s_0+mn}, A_0) \subset \text{im } H^n(f'_n o T^{s_0}, A_0)$.

Extending (5.8.2) as follows

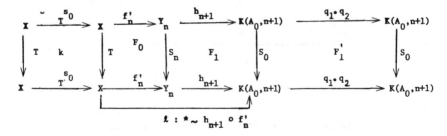

one obtains $\alpha(\mathcal{L}(q_1 q_2 \cdot 1) o \ell o T^{s_0}, F_0 * k, F'_1 * F_1) = q_1 q_2 [H^n(T^{s_0}, A) \alpha(\ell, F_0, F_1)] \in \text{im } H^n(f'_n o T^{s_0, A_0})$ and one can alter $F'_1 * F_1$ by $\omega : Y_n \to \Omega K(A_0, n+1) \cdot H^n(f_n o T^{s_0}, A_0)\omega = \alpha(\mathcal{L}(q_1 q_2 1) o \ell o T^{s_0}, F_0 * k, F'_1 * F_1)$ to obtain $\hat{F}_1$ with $\alpha(\mathcal{L}(q_1 q_2 1) o \ell o T^{s_0}, F_0 * k, \hat{F}_1) = 0$. This completes the inductive construction of Y_n, f'_n, f''_n. Obviously one has $H^m(f'_n, Z_p)$ injective, $\text{im } H^m(f'_n, Z_p) = \text{im } H^m(T^{mn}, Z_p) \, m \leq n$ and consequently $H^m(f''_n, Z_p)$ is injective, $m \leq n$.

Using the 5.8 construction, one can pass to a limit $Y = \lim_{\leftarrow} Y_n$ $\quad g : Y \to X$ induced by g_n. Obviously $\ker H^m(g, Z_p) = \bigcup_t \ker H^m(T^t, Z_p)$ and putting 5.8.4 and 5.8.6 together one obtains

5.9. COROLLARY. *Let $T : X \to X$, $\mathrm{Coker}[H^*(T, Z)/\text{torsion}]$ p torsion free (X not necessarily restricted to have finite dimensional homology or homotopy groups.) Then $\bigcap_t \mathrm{im}\, H^*(T^t, Z_p)$ is realizable by a mod p retract: $\tilde{Y} \xrightarrow{g} X \xrightarrow{f} Y$ g of 5.8, f of 5.7.4 $H^*(f \circ g, Z_p)$ an isomorphism.*

5.10. Remarks: In general one cannot expect the existence of a mod p equivalence $Y \to \tilde{Y}$. After localization, an inverse obviously exists but one cannot be certain of the existence of a diagram

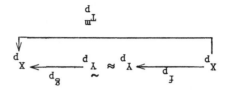

unless Y_p satisfies some finiteness conditions.

If X is finite dimensional (or having finite dimensional homotopy graded group) and if it rationally an H or a co-H space not only a (5.10.1) diagram exists, but also restriction on T is inessential as T^m can be replaced by $\hat{T}$ satisfying the 5.7.4, 5.7.5-5.8 -5.9 restrictions and $H^*(\hat{T}, Z_p) = H^*(T^m, Z_p)$.

REFERENCES

[1] E. Dror and A. Zabrodsky, *Unipotency and nilpotency in homotopy equivalences*, Topology **18**, (1979), 187–197.

[2] D. Sullivan, *Geometric topology*, MIT Notes.

[3-1] C. Wilkerson, *Applications of minimal simplical groups*, Topology **15**, (1976), pp. 111–130.

[3-2] C. Wilkerson, *Genus and Cancellation*, Topology **17**, (1975), pp. 29–36.

[4-1] A. Zabrodsky, *Hopf spaces*, North Holland Math Study **22**, (1976).

[4-2] A. Zabrodsky, *Power spaces*, (Mimeographed).

[4-3] A. Zabrodsky, *On rank 2 mod odd H-spaces*, London Math. Soc., Lecture Notes **12**, (1974), pp. 119–128.

[4-4] A. Zabrodsky, *P-equivalences and homotopy type*, Springer Lecture notes **418**.

[4-5] A. Zabrodsky, *Secondary operations in the module of indecomposables*, Proc. Adv. Study Inst. on Algebraic topology, Aarhus, (1970), pp. 657–672.

HEBREW UNIVERSITY

FOUR MANIFOLDS

Contemporary Mathematics
Volume **44**, 1985

ON FAKE $S^3 \tilde{\times} S^1 \# S^2 \times S^2$

SELMAN AKBULUT*

In [A] we have constructed a fake $S^3 \tilde{\times} S^1 \# S^2 \times S^2$, that is a closed smooth manifold M^4 which is simple homotopy equivalent to $S^3 \tilde{\times} S^1 \# S^2 \times S^2$ but not diffeomorphic to it. In fact, they are not even normally cobordant to each other. Here $S^3 \tilde{\times} S^1$ denotes the twisted S^3 bundler over S^1. It is interesting to note that by [FQ] M^4 is topologically standard. Here we announce a curious property of M^4; i.e. M^4 is obtained by twisting $S^3 \tilde{\times} S^1 \# S^2 \times S^2$ along an imbedded 2-sphere (Gluck construction). This implies that $M \# \mathbb{C}P^2$ is standard. The proof of this fact is long; it will appear elsewhere. Instead here we specifically identify this imbedding of 2-sphere by drawing a picture. We end the paper by giving a quick alternative construction of a fake $S^3 \tilde{\times} S^1 \# S^2 \times S^2$. We would like to thank Larry Taylor for useful conversations on surgery theory.

§1. Notations

Recall that we can visualize the handlebody structures of connected smooth 4-manifolds by positioning ourselves at the boundary of the zero handle $(= S^3)$. We see the attaching $S^0 \times B^3$ of a 1-handle $B^1 \times B^3$ as a pair of balls, and the attaching $S^1 \times B^2$ of a 2-handle $B^2 \times B^2$ as a framed knot. The framed knots are allowed to go over the 1-handles, hence, we only see the part of them which lie in S^3, i.e., arcs entering one of the balls and leaving from the others as in Figure 1. The framings are determined by the vector field on the plane of the paper (plus the number of twists added when indicated as in Figure 1.)

There are many ways to attach 1-handles, here we use only one of them; namely, if we visualize coordinate axis at the centers of the balls, they are identified via $B^1 \times B^3$ by the map $(x, y, z) \to (x, -y, -z)$. Since 3 and 4- handles are attached the standard way, we don't need to visualize them. For more systematic description of 4-dimensional handlebodies, we refer the reader to [A].

*Sloan Fellow, and supported in part by N.S.F.

§2. Gluck construction

We start with the handlebody picture of $S^3 \tilde{\times} S^1 \#\ S^2 \times S^2$ which is $B^3_+ \tilde{\times} S^1 \# S^2 \times S^2 \underset{\partial}{\cup} B^3_- \tilde{\times} S^1$. We draw $B^3_+ \tilde{\times} S^1 \# S^2 \times S^2$ in Figure 2. Now, we describe an imbedding of S^2 into $S^3 \tilde{\times} S^1 \# S^2 \times S^2$ as follows. Figure 3 is a picture of an imbedding $D^2_+ \hookrightarrow B^3_+ \tilde{\times} S^1 \# S^2 \times S^2$ (the shaded disc) such that $\gamma = \partial D^2_+$ is an unknot on the boundary $(= S^2 \tilde{\times} S^1)$. One can verify the last claim by simply

cancelling $\overset{\circ}{\mathcal{C}}\overset{\circ}{\mathcal{C}}$ $(= S^2 \times S^2)$ by surgering one of the 2-spheres and tracing γ in the picture. Since γ is an unknot, it bounds a trivial 2-disc D^2_- in $B^3_- \tilde{\times} S^1$. Then let $K^2 = D^2_+ \underset{\partial}{\cup} D^2_-$, K^2 is an imbedded 2-sphere in $S^3 \tilde{\times} S^1 \# S^2 \times S^2$.

THEOREM 1. M^4 is obtained by doing the Gluck construction to $S^3 \tilde{\times} S^1 \# S^2 \times S^2$ along K. Namely,

$$M^4 = (S^3 \tilde{\times} S^1 \# S^2 \times S^2 - \text{int}(K \times D^2)) \underset{\varphi}{\cup} K \times D^2$$

where $K \times D^2$ is the closed tubular neighborhood of K, and φ is the self diffeomorphism of $\partial(K \times D^2) = S^2 \times S^1$ given by $(x,y) \to (\alpha(y)x, y)$ with $\alpha \in \pi_1 SO_3$ is the nontrivial element.

Figure 4 is the handlebody picture of M^4.

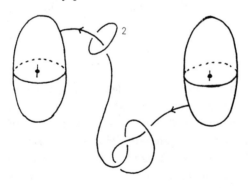

FIGURE 1

FIGURE 2

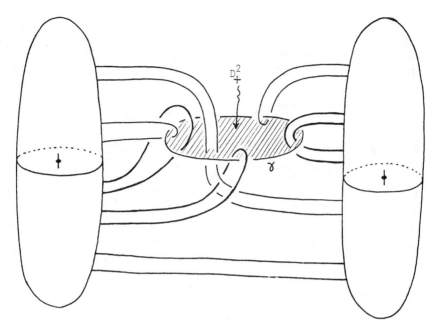

FIGURE 3

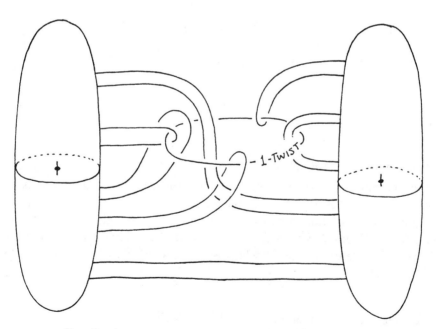

Here the above notation means that all the arcs going through γ are twisted once with a full right handed twist.

FIGURE 4

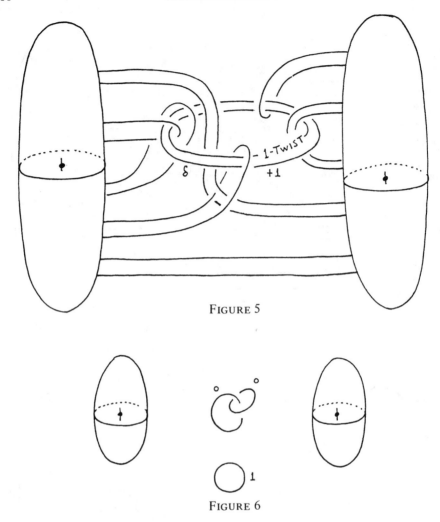

FIGURE 5

FIGURE 6

COROLLARY 2. $M\#\mathbb{C}P^2 = S^3\tilde{\times}S^1\#S^2\times S^2\#\mathbb{C}P^2$.

PROOF. Since ∂D_+ is an unknot in $S^2\tilde{\times}S^1$ the circle δ in Figure 5 is also an unknot in $S^2\tilde{\times}S^1$. $M\#\mathbb{C}P^2$ is obtained by attaching a 2-handle to an unknot of Figure 4 with $+1$ framing. Hence, if we attach a 2-handle with $+1$ framing along δ as in Figure 5, we get $M\#\mathbb{C}P^2$; but then by sliding all 2-handles over this new 2-handle gives Figure 6 which is (along with $B^3_-\tilde{\times}S^1$) is $S^3\tilde{\times}S^1\#S^2\times S^2\#\mathbb{C}P^2$. $\square$

§3. A quick construction of a fake $S^3\tilde{\times}S^1\#S^2\times S^2$

By attaching two 2-handles to B^4 we can obtain a smooth W^4, with $W\simeq S^2\times S^2 - B^4$ and $\partial W = \Sigma^3\#\Sigma^3$, where Σ^3 is a Rochlin invariant 8 homology sphere. To see this, start with the 4-manifold N^4 of Figure 7. $\partial N = \Sigma^3\#\Sigma^3$, where $\Sigma^3 = \Sigma(2,3,7)$ is the Rochlin invariant 8 homology sphere which is the link of the singularity $z_1^2 + z_2^3 + z_3^7 = 0$ in $\mathbb{C}^3$. Attach 2-handles to N^4 as in Figure 8 to obtain $Q^4 = N^4 \cup h_1^2 \cup h_2^2$ with $\partial Q = S^3$. The fact that $\partial Q = S^3$ can be seen by the handle slides indicated in Figure 9. To obtain W^4, we start with B^4 and attach the dual 2-handles of h_1^2 and h_2^2 to ∂B^4, then clearly $\partial W = \partial N = \Sigma \# \Sigma$ and $W \simeq S^2 \times S^2 - B^4$ (check framings).

By attaching a 3-handle H^3 to W along the obvious S^2 of ∂W, we get $W_1 = W \cup h^3$ with $\partial W_1 = \Sigma^3 \cup \Sigma^3$. Let Q^4 be the closed manifold obtained by identifying the two boundary components of W_1.

THEOREM 3. Q^4 is a fake $S^3 \tilde{\times} S^1 \# S^2 \times S^2$.

PROOF. From the obvious homotopy equivalence $W_1 \to S^3 \times I \# S^2 \times S^2$, we get a simple homotopy equivalence $f : Q^4 \to S^3 \tilde{\times} S^1 \# S^2 \times S^2$ with the property that Σ^3 is the transverse inverse image of a fibre $S^3 \subset S^3 \tilde{\times} S^1 \# S^2 \times S^2$. Consider the map $\overline{f} : Q \to S^3 \tilde{\times} S^1$ obtained by pinching $S^2 \times S^2$ to a point; then $(Q, \overline{f}) \in [S^3 \tilde{\times} S^1; G/0]$ which is the normal cobordism classes of degree 1 normal maps to $S^3 \tilde{\times} S^1$. According to [CS] $\alpha(Q, \overline{f}) = 2\mu(\Sigma) -$

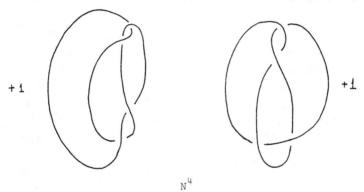

$+1$ $+1$

N^4

FIGURE 7

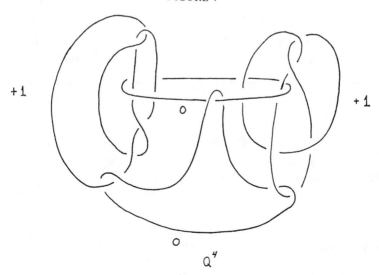

$+1$ $+1$

0

Q^4

FIGURE 8

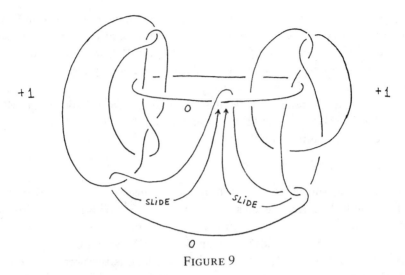

+1   +1

SLIDE   SLIDE

0

FIGURE 9

$\text{Sign}(W_1)(\text{mod } 32) = 16(\text{mod } 32)$ is an invariant of $[S^3 \widetilde{\times} S^1; G/0]$, where $\mu(\Sigma)$ denotes the Rochlin invariant of Σ, and $\text{Sign}(W)$ denotes the signature of W.

We claim that Q cannot be s-cobordant to $S^3 \widetilde{\times} S^1 \# S^2 \times S^2$. If there was an s-cobordism H^5 with $\partial H = H_+ \cup H_-$, $H_- = Q$ and $H_+ = S^3 \widetilde{\times} S^1 \# S^2 \times S^2$; by attaching a 3-handle to H_+, we get a cobordism $\overline{H}$ with $\partial \overline{H} = Q \cup S^3 \widetilde{\times} S^1$. Since $\pi_2(S^3 \widetilde{\times} S^1) = 0$ the map $\overline{f}$ extends to a degree 1 normal map $F : \overline{H} \to S^3 \widetilde{\times} S^1$. Let $F|S^3 \widetilde{\times} S^1 = h$, then $\alpha(S^3 \widetilde{\times} S^1, h) = \alpha(Q, \overline{f}) \neq 0$. Let L^3 be the framed manifold which is a transverse inverse image of S^3 under h, let C be a closed tubular neighborhood of L^3 and let $Z^4 = S^3 \widetilde{\times} S^1 - \text{int } (C)$. By Mayer-Vietoris sequence of Z and C we get $H_2(\partial Z) \to H_2(Z)$ onto which implies that $\text{Sign } (Z) = 0$. By taking the universal cover $S^3 \times \mathbb{R}$ of $S^3 \widetilde{\times} S^1$ we get an imbedding of L^3 into $S^3 \times I$, L^3 separates $S^3 \times I$ into two parts N_1 and N_2. Then by applying Mayer-Vietoris to N_1, N_2 we get $H_2(\partial N_1) \to H_2(N_1)$ onto, hence $\text{Sign } (N_1) = 0$. Therefore $\mu(L^3) = \text{Sign}(N_1) (\text{mod} 16) = 0$. By [CS] we compute

$$\alpha(S^3 \widetilde{\times} S^1, h) = 2\mu(L^3) - \text{Sign } (N) \quad (\text{mod } 32)$$
$$= 0 \qquad\qquad\qquad \text{contradiction} \quad \square$$

Remark: The proof also shows that $(Q, \overline{f})$ cannot be normally cobordant to $(S^3 \widetilde{\times} S^1 \# S^2 \times S^2, \overline{id})$.

REFERENCES

[A]   S. Akbulut, *A fake 4-manifold*, to appear, Proceedings on 4-manifolds, *A.M.S. Summer Research Conference series* (1982).

[CS] S. Cappel and J. Shaneson, *Some new 4-manifolds*, Ann. of Math., **104** (1976), 61–72.

[FQ] M. Freedman and F. Quinn, *Topology of 4-manifolds*, to appear, Ann. of Math. series.

MICHIGAN STATE UNIVERSITY

Contemporary Mathematics
Volume 44, 1985

MANIFOLD HAVING NON-AMPLE NORMAL BUNDLES IN QUADRICS

Norman Goldstein

Let $Z \subset \mathbb{P}_{\mathbb{C}}^{r+1}$ be a smooth r-dimensional quadric;

$$Z = \{z \in \mathbb{P}^{r+1} : zQz = 0\}$$

for some matrix $Q \in GL(r+2, \mathbb{C})$. For $a, b \in \mathbb{C}^{r+2}, Q$ defines a pairing $aQb = \Sigma Q_{ij} a_i b_j$.

The orthogonal group,

$$G = O(r+2, \mathbb{C}) = \{g \in GL(r+2, \mathbb{C}) : g^t Q g = Q\}$$

acts transitively on Z. This action determines a finite number of holomorphic vector fields $v_0, \ldots, v_N$ that span TZ, i.e., $v_0(z), \ldots, v_N(z)$ span $T_z Z$ for each $z \in Z$. These induce a map

$$\varphi : T^* Z \to \mathbb{C}^{N+1}$$

$$T_z^* Z \quad \alpha \mapsto (\alpha(v_0(z)), \ldots, \alpha(v_N(z)))$$

or, also, $\varphi : \mathbb{P}(T^* Z) \to \mathbb{P}^N$, which we call the ampleness map. In other guises, and for various (homogenous spaces), Z, this map has been studied for many years; e.g. Springer [10], Steinberg [11], Borho-Kraft [1], Guillemin-Sternberg [5], and a paper of mine [2].

Let $Y \subset Z$ be a smooth, connected, complex submanifold. The normal bundle, NY, of Y is defined by the sequence

$$0 \to TY \to TZ|_Y \to NY \to 0 \,.$$

Let φ' be the restriction of φ to $\mathbb{P}(N^* Y) \subset \mathbb{P}(T^* Z)$. It is a fact (e.g. Sommese [8, remark 1.4.1]) that the vector bundle NY is ample precisely when φ' is finite to 1.

I won't elaborate as to why one *cares* whether NY is ample. Let me just say that the hypothesis of ampleness is used to obtain results on, for example, Lefschetz theorems (Sommese [9,§3]), and bounds on cohomological dimension, Hartshorne [6, Ch. 7]. In short, one would like to know when NY is not ample.

Example. Let V be a subvariety of $\mathbb{P}^{r+1}$ that meets Z transversely (in particular, $Z \cap \text{sing}(V) = \varnothing$), and let Y be a connected component of $V \cap Z$.

Then $NY = N_Z Y$ is ample; indeed, $N_Z Y \simeq (N_{\mathbb{P}^{r+1}} V)|_Y$ which is ample (in fact, *very* ample).

Let A be any subset of $\mathbb{P}^{r+1}$, and put $A^\perp = \{x \in \mathbb{P}^{r+1} : xQa = 0 \ \forall a \in A\}$. Concretely, $A^\perp$ is a linear $\mathbb{P}^{r-k}$ when A spans a $\mathbb{P}^k$. Also, for any submanifold $Y^m \subset \mathbb{P}_{\mathbb{C}}$, let $\mathbb{P}_y Y$ denote the linear $\mathbb{P}^m$ tangent to Y at y. A line in $\mathbb{P}_{\mathbb{C}}$ refers to a linear P^1.

PROPOSITION. *Let $Y^m \subset Z \subset \mathbb{P}^{r+1}$ be a smooth subvariety of the quadric Z. Then NY is not ample if and only if there is a line $\ell \subset Y$ satisfying*

$$\mathbb{P}_y Y \subset \ell^\perp \text{ for each } y \in \ell.$$

Call such a line an ampleness line for Y.

Example. Let Y be a linear $\mathbb{P}^m, m \geq 1$, in Z. Then NY is *not* ample. In fact, any line $\ell \subset Y$ is an ampleness line for Y.

Next, let's "reduce" the search for Y's with non-ample normal bundles, to considering only surfaces. Let $Y^m, m \geq 3$, be a submanifold of Z^r having a non-ample normal bundle. Let $\ell \subset Y$ be an ampleness line. Then, one can choose a hyperplane H in $\mathbb{P}^{r+1}$ that contains ℓ, and which meets both Y and Z transversely (because $m \geq 3$). Now, $Y \cap H$ is smooth, and has a non-ample normal bundle in the quadric $Z \cap H$; in fact, ℓ is, also, an ampleness line for $Y \cap H$. By repeating this process, one is led to considering surfaces in Z^{r-m+2}. Given the strong geometric relationship between Y and $Y \cap H$ (e.g. the Lefschetz theorem on hyperplane sections), there is every reason to expect that if one could describe all surfaces in Z^{r-m+2} having a non-ample normal bundle, then one could, also, describe all submanifolds Y^m of Z^r having a non-ample normal bundle.

The smallest non-trivial case to consider is of surfaces in the 4-quadric. As explained earlier, any linear $\mathbb{P}^2 \subset Z^4$ has a non-ample normal bundle. Panantonopoulou [7] states that these are the only examples for Z^4. However, below, I would like to describe some other surfaces in Z^4 having non-ample normal bundles.

Let $\mathcal{Q}^1 \subset \mathbb{P}^3$ be a smooth quadric curve, and $\kappa \in \mathbb{P}^3$, not lying in the plane containing $\mathcal{Q}$. For each $t \in \mathcal{Q}$, let $\mathbb{P}_t^2$ be the linear $\mathbb{P}^2 \subset \mathbb{P}^3$, spanned by $\mathbb{P}_t \mathcal{Q}$ and κ. We view $Z = Gr(1, \mathbb{P}^3)$, via the Plücker embedding. For each $t \in \mathcal{Q}$ let $\ell_t = \{\mathbb{P}^1 \subset \mathbb{P}^3 : t \in \mathbb{P}^1 \subset \mathbb{P}_t^2\}$, which is a line in $\mathbb{P}^5$, and a Schubert variety in Z. Let $X_4 = \cup \{\ell_t : t \in \mathcal{Q}\}$. Then $X_4 \simeq \mathbb{P}^1 \times \mathbb{P}^1$ has degree 4 in $\mathbb{P}^5$, and each line ℓ_t is an ampleness line for X_4. In fact, if X is any surface in Z having an infinity of ampleness lines, then either X is a linear $\mathbb{P}^2$, or $X = g(X_4)$ for some $g \in SO(6, \mathbb{C})$; a proof of this fact is in my paper [3].

Another example is $X_3 \simeq \mathbb{P}^2$ with one point blown up, and is a surface of degree 3. There is only one ampleness line; it is one of the lines of the ruling of X_3, and is *not* the exceptional divisor.

The last example that I will describe is $X_6 \simeq \mathbb{P}^2$ with 3 general points blown up, a surface of degree 6. There are 6 ampleness lines; they form the hexagon that arises in the blowing up.

I know of one other example, which I found after discussing torus embeddings with Jonathan Fine. I do not have a list of all possible examples, but I would like to finish this talk by explaining where one can look in order to find them all.

An irreducible hypersurface $Y^3 \subset Z = Gr(1, \mathbb{P}^3) \subset \mathbb{P}^5$ is given by a homogeneous polynomial $f(z_0, \ldots, z_5) = 0$. Following Green and Morrison [4], we call Y a Chow form if

$$\frac{\partial f}{\partial z_0} \frac{\partial f}{\partial z_5} - \frac{\partial f}{\partial z_1} \frac{\partial f}{\partial z_4} + \frac{\partial f}{\partial z_2} \frac{\partial f}{\partial z_3} = 0$$

identically, on Y. Equivalently, either there is a surface $S \subset \mathbb{P}^3$ such that Y consists closure of $\{\mathbb{P}^1 \subset \mathbb{P}^3 : \mathbb{P}^1$ is tangent to S at a smooth point of $S\}$, or else there is a curve $C \subset \mathbb{P}^3$ such that Y is the closure of $\{\mathbb{P}^1 \subset \mathbb{P}^3 : \mathbb{P}^1$ meets $C\}$. From this geometric characterization of Chow forms, we see that Y contains an infinity of lines; lets call them *special lines* for Y. A surface $X \subset Z$ with a non-ample normal bundle can arise in 2 ways:

A) X is a component of the singular set of some Chow form Y, or

B) X is a component of the intersection of two Chow forms Y_1 and Y_2.

In A), X contains a special line of Y, and in B), X contains a line which is special for Y_1 and for Y_2. In each case, this is an ampleness line for X. An example of A) is X_6; and X_3 is an example of B). In fact, except for linear $\mathbb{P}^2$'s and X_4, this expression for X in terms of Chow forms is unique.

REFERENCES

[1] Borho, W., Kraft, H.: *Über Bahnen und deren Deformationen bei linearen Aktionen reductiver Gruppen*, Comment. Math. Helvetici **54**, 61–104 (1979).

[2] Goldstein, N.: *Ampleness and connectedness in complex G/P*. Trans. Amer. Math. Soc., 274, 367–373 (1982).

[3] Goldstein, N.: *A special surface in the 4-quadric*. Preprint.

[4] Green, M., Morrison, I.: *The equations defining Chow varieties*. Preprint.

[5] Guillemin, V., Sternberg, S.: *Convexity properties of the moment mapping*. Inv. Math. **67**, 491–513 (1982).

[6] Hartshorne, R.: *Cohomological dimension of algebraic varieties*. Ann. of Math. **88**, 403–450 (1968).

[7] Papantonopoulou, A.: *Surfaces in the Grassman variety G(1,3)*. Proc. Amer. Math. Soc. **77**, 15–18 (1979).

[8] Sommese, A.: *Submanifolds of Abelian varieties*. Math. Ann. **233**, 229–256 (1978).

[9] Sommese, A.: *Complex subspaces of homogeneous complex manifolds II—homotopy results.* Nagoya Math. J. **86**, 101–129 (1982).

[10] Springer, T.: *The unipotent variety of a semisimple group.* Algebraic Geometry *(papers presented at the Bombay Colloquium 1968)* 373–391, Tata Institute, 1969.

[11] Steinberg, R.: *On the desingularization of the unipotent variety.* Inv. Math. **36**, 209–224 (1976).

PURDUE UNIVERSITY

Contemporary Mathematics
Volume 44, 1985

LEFSCHETZ FIBRATIONS OF RIEMANN SURFACES
AND DECOMPOSITIONS OF COMPLEX ELLIPTIC SURFACES

RICHARD MANDELBAUM*

Introduction

Let M be a simply connected smooth compact 4-manifold. Then as a consequence of results of Whitehead [Wh] and Milnor [Mil] we know that the congruence class of the intersection form L_M of M determines the homotopy type of M. Thus such manifolds M_1, M_2 are homotopy equivalent if and only if L_{M_1} is congruent to L_{M_2}. These results were extended by Wall [W] to show that: (1) for simply connected 4-manifolds homotopy equivalent implies h-cobordant, and (2) M_1 h-cobordant to M_2 implies that for some integer $k \geq 0$, $M_1 \# k(S^2 \times S^2) = M_2 \# k(S^2 \times S^2)$, where $\#$ is 'connected sum' and $=$ is diffeomorphic (of course if the smooth h-cobordism theorem were true in dimension 4 then we could conclude that $k = 0$ above.)

For our purposes it is convenient to consider a slight variation of Wall's result (2) above. It is well-known that $S^2 \times S^2 \# P = 2P \# Q$ where P is CP^2 with its usual orientation and Q is CP^2 with reversal orientation. Thus we deduce that if M_1 is h-cobordant to M_2 then there exist integers $k_1 \geq 0$, $k_2 \geq 0$ such that

$$(3) \qquad M_1 \# k_1 P \# k_2 Q = M_2 \# k_1 P \# k_2 Q .$$

Now although it is not true that any simply-connected 4-manifold X is even homotopy equivalent to a connected sum of P's and Q's (for example neither $S^2 \times S^2$, nor the $K3$ surface is homotopy equivalent to such a connected sum) it follows readily from the classification theory of quadratic forms [Ser] over $\mathbb{Z}$ that either $X \# P$ or $X \# Q$ *is always* homotopy equivalent to such a connected sum. Thus we can combine this classification theory with (3) above to obtain:

Proposition

Let M be a smooth simply connected compact 4-manifold. Then there exists $k_1 \geq 0$, $k_2 \geq 0$, $a \geq 0$, $b \geq 0$ such that

$$(4) \qquad M_1 \# k_1 P \# k_2 Q = aP \# bQ .$$

If we can take $k_1 = k_2 = 0$ in (4) above we say that M is completely decomposable. In the papers [Man 1,2; MM 1,2] we have shown that for a wide variety

---

*Author partially supported by NSF grant MCS77-04165.

of complex surfaces V, $V \# P$ is completely decomposable. (We call such a V almost completely decomposable (ACD).)

What happens if M is not simply connected. Things, naturally, become more complicated. For example, it is no longer necessarily true that the homotopy type of the manifold is determined by its cohomology ring nor that homotopy equivalent manifolds are h-cobordant. For example in [CS] a smooth manifold Q is exhibited which is homotopy equivalent to RP^4 but not smoothly h-cobordant to it. Despite all this in many significant cases $M \# P$ still decomposes nicely.

We recall that among the most interesting class of 4-manifolds are the elliptic surfaces. (See Definition in §2). In the simply-connected case they provide us with examples of infinite collections of homotopy equivalent 4-manifolds no two of which are known to be diffeomorphic! In [Man 2] all the simply connected elliptic surfces are nevertheless shown to be almost completely decomposable. In the present paper we consider the question of the decomposability of $V \# P$ for V a non-simply connected elliptic surface (or its close relative, the Lefschetz Torus fibration, (see §2 for definition)). Our principal result on such decompositions is

THEOREM 3.7. *Let $V \xrightarrow{\pi} R$ be a minimally elliptic surface with genus $(R) = g$ and suppose $\pi_* : \pi_1(V) \to \pi_1(R)$ is an isomorphism.*
Then
1) $V \# P = S^2 \times R \# (2g + 2P_a(V) - 1)P \# (2g + 10P_a(V) - 2)Q$.
2) If $g > 0$ then

$$V \approx T^2 \times R \oplus \bigoplus_{i=1}^{P_a(V)} (V_0)_i$$

An outline of the paper is as follows: In §1 we define elliptic surfaces and elliptic surfaces of standard type and state the main theorems concerning them. In §2 we define Lefschetz fibrations and relate them to elliptic surfaces and in §3 we discuss surgeries on elliptic surfaces and proves our main theorems. Lastly we conclude with an appendix proving Lemma 3.2-1).

We wish to thank D. Cox, A. Kas and B. Moishezon for useful discussions on elliptic surfaces.

We also wish to thank S. Akbulut for help with the Calculus of links and Lemma 3.2-1.

§1. Elliptic Surfaces.

Let V be a complete analytic surface. We shall say that V is an elliptic surface if there exists a proper holomorphic map $\Phi : V \to R$ of V onto a non-singular compact Riemann Surface such that for generic $x \in R$, $f^{-1}(x)$ is a (non-singular) elliptic curve. We consider V to be an analytic fibre space over R and call R the base curve and Φ the projection map of V. We define the fibre $\Phi^*(x)$ over x in R as the divisor of the holomorphic function $J \circ \psi - J(x)$, where J is any local uniformization variable on a neighborhood of x in R with value $J(x)$ at x.

We recall that a curve θ on V is called an exceptional curve if θ is a non-singular rational curve with $\theta \cdot \theta = -1$. We shall assume that any fiber of V is free from exceptional curves. (Such a surface is technically called a minimally elliptic surface and we are assuming that all elliptic surfaces considered henceforth are minimally elliptic. Note that by [Kod] if $\tilde{V}$ is any elliptic surface then there exists a proper holomorphic map $f : \tilde{V} \to V$ of V onto a minimal elliptic surface which is an analytic isomorphism of $\tilde{V} - f^{-1}(p_1,...,p_n) \xrightarrow{f} V - \{p_1,...,p_n\}$ for some finite set of points $p_1,...,p_n$. Furthermore, topologically $\tilde{V} = V \# nQ$ so for topological purposes it certainly suffices to restrict attention to the minimally elliptic surfaces).

We shall say $\Phi^*(x)$ is a regular fiber if $d(J \circ \psi)$ is nowhere zero on $\Phi^{-1}(x)$. Having assumed that V was minimally elliptic we see that if $\Phi^*(x)$ is a regular fiber then $\Phi^*(x) = \Phi^{-1}(x)$ is a non-singular elliptic curve.

Clearly an elliptic surface can have at most a finite number of singular fibers $C_{a_i} = \Phi^*(a_i)$, $a_1,...,a_n \in R$. We write each such singular fiber in the form $C_{a_i} = \Sigma_j n_{ij} \Gamma_{ij}$ where the Γ_{ij} are irreducible curves and the n_{ij} are positive integers. We let $m_i = g.c.d(n_{i1}, n_{i2}...)$ and call m_i the multiplicity of the fiber C_{a_i} at a_i. We note that even if $\Phi^{-1}(a_i)$ is non-singular, C_{a_i} may still be a singular-fiber by virtue of having multiplicity > 1.

If $\Phi^{-1}(a_i)$ has singular points we shall call the fiber C_{a_i} a degenerate singular fiber or sometimes simply a degenerate fiber.

Examples

1) The simplest example of an elliptic surface is given by the rational elliptic surface V_R which we can construct as follows:

Let C_1, C_2 be non-singular elliptic curves in CP^2 with transversal intersection (for example let C_1 be given by $Z_2^3 + Z_1^3 + Z_0^3 = 0$ and C_2 by $Z_2^3 + 2Z_1^3 + 3Z_0^3 = 0$). Let $V_R \subset CP^1 \times CP^2$ be given by $\lambda C_1 + \mu C_2 = 0$. (In our case by $(\lambda + \mu)Z_2^3 + (\lambda + 2\mu)Z_1^3 + (\lambda + 3\mu)Z_0^3 = 0$) and let $\Phi : V_R \to CP^1$ be the restriction to V_0 of the projective map $\pi_1 : CP^1 \times CP^2 \to CP^1$. Then it can be verified that V_R is a minimally elliptic surface. (V_R is not a minimal surface. It has exceptional curves of the first kind, however, each of them are transverse to all the fibers and so do not lie in a fiber. Thus V_R is nevertheless minimally elliptic. It is quite easy to see that V_R is analytically just CP^2 blown up at the 9 distinct points of intersection in $C_1 \cap C_2$. In particular, if $\pi : V_R \to CP^2$ is the restriction of the projection map $\pi_2 : CP^1 \times CP^2 \to CP^2$, then π is the blowing down map)

2) Let X be the locus $Z_0^4 - Z_1^4 - Z_2^4 + Z_3^4 = 0$ in CP^3. Then X is called a Fermat $K3$ surface. We describe an elliptic structure on X as follows:

Let $\tilde{X}$ be the intersection of the hypersurfaces

$$\lambda(Z_0^2 + Z_1^2) - \mu(Z_2^2 - Z_3^2) = 0$$
$$\lambda(Z_2^2 + Z_3^2) - \mu(Z_0^2 - Z_1^2) = 0$$

in $CP^1 \times CP^3$.

Let f be the restriction of projection on the 2nd factor to $\tilde{X}$ and π the restriction of projection on the first factor. Then a straightforward calculation shows that $f : \tilde{X} \to X$ is an analytic isomorphism and that $\pi : \tilde{X} \to CP^1$ is an analytic elliptic fibration of $\tilde{X}$. In fact, $\tilde{X}$ has precisely six singular fibers, all degenerate and non-multiple, lying over $[1,0]$, $[0,1]$, $[1,1]$, $[-1,1]$, $[i,1]$ and $[-i,1]$.

We note that both of the examples above were simply connected. In the case of the simply-connected elliptic surfaces the following is known

THEOREM 1.1. *[Man 2, Msh] Let $\Phi : V \to R$ be an elliptic surface. Then V is simply connected if and only if*
1) $R = S^2$
2) *V has at least one degenerate fiber*
3) *V has no more than two multiple fibers and their multiplicities are relatively prime.*
Furthermore, if V is simply connected it is almost completely decomposable.

Thus for a simply connected elliptic surface V, $V \# P$ decomposes as a simple connected sum.

Now suppose $V \overset{\Phi}{\to} R$ is a compact elliptic surface and $f : S \to R$ is a holomorphic map of the non-singular Riemann Surface S on to R. Then there exists a unique elliptic surface $f^*V \overset{f^*\Phi}{\to} S$ making the following diagram commute:

$$
\begin{array}{ccc}
f^*V & \overset{\tilde{f}}{\to} & V \\
f^*\Phi \downarrow & & \downarrow \Phi \\
S & \overset{f}{\to} & R
\end{array}
$$

We call f^*V the pullback of $V \overset{\Phi}{\to} R$ by f. For simply-connected elliptic surfaces without multiple fibers one has:

THEOREM 1.2 [MSH]. *There exists a compact elliptic surface $V_0 \overset{\pi}{\to} S^2$ such that if $X \to S^2$ is any simply connected elliptic surface without multiple fibers then for some map $f : S^2 \to S^2$, X is diffeomorphic to (in fact a complex analytic deformation of) the pullback of $V_0 \overset{\phi}{\to} S^2$ by f. Furthermore, V_0 is a complex deformation of the surface V_R of example 1 and thus diffeomorphic to $P \# 9Q$.*

We introduce some more terminology. If the elliptic surface X is diffeomorphic to the pullback of $V_0 \overset{\pi}{\to} S^2$ by some map $f : R \to S^2$, where R is a compact Riemann surface and the Branch locus $B(f) \subset S^2$ of f is disjoint from the critical values $\{a_i\}$ of π, we shall say that it is an elliptic surface of standard type. In order to adequately investigate the topology of standard elliptic surfaces we must first extend our category of interest to C^∞ manifolds of Lefschetz type which we define in the next section.

§2: Lefschetz Fibrations

Definition 2.1 (Moishezon, [Msh, pg. 162])

Suppose $f : M^4 \to R^2$ is a proper map of a connected 4-manifold onto a connected 2-manifold with $\partial M = f^{-1}(\partial R)$.

Then (M, f, R) is a Lefschetz-fibration of genus g if and only if

1) There is a finite set of points $a_1, \ldots, a_n \in R - \partial R$ such that $M - f^{-1}\{a_i\} \to R - \{a_1\}$ is a fiber bundle with connected fibre. (We call the a_i the critical values of (M, f, R)).

2) $rk \ H_1(f^{-1}(a_i); \mathbb{Z}) = rk \ H_1(f^{-1}(x); \mathbb{Z}) - 1$ for any $i \in \{1, \ldots, n\}$ and $x \in R - \{a_i\}$.

3) For any a_i there exists a single point $c_i \in f^{-1}(a_i)$ such that c_i is the only critical point of f on $f^{-1}(a_i)$ and it is possible to write f locally as $y_i = x_{i1}^2 + x_{i2}^2$ for some choice of local complex coordinates y_i around a_i and (x_{i1}, x_{i2}) around c_i.

4) Genus $f^{-1}(x) = g$ for $x \notin \{a_1, \ldots, a_n\}$.

Note that the genus of $f^{-1}(x)$ for $x \notin \{a_i\}$ does not depend on x.)

We shall call a Lefschetz fibration of genus 1 a Lefschetz fibration of Tori. If M is compact and $\partial M = \varnothing$ we shall say our fibration is compact.

We relate our Lefschetz fibrations to general elliptic surfaces by means of the following theorem.

THEOREM 2.2. *[Msh, p.155]*

Let V be an elliptic surface without multiple fibers. Then there exists a complex analytic deformation $\tilde{V}$ of V such that $\tilde{V}$ has the structure of a Lefschetz fibration of Tori. Furthermore, the surface $V_0 \overset{\pi}{\to} S^2$ is a compact Lefschetz fibration of Tori.

Note that from the point of view of diffeomorphy type of elliptic surfaces without multiple fibers it suffices to limit attention to those which are Lefschetz fibrations.

Now let $F : M^4 \to R^2$ be a Lefschetz fibration of Torii and suppose γ is an arc in R containing all the critical values a_i of f. Then $M - f^{-1}\{\gamma\} \to R - \{\gamma\}$ is a Torus bundle. If this bundle is a product we shall call our fibration a product fibration. We shall define one operation, that of direct sum, on Lefschetz fibrations. To do this properly, we recall some more results of Moishezon [Msh].

Definition 2.3

Let $f_i : M_i \to R_i$, $i = 1, 2$ be Lefschetz fibrations of genus g. Then f_1 is Lefschetz isomorphic to f_2 if and only if there exists orientation preserving diffeomorphisms $\overline{\alpha} : M_1 \to M_2$ and $\alpha : R_1 \to R_2$ with $\alpha f_1 = f_2 \overline{\alpha}$.

Definition 2.4

Let $f : M \to R$ be a Lefschetz fibration with non-empty critical set $\{a_1, \ldots, a_n\}$. Let $a \in R - \{a_1, \ldots, a_n\}$. Then the global monodromy of f (at a) is the canonical homomorphism

$$\pi_1(R - \{a, \ldots, a_n\}, a) \overset{A_{f,a}}{\to} \mathrm{Aut}_+(H_1(f^{-1}(a), \mathbb{Z})$$

We say f has surjective global monodromy if A_f is onto.

LEMMA 2.5. *[Msh, p.169]* Let $f_i : M_i \to R$ $i = 1, 2$ be Lefschetz fibrations of genus g with equal and non-empty critical set $\{a_1, \ldots, a_n\}$. Let $a \in R - \{a_1, \ldots, a_n\}$ and identify $f_1^{-1}(a)$ with $f_2^{-1}(a)$. Then $A_{f_1,a} = A_{f_2,a}$ and $A_{f,a}$ is onto implies f_1 is Lefschetz isomorphic to f_2.

Definition 2.6

Let $f_i : M_i \to R_i$ $i = 1, 2$ be two Lefschetz fibrations of genus g with $\partial R_i = \varnothing$. Suppose D_i are closed discs in R_i such that $M_i|_{D_i} = D_i \times F_g$, where F_g is an oriented 2-manifold of genus g.

Suppose $\eta : \partial D_1 \to \partial D_2$ is an orientation reversing diffeomorphism and let $\bar{\eta} : M_1|_{D_1} \to M_1|_{D_2}$ be the diffeomorphism corresponding to $\eta \times Id$ under the identification of $M_i|_{D_i}$ with $D_i \times F_g$.

Then set

$$M = M_1 \oplus M_2 = \overline{(M_1 - f_1^{-1}(D_1)} \cup_{\bar{\eta}} \overline{(M_2 - f_2^{-1}(D_2))}$$

and

$$R = R_1 \# R_2 = \overline{R_1 - D_1} \cup_\eta \overline{(R_2 - D_2}$$

and define

$$f : M \to R \text{ by } f|\overline{M_i - f^{-1}(D_i)} = f_1|\overline{M_i - f_i^{-1}(D_i)}$$

We note that (M, f, R) is clearly a Lefschetz fibration and by Lemma 2.5 is independent of η. We call it the direct sum of the Lefschetz fibrations (M_i, f_i, R_1) and also write $f = f_1 \oplus f_2$.

We now consider the relation between the topology of M_1, M_2 and M in the direct sum construction $M = M_1 \oplus M_2$ above.

We begin with a Lemma showing what direct summing with V_0 does.

LEMMA 2.7. Let $f : M \to R$ be a compact Lefschetz fibration of Torii. Let $\pi : V_0 \to S^2$ be as in Theorem 1.2 and let $X = M \oplus V_0$.

Let a be a non-critical value of f and let K be a bouquet of two 1-spheres α_1, α_2 embedded as a spine of $f^{-1}(a)$.

Then

$$X \# P = P \# 9Q \# \chi_K(M),$$

where $\chi_K(M)$ is M surgered successively along 2 disjoint 1-spheres β_1, β_2 isotopic to α_1, α_2 respectively.

PROOF. We recall by our discussion of V_R and V_0 in §1 that V_0 is diffeomorphic to $(P \# 8Q)$ blown up by a σ-process at one point. But then our conclusion follows immediately from Lemma 3.3 and Proposition 2.5 of [Man 1]. □

COROLLARY 2.8. *Let $f_1 : M \to R$ be a compact Lefschetz fibration of Torii. Suppose $g : V \to S^2$ is a compact Lefschetz fibration of Torii with V simply connected. Let K be as above and let $X = M \oplus V$.*
Then

$$X \# P = (2n - 1)P \# (10n - 2)Q \# \chi_K(M)$$

where $12n = c_2(V) = $ the Euler number of V.

PROOF. By theorem 9 of [Msh] we have that $c_2(V) = 12n$ for some $n > 0$ and that $V = \bigoplus_{i=1}^n (V_0)_i$.

Let $X_{n-j} = \bigoplus_{i=1}^j (V_o)_i \oplus M$ with $X = X_0$. Thus

$$X = V_0 \oplus X_1 \text{ so by our Theorem } X \# P = 9Q \# P \# \chi_K(X_1)$$

But if $n = 1$ we are finished while if $n > 1$ we see that K is homotopic to zero in X_1 so that by [W] we have $P \# \chi_K(X_1) = 3P \# 2Q \# X_1$. Thus

$$
\begin{aligned}
X \# P &= 8Q \# (P \# \chi_K(X_1)) \\
&= 8Q \# (2Q \# 2P) \# (P \# X_1) \\
&= 10Q \# 2P \# (8Q \# P \# \chi_K(X_2)) \\
&= 3P \# 180 \# \chi_K(X_2)
\end{aligned}
$$

Continuing inductively we obtain the desired result. $\square$

We are thus left with the problem of determining what $\chi_K(M)$ is for a general compact Lefschetz fibration $f : M \to R$. Although we can't do this in all cases, there are certain suggestive ones in which we obtain success. We thus consider the problem of surgering Lefschetz fibrations in the next section.

§3. Surgering Lefschetz fibrations

LEMMA 3.1. *Let $f_i : M_i \to R_i$ be compact Lefschetz fibrations of genus g and suppose K is a bouqut of $2g$ 1-spheres embedded as a spine of the compact oriented manifold of genus g, F_g. Let $a \in R$ and identify $f_1^{-1}(a)$ with F_g.*
Then

$$\chi_K(M_1 \oplus M_2) = \overline{\chi_K(M_1) - D_1^2 \times S} \cup_\eta \overline{\chi_K(M_2) - D_2^2 \times S^2}$$

for some diffeomorphism $\eta : \partial(D_1^2 \times S^2) \to \partial(D_2^2 \times S^2)$.

PROOF. Using the notation of Definition 2.6, we have

$$M_1 \oplus M_2 = \overline{M_1 - f_1^{-1}(D_1)} \cup_{\overline{\eta}} \overline{M_2 - f_2^{-1}(D_2)}$$

which we can identify with

$$\overline{M_1 - D_1^2 \times F_g} \cup_{\overline{\eta}} \overline{M_2 - D_1^2 \times F_2}$$

Let e_α, $\alpha = 1, \ldots 2g$ be disjoint 1-spheres in $\partial(D_1^2 \times F_g) = \partial(D_2^2 \times F_g)$ which correspond to the bouquet K. Let $M_i' = M_i - D_i^2 \times F_g$ for $i = 1, 2$. Then by an argument identical to that of Lemma 2.4 of [Man 1] we see one can identify

$$\chi_K(M_1 \oplus M_2) \text{ with } M_1' \cup \bigcup_{\alpha=1}^{2g} B_{1,\alpha} \cup M_2' \cup \bigcup_{\alpha=1}^{2g} B_{2,\alpha}$$

where the $B_{i,\alpha}$ are 2-handles attached to M_1' along some neighborhood $e_\alpha \times D^2$ of e_α considered as lying in the boundary of M_1' and M_2'. But clearly $D_i^2 \times F_g \cup \bigcup_{\alpha=1}^{2g} B_{i,\alpha}$ can, as in Lemma 2.3 of [Man 1], be seen to be diffeomorphic to a $D^2 \times S^2$ and thus applying the argument of Lemma 2.4 of [Man 1] we obtain

$$M_i' \cup \bigcup_{\alpha=1}^{2g} B_{i,\alpha} \approx \chi_{\cup e_\alpha} M_i - D^2 \times S^2$$

Thus

$$\chi_K(M_1 \oplus M_2) = \overline{\chi_K(M_1) - D^2 \times S^2} \cup_\eta \overline{\chi_K(M_2) - D^2 \times S^2}$$

for some autodiffeomorphism $\eta : S^1 \times S^2$. By a slight abuse of notation we shall let

$$\chi_K(M_1) \oplus \chi_K(M_2) \text{ represent } \overline{\chi_K(M_1) - D^2 \times S^2} \cup \overline{\chi_K(M_2) - D^2 \times S^2}.$$

We now investigate $\chi_K(M)$ for specific fibrations $M \to R$. The easiest fibrations to handle are those without critical values. Then M is simply a F_g-bundle over R. We shall use the notation $M = F_g \tilde{\times} R$ to denote that M is some (we don't necessarily know exactly which) F_g bundle over R. For the case of product bundles we have

LEMMA 3.2. *(Akbulut) Let K, F_g be as above. Then*
1) $\chi_K(F_g \times T^2) = S^2 \times T^2 \# 2g(S^2 \times S^2)$
2) $\chi_K(F_g \times F_p) = S^2 S^2 \times F_p F_p \# 2\tilde{g}p(S^2 \times S^2)$

PROOF. We defer the proof of 1) to the appendix. To prove 2) we note that

$$F_g \times F_p = F_g \times T^2 \oplus F_g \times F_{p-1}.$$

Thus by Lemma 3.1   1), we obtain

$$\chi_K(F_g \times T^2) = \chi_K(F_g \times T^2) \oplus \chi_K(F_g \times F_{p-1})$$
$$+ 2g(S^2 \tilde{\times} S^2) \# [S^2 \times T^2 \oplus \chi_K(F_g \times F_{p-1})]$$

which by induction is then

$$2g(S^2 \tilde{\times} S^2) \# [S^2 \times T^2 \oplus S^2 \times F_{p-1}] \# 2g(p-1)(S^2 \tilde{\times} S^2)$$

as desired. □

(Note that we use the fact that if X and Y are S^2 bundles over S^2. Then $X \# Y$ is either $X \# X$ or $Y \# Y$. See [W] for a proof.)

Note:

If M is an F_g bundle over T^2 it would again appear that $\chi_K(F_g \tilde{\times} T^2) = S^2 \tilde{\times} T^2 \# 2g(S^2 \tilde{\times} S^2)$. However, we can at the moment prove this only for specific examples.

We can now obtain some information on the problem of decomposing Lefschetz fibrations of Torii. We have

THEOREM 3.3. *Let $f : M \to R$ be a compct Lefschetz Torus product fibration with*

$f_* : \pi_1(M) \to \pi_1(R)$ *an isomorphism. Let $g = genus(R)$.*

Then for some $N > 0$, f has $12n$ critical values and

$$M \# P = S^2 \times R \#(2g + 2n - 1)P \#(2g + 10n - 2)Q$$

PROOF. Since $f_* : \pi_1(M) \to \pi_1(R)$ is an isomorphism, it is clear that f must have at least one critical value. Let $A \subset R$ be the set of its critical values. Since A is a finite set of points on R we can find a 2-disc $D^2 \subset R$ with $A \subset$ interior(D^2). Since f is a product fibration we have that $M|_{\overline{R-D^2}}$ is trivial. Thus $M - f^{-1}(D^2) = (R - D^2) \times T^2$. In particular $M|f-1(\partial D^2) = S^1 \times T^2$.

We can thus find Lefschetz fibrations $f_1 : N \to R$ and $f_2 : V \to S^2$ with $M = N \oplus V$ and such that $N = R \times T^2$. But f_2 has at least one critical value so by Theorem 9 of [Msh], V is the direct sum $\bigoplus_{i=1}^{n}(V_0)_i$ where

$$12n = \# \text{ of critical points of } f_2$$
$$= \# \text{ of critical points of } f$$
$$= \text{ Euler number of } V.$$

But then by Corollary 2.8 we obtain that

$$M \# P = (2n - 1)P \#(10n - 2)Q \# \chi_K(N)$$

with $n > 0$ as above and K as in Corollary 2.8. But by Lemma 3.2 we have

$$\chi_K(N) = \chi_K(T^2 \times R) = (S^2 \times R) \# 2g(S^2 \times S^2)$$

But $n > 0$ and $P \# 2g(S^2 \times S^2) = (2g + 1)P \# 2gQ$. Thus

$$M \# P = (2n + 2g - 1)P \#(10n + 2g - 2)Q \#(S^2 \times R)$$

as desired. □

We recall that Lefschetz fibrations were introduced to help us understand the topology of elliptic surfaces.

THEOREM 3.4. *Let V be a compact complex elliptic surface of standard type. Let $g = \frac{1}{2} rank H_1(V, \mathbb{Z})$ and let $n = p_a(V)$. Then*

$$V \# P = S^2 \times F_g \#(2g + 2n - 1)P \#(2g + 10n - 2)Q.$$

PROOF. Since V is of standard type we see there exists a commutative diagram

$$\begin{array}{ccc} V & \xrightarrow{\bar{f}} & V_0 \\ \pi \downarrow & & \downarrow \pi \\ R & \xrightarrow{f} & S^2 \end{array}$$

for some map $f : R \to S^2$ of the Riemann Surface R.

It is a straightforward calculation to see that $rk\, H_1(V, \mathbb{Z}) = 2$ genus (R) and that $\bar{\pi}$ has 12 (deg f) critical values. Furthermore, a straightforward calculation shows that $(\deg f) = p_g(V)$ and that $\pi_1(V) \xrightarrow{\bar{\pi}_*} \pi_1(R)$ is an isomorphism. Since V_0 is of Lefschetz type, we can assume that up to complex analytic deformation so is V. Lastly, to see that $V \xrightarrow{\bar{\pi}} R$ is a product fibration, it suffices to show that there exists a basis of 1-cycles $\alpha_1, \ldots, \alpha_{2g}$ for $H_1(R, \mathbb{Z})$ such that the monodromy of $V \xrightarrow{\bar{\pi}} R$ around each such curve is the identity. But this is clearly true by the definition of standard type. Thus V satisfies the hypothesis of Theorem 3.3 and we are done.

Note: Clearly we used the hypothesis of standard type only to guarantee that our surface was diffeomorphic to a product Lefschetz fibration with $\pi_1(V) \to \pi_1(R)$ an isomorphism and our theorem holds 'mutatis mutandis' for all such surfaces.

Before proceeding further, we therefore recall some facts about the topology of such surfaces. [Man 2, Sh]. So suppose V is a minimally elliptic surface with projection $f : V \to R$ and let $p_1 \ldots p_n \in \mathbb{R}$ be the critical values of f. Then

$$\chi(V) = \text{Euler Number of } V$$
$$= \Sigma \chi(f^{-1}(p))$$
$$= 12 \, p_a(V)$$
$$= 12(1 - q(V) + p_g(V))$$

where

$p_a(V) = $ arithmetic genus of $V = 1 - q(V) + p_g(V)$

$q(V) = $ irregularity of V

$\quad = $ dimension of the space of holomorphic 1-forms on V

$g(R) = $ genus (R)

$p_g(V) = $ geometric genus of V

$\quad = $ dimension of the space of holomorphic 2-forms on V

We can now ask, when is an Elliptic Surface of standard type? More generally, when is it diffeomorphic to a product Lefschetz fibration with $\pi_1(V) \to \pi_1(R)$ an isomorphism.

THEOREM 3.5.   *Let $V \xrightarrow{f} R$ be a minimally elliptic surface without multiple fibers. Suppose f has at least one critical value. Then $p_a(V) \geq 2$ implies V is of standard type.*

THEOREM 3.6.   *Let $V \xrightarrow{f} R$ be a minimally elliptic surface without multiple fibers. Suppose f has at least one critical value. Then V is diffeomorphic to a product Lefschetz fibration.*

PROOF OF THEOREMS 3.5 AND 3.6.

We note that by the Noetler formula $\chi(V) = 12 p_a(V)$. Furthermore, by [Msh] we can assume without loss of generality that $V \xrightarrow{f} R$ is already a Lefschetz Fibration. Then $\chi(V) = \# \{\text{critical values of } f\} > 0$. Thus by the remarks after Theorem 3.4 we see that Theorem 3.5 implies 3.6 for $p_a(V) \geq 2$ and so in Theorem 3.6 we must simply analyze the case when $p_a(V) = 1$. Our proof(s) then follow from the following three lemmas.

LEMMA 1. *Let* $V_1 \to R$, $V_1 \to S$ *be elliptic surfaces with no multiple fibers and* $\chi(V_1) \neq 0$. *Suppose genus* $(R)=genus(S)$ *and* $\chi(V_1) = \chi(V_2)$.
   Then V_1 *is diffeomorphic to* V_2.

PROOF. By a straightforward adaptation of the arguments in Lemma 1 and Lemma 2 of [Kas 3] we can deduce that every elliptic surface of the above type is a complex analytic deformation of a basic elliptic surface. However, in his thesis [Sei], Seiler shows that for every pair (g,p) of non-negative integers

1)   there exists a coarse moduli scheme $E_{(g,p)}$ of elliptic surfaces $V \to B$ with genus $(B) = g$ and $p_a(V) = p$

2) $E_{(g,p)}$ is connected.

Thus any two basic elliptic surfaces with the same numerical invariants are deformations of one another. But $\chi(V_1) = (V_2)$ implies $p_a(V_1)p(V_2)$ so that V_1 and V_2 are diffeomorphic.   □

LEMMA 2. *For any pair of non-negative integers* (g,p) *with* $p \geq 2$ *there exists an Elliptic Surface of standard type* $V \to F_g$, *where* F_g *is a Riemann Surface of genus* g *and* $p_a(V) = p$.

PROOF. Suppose $g > 0$ Let $n = 2(g + p - 1)$. Thus $n - 2p + 2 = 2g > 0$ so that by [HL] or [Hur] there exists a p-sheeted branched cover $F_g \xrightarrow{f} S^2$ of S^2 having exactly n branch points each of order 2. We can assume without loss of generality that the branch locus $B(f) \subset S^2$ of f is disjoint from the set of critical values of $\pi : V_0 \to S^2$, V_0 as in §2. Let V be the pullback of $V_0 \to S^2$ by f. Then V is certainly standard and $p_a(V) = \chi(V)/12 = p \cdot \chi(V_0)/12 = p$ since $\chi(V_0) = 12$.
   Now if $g = 0$ our result is a direct consequence of Theorem 1.2.   □
   Note that if $g = 0$ then Theorem 3.5 is the even if $p_a(V) = 2$ as a consequence of 1.2.

LEMMA 3. *Let* g *be a non-negative integer. Then there exists an Elliptic Surface* $V \to F_g$ *with* $p_a(V) = 1$ *and such that* V *is diffeomorphic to a product Lefschetz Fibration*.

PROOF.
Claim 1: There exists an elliptic surface $E \to T^2$, fibred over a torus which satisfies
   1) $p_a(E) = 1$
   2) E has no multiple fibers
   3) E is diffeomorphic to $T^2 \times T^2 \oplus V_0$
   4) E has exactly one singular fiber of type I_6^*
Assuming the above claim, we proceed as follows:
   Let $x \in T^2$ be the point above which I_6^* lies. For any integer $r \geq 0$ we can construct a 2-fold covering of T^2 of genus $r + 1$ with exactly $2r$ branch points, one of which we can assume to be x. Let $F_{r+1} \xrightarrow{f} T^2$ be that covering and set $y = f^{-1}(x)$. Denote f^*E by V.

Then it is clear that

1) V is an elliptic surface with exactly one singular fiber of type I_{12}
2) V is diffeomorphic to $T^2 \times F_{r+1} \oplus V_0$ and so $p_a(V) = \chi(V)/12 = 1$ as desired.

We are thus left to prove our claim. To do this, we first review the theory of elliptic modular surfaces as discussed in [Sh, §4,§5].

Suppose Γ is a subgroup of finite index in $SL(2, \mathbb{Z})$ and let Γ act on the upper half plane H by $\gamma = \left(\begin{smallmatrix} a & b \\ c & d \end{smallmatrix}\right) \in \Gamma$, $z \in H$ goes to $\left(\frac{az+b}{cz+d}\right)$. Then H/Γ can be compactified by the addition of a finite number of cusps to obtain a Riemann Surface $R_\Gamma = \overline{H/\Gamma}$. For $\Gamma \subset \Gamma'$, we can obtain a holomorphic map of R_Γ onto $R_{\Gamma'}$ by a straightforward extension of the canonical map $H/\Gamma \to H/\Gamma'$. Thus there always exists a canonical map $j_\Gamma : R_\Gamma \to CP^1$ of R_Γ onto CP^1, which is obtained by taking $\Gamma' = SL(2, \mathbb{Z})$ and noting that $H/SL(2, \mathbb{Z}) \cong CP^1$.

Now suppose Γ acts effectively on H (i.e. $-1 \notin \Gamma$). Let $\mu = [SL(2,\mathbb{Z}) : \Gamma \cdot \{\pm 1\}]$ and set $c = \#$ of cusps in R_Γ and $e = \#$ of elliptic points in R_Γ and $s = c + e$. Let $\Sigma_\Gamma = $ set of cusps and elliptic points and let $R'_\Gamma = R_\Gamma - \Sigma_\Gamma$. Then it can be shown that R'_Γ and j_Γ determine a unique representation of $\phi_\Gamma : \pi_1(R'_\Gamma) \to \Gamma \subset SL(2, \mathbb{Z})$ and thus by [Kod] an elliptic surface E_Γ over R_Γ with functional invariant j_Γ and homological invariant $G_\Gamma = $ sheaf over R_Γ determined by ϕ_Γ.

E_Γ is known as the elliptic modular surface attached to Γ. Its numerical characteristics and singular fibers can be shown to be: [Sh, Prop. 4.2, 4.3].

Fact 1

Let E_Γ be the elliptic modular surface associated to Γ.

$$\text{Then } E_\Gamma \text{ has } c_1 \text{ singular fibres of type } I_b(b \geq 1)$$
$$c_2 \text{ singular fibres of type } I_b^+ (b \geq 1)$$
$$\text{and} \quad e \text{ singular fibres of type } IV$$

where
$$c = 1 = \# \text{ of cusps of type I of } R_\Gamma$$
$$c_2 = \# \text{ of cusps of type II of } R_\Gamma$$

(where a cusp is of type I if the stabilizer of its representative in $\mathbb{Q} \cup \{\infty\}$ has a generator which is conjugate to $\left(\begin{smallmatrix} 1 & b \\ 0 & 1 \end{smallmatrix}\right)$ and of type II if it is conjugate to $\left(\begin{smallmatrix} -1 & -b \\ 0 & -1 \end{smallmatrix}\right)$.)

$$e = \# \text{ of elliptic points of } R_\Gamma$$

Furthermore, the genus of Γ is given by

$$g = \text{ genus of } R_\Gamma = \frac{1}{2}\left[\mu/6 - c - \frac{2}{3}e\right] + 1$$

and

$$p_g(E_\Gamma) = \text{ geometric genus of } E_\Gamma$$
$$= 2g - 2 + \frac{1}{2}c_1 + c_2 + e$$
$$q(E_\Gamma) = \text{ irregularity of } E_\Gamma = g.$$

Now suppose $\Gamma =$ Commutator Subgroup of $L(2, \mathbb{Z})$. Then one can check that:

$$\mu_\Gamma = 6, c_1 = 0, \ c_2 = 1, \ \text{and} \ e = 0 \text{ and therefore } p_g = 0 \text{ and } q = g = 1.$$

Thus E_Γ has only one singular fibre of type I_6^* and is fibered over a Torus.

By Theorem 8a of [Msh] we can slightly deform $E_\Gamma \to T^2$ to a diffeomorphic surface $\tilde{E} \to T^2$ such that $\tilde{E}$ has exactly 12 fibers of type I_1. But by the main theorem of [MH], $\tilde{E}$ must then be diffeomorphic to $T^2 \times T^2 \oplus V_0$ as desired! Thus our claim is proven. $\square$

We note that Kas as observed ([Kas 2]) that even without resorting to [Sec] one can prove Lemma 1 provided that $p_a(V) \geq g(R)14$ by using a straightforward modification of the arguments of [Kas 2]. We leave details to the interested reader!

We recall that if $V \overset{f}{\to} R$ is an elliptic surface with at least one degenerate fiber then as in [CZ]

$$\pi_1(V) = \{(x_i, y_i, z_j\} \ L = 1 \ldots g \ \ j = 1 \ldots n \ \ \prod_{i=1}^{g} [x_i, y_i] \prod_{i=1}^{n} z_j = 1, \ z_j^{m_j} = 1$$

where $g =$ genus (R) and $V \to R$ has n multiple fibers of multiplicities $m_1 \ldots m_n$. Thus if $f_* : \pi_1(V) \to \pi_1(R)$ is an isomorphism then either V has no multiple fibers or $g(R) = 0$ and V has at most 2 multiple fibers, whose multiplicities are relatively prime. Thus noting Theorem 1.1 and [Msh, Man 2] we can combine Theorem's 3.4, 3.5, and 3.6 to obtain:

THEOREM 3.7. Let $V \overset{f}{\to} R$ be an elliptic surface with genus $(R) = g$ and suppose $f_* : \pi_1(V) \to \pi_1(R)$ is an isomorphism.
Let $n = p_a(V)$ and $m = |C_1^2(V)|$. Then

$$V \# P = S^2 \times F_g \#(2g + 2n - 1)P\#(2g + 10n + m - 2)Q$$

Appendix

This appendix proves the following lemma first discovered by S. Akbulut.

LEMMA. (Akbulut) Let F_g be a compact orientable surface of genus g. Let K be a bouquet of $2g$ 1-spheres $\alpha_1, \ldots \alpha_{2g}$ embedded as a spine of F_g. Let $\beta_1, \ldots \beta_{2g}$ be disjoint 1-spheres in $M = F_g \times T^2$ isotopic to $\alpha_1, \ldots, \alpha_{2g}$ in $F_g \times * \to M$. Then

$$\chi_K(M) = M \text{ surgered along } \beta_1, \ldots, \beta_{2g}$$
$$= S^2 \times T^2 \# 2g(S^2 \underset{\sim}{\times} S^2)$$

with $S^2 \underset{\sim}{\times} S^2$ being either $S^2 \times S^2$ or $P\#Q$ depending on the framings chosen for the surgery.

PROOF. We shall use the link calculus as developed by Kirby in [Kirb] and expanded in [AK] to illustrate our surgeries. For the sake of simplicity we shall work only with $T^2 \times T^2$ and surger along a canonical basis α_1, α_2 of $T^2 \times *$. The proof is identical in the general case, however, the necessary picture are harder to draw. (Actually, the only case of the Lemma we use is the $T^2 \times T^2$ case so as far as our paper is concerned that case is sufficient.)

We begin with a handlebody picture of T^4. All 2−handles have zero-framings and for the sake of simplicity, we shall do zero-framed surgery throughout. Thus we suppress the 0's on all the 2-handle. We now surger T^4 along the core of the 1-handle represented by the $S^0 \times D^3$ pair centered at zero and infinity in Fig. 0. The result is the 2-handle with 0-framing labelled W in Fig. 1. We have also labelled 2 of the remaining 4 2-handles as α and A in Fig. 1.

We now push off a copy of α and add it to W to get W' as shown in Figure 2. We now push the 2-handle A off the 1-handle X in Fig. 2 to get Fig. 3 and then continue to isotop A until we get to the situation in Fig. 4. We now push off a copy of $-\alpha$ and add it to W' to get W'' as in Fig. 5.

Next we will pull the 2-handle B off the 1-handle Y to get Figure 6 in which we have also isotoped W'' and α slightly for greater clarity. We redraw Figure 6 as Figure 6a to indicate a further isotop of B. We now push off a copy of β and $-\beta$ and add them onto W''. This gives Figure 7 with $\overline{W} = W'' + \beta - \beta$. We can now pull part of $\overline{W}$ off the 1-handles X and Z of Figure 7. We also pull the 2-handle C off of Z. With a bit of isotoping we get Figure 8.

In the series of Figures 8A, 8B, 8C, we push off a copy of $+\gamma$ and $-\gamma$, add them to $\overline{W}$ to get $\hat{W}$ and then push $\hat{W}$ off the 1-handle Y and X. Finally we push $\hat{W}$ completely off Z to get Figure 8D.

We now surger Z to get the 2-handle in Figure 9. Then pushing off copies of A and B and adding them to ς we get $\overline{\varsigma}$ as in Figure 9A. We can now pull α off Y and β off X to get Figure 10.

Figure 10, however, is simply a handlebody picture of $S^2 \times T^2 \# 2(S^2 \times S^2)$. (The 2-handles $\overline{\varsigma}$ and $\hat{W}$ are cancelled by 3-handles (not drawn) in Figure 0. (Recall that our picture of T^4 has 4 undrawn 3-handles)). Had we used different framings when doing our surgeries, we would have obtained $S^2 \times T^2 \# 2(P \# Q)$ instead. Thus our result is obtained as desired.

REFERENCES

[AK] S. Akbulut and R. Kirby, *An exotic involution of* S^4 (to appear).

[CS] S. Cappel and J. Shaneson, *Some new four manifolds*, Annals of Math. **104** (1976), 61–72.

[CZ] D. Cox and S. Zucker, preprint.

[HL] Hensel K. and Landsberg, G., *Theorie der Algebraischen Funktionen einer Variabelen* (Chelsea reprint, 1968).

[Hur] Hurwitz, A., *Über Riemansche Flöchen mit gegebenen Verzweigungspunkten.* Math. Ann. **39** (1891), 1–61.

[Kas 1] Kas, A., *Weerstrass normal forms and invariants of elliptic surfaces*, Trans. Amer. Math. Soc. **225** (1977), 259–266.

[Kas 2] Kas, A., Private communication.

[Kas 3] Kas, A., *On the deformation types of regular elliptic surfaces*, Complex Analysis and Algebraic Geometry, Cambridge University Press, 1977, 107–112.

[Kirb] R. Kirby, *A calculus for framed links in S^3*, Invent. Math. **45** (1978), 35–36.

[Kod] K. Kodaira, *On Compact Analytic Surfaces II*, Annals of Math 77 (163), 111–152.

[Man1] R. Mandelbaum, *Algebraic Surfaces and the Irrational Connected Sum of Four Manifolds*, Trans. Amer. Math. Soc. **247** (1979) 137–156.

[Man2] R. Mandelbaum, *On the topology of Elliptic Surfaces*, Studies in Algebraic Topology, Advances in Mathematics, Supplementary Studies, Vol. 5, 143–166.

[Man3] R. Mandelbaum, *Four dimensional topology*, Bull. Amer. Math. Soc. **2** (1980), 1–159.

[MH] R. Mandelbaum and J. R. Harper, *Global monodromy of elliptic Lefschetz fibrations*, Canadian Math. Soc. Conference Proceedings **2** (1982), 35–41.

[MM1] R. Mandelbaum and B. Moishezon, *On the topological structure of non-singular algebraic surfaces in CP^3*, Topology, Vol 15 (1976), 23–40.

[MM2] R. Mandelbaum and B. Moishezon, *On the topology of simply-connected Algebraic Surfaces*, Trans. Am. Math. Soc. **460** (1980), 195–222.

[Mil] J. Milnor, *On simply connected 4-manifold*, Symposium International de Topologla Algebrica, Mexico (1958), 122–128.

[Msh] B. Moishezon, *Complex Surfaces and Connected Sums of Complex Projective Planes*, Lecture Notes in Math **603**, Springer-Verlag, 1977.

[Sei] W. Seiler, *Moduln Elliptischen Flächen mit Schnitt*, Ph.D. dissertation, July 1982, Universität Karlsrche. Bull. AMS **68** (1962).

[Ser] J. P. Serre, *Cours D'Arithmetique*, Presses Univesitaire De France (1970).

[Sh] T. Shoida, *On Elliptic Modular Surfaces*, J. Math. Soc. Japan. Vol. 24 **1**, (1972), 20–59.

[W] C. T. C. Wall, *On simply connected 4-manifolds*, J. Lond. Math. Soc. **39** (1964), 141–149.

[Wh] J. H. C. Whitehead, *On simply connected 4-dimensional polyhedra*, Comm. Math. Lelv. **22** (1949), 48–92.

UNIVERSITY OF ROCHESTER

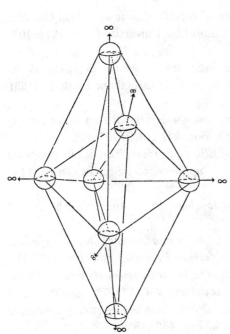

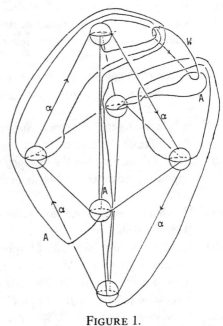

FIGURE 1.

T^4 surgered along the core of the 1-handle represented by the $S^0 \times 0^3$ centered at zero and infinity in Figure 0. The result is the 2-handle labelled W in the picture.

FIGURE 0.

A handlebody picture of the 0,1 and 2 handles of T^4.

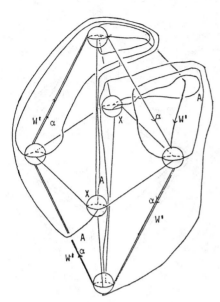

FIGURE 2.
We have pushed off a copy of α and added it to W to get W'.

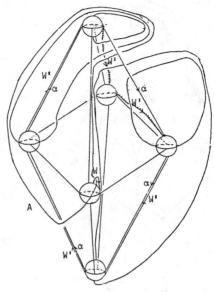

FIGURE 3.
We have slid the 2-handle represented by A in Figure 1 off the 1-handle X. In doing so we have pulled W' over X.

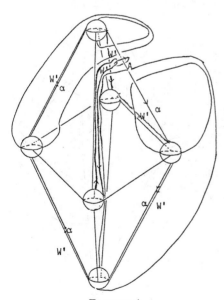

FIGURE 4.
We have isotoped A to its new position. Again W' gets pulled along.

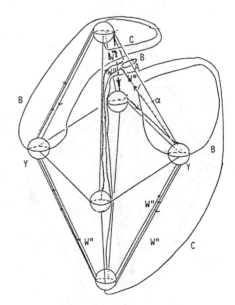

FIGURE 5.
We now push off a copy of $-\alpha$ and add it to W' to get W''.

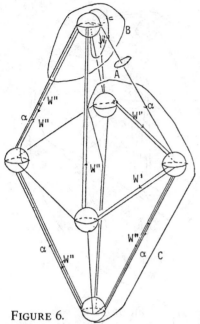

FIGURE 6.

We can now pull B off the 1-handle Y and isotop it to the position illustrated. We have also isotopoed W'' and α somewhat to simplify the picture.

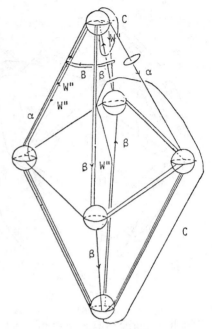

FIGURE 6A.

A further isotopy of B.

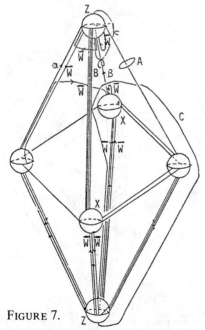

FIGURE 7.

We have pushed off a copy of $+\beta$ and $-\beta$ and added them onto W''. We call the result $\overline{W}$.

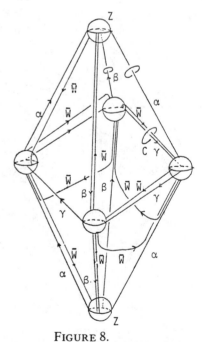

FIGURE 8.

We have now pulled part of $\overline{W}$ off the 1-handles X and Z and isotopoed the result a bit. We also pulled C off Z and isotoped it as well.

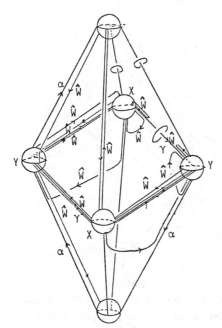

FIGURE 8A.
We have now pushed off a copy of $+\gamma$ *and* $-\gamma$ and added them onto $\bar{W}$. We call the result $\hat{W}$.

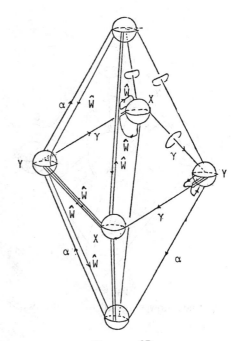

FIGURE 8B.
We pull W off Y and partially off X and isotop.

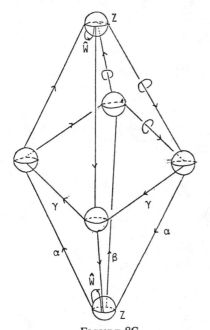

FIGURE 8C.
We continue pulling W off Y and X and isotop.

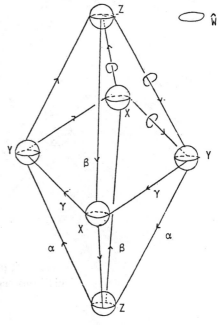

FIGURE 8D.
We now pulled W completely off Z.

310 RICHARD MANDELBAUM

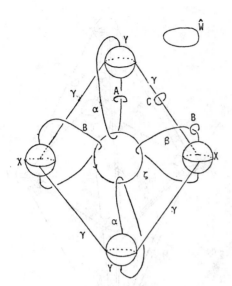

FIGURE 9.
We have now surgered to get the two
handle ζ.

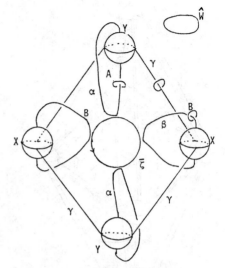

FIGURE 9A.
We have pushed off copies of A and B
and added them to ζ to get $\bar{\zeta} = \zeta + A + B$.

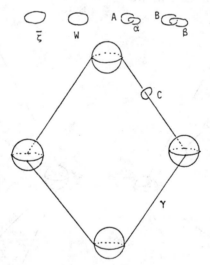

FIGURE 10.
The result of pulling α off Y and β off
X and isotoping a little.

Contemporary Mathematics
Volume 44, 1985

ALGEBRAIC SURFACES AND
THE ARITHMETIC OF BRAIDS, II

B. MOISHEZON

§1. Introduction.

In ([2]) we outlined a program for studying branch curves of algebraic surfaces by their "braid monodromys" (for Definitions see [1]). The most interesting cases correspond to branch curves of generic projections of pluricanonically embedded algebraic surfaces of general type. In ([3]) U. Persson constructed large classes of surfaces of general type. The basic objects in the constructions of ([3]) are double coverings of rational surfaces and especially double coverings of quadrics (e.g. $CP^1 \times CP^1$). If we want to understand braid monodromys for U. Persson's pluricanonically embedded surfaces we must first study braid monodromies for some projective embeddings of the corresponding rational surfaces, and especially for projective planes and quadrics. Such a study is the main subject of the present paper.

We denote by V_n a projective surface corresponding to a Veronese embedding of degree n of CP^2. By $F_0(m,n)$ we denote a projective surface obtained by a projective embedding of F_0 corresponding to the linear system $|m\ell_1 + n\ell_2|$, where ℓ_1, ℓ_2 are generators of F_0, $(F_0 \simeq CP^1 \times CP^1)$.

Let X be a projective surface which is either V_n or $F_0(m,n)$, $\pi : X \to CP^2$ a generic projection of X to CP^2 and $S \subset CP^2$ the branch curve of π. Our aim is to write explicit formulae describing the braid monodromy of S (see [1]). Such formulae immediately give finite presentations for $\pi_1(CP^2 - S, *)$ and explicit descriptions for the mapping class group monodromy (system of Lefshetz vanishing cycles) for a Lefshetz pencil of hyperplane sections of X. When $X = V_n$ we just get a generic pencil of plane curves of degree n. So from our braid monodromy formulae we can explicitly write a system of Lefshetz vanishing cycles for such a pencil.

Another application of our results is the construction of the first examples of simply-connected algebraic surfaces of positive index (joint paper with M. Teicher (see [4])).

§2. Construction of projective degenerations.

Let Z be a complex algebraic variety, $Z = A \cup B$, A, B subvarieties of Z, and let $g : Z \to \mathbb{C}P^{N-1}$ be a projective embedding. We say that (Z, A, B, g) satisfies *Assumption* (A_1) if the following is true:

(1) B is defined by a sheaf of principal ideals,

(2) the ideal of $C = A \cap B$ on A is equal to the restriction to A of the ideal of B on Z and

(3) C does not have common irreducible components with A.

Clearly C is a Cartier divisor on A. Let $\{U_r\}$ be an affine open covering of Z such that in each U_r there is a regular function τ_r generating the ideal of $B \cap U_r$ in U_r. Set $W = Z \times \mathbf{C}^1$, $Z_0 = Z \times 0$, $A_0 = A \times 0$, $B_0 = B \times 0$, $C_0 = C \times 0$. Let $p : W \to \mathbf{C}^1$, $q : W \to Z$ be canonical projections, $\hat{B} = q^{-1}(B)$, $\hat{U}_r = q^{-1}(U_r)$, $\hat{\tau}_r = q^*(\tau_r)$, and let t be a complex coordinate in $\mathbf{C}^1$. It is clear that $B_0 = \hat{B} \cap Z_0$ and that the ideal of $B_0 \cap \hat{U}_r$ in $\hat{U}_r$ is generated by t and τ_r. Denote by $\phi : \tilde{W} \to W$ the monoidal transformation of W with the center B_0, $\tilde{B}_0 = \phi^{-1}(B_0)$, $\tilde{C}_0 = \phi^{-1}(C_0)$, and let $\overline{A}_0$, $\overline{\hat{B}}$ be the strict images of A_0 and $\hat{B}$ in $\tilde{W}$, $\overline{B}_0 = \overline{\hat{B}}_0 \cap \tilde{B}_0$, $U_{r_0} = \phi^{-1}(U_r) - \overline{A}_0 \cap \phi^{-1}(U_r)$, $U_{r_1} = \phi^{-1}(U_r) - \overline{\hat{B}} \cap \phi^{-1}(U_r)$, $u_r = \frac{\hat{\tau}_r}{t}$. It is easy to see that $\overline{A}_0$ and $\overline{\hat{B}}$ are isomorphic respectively to A_0 and $\hat{B}$, $\hat{A}_0 \cap \overline{\hat{B}} = \emptyset$ and $\phi | \tilde{C}_0 : \tilde{C}_0 \to C_0$ is a projective line bundle over C_0.

Let G be the line bundle corresponding to a hyperplane section of Z and $\alpha \in \mathcal{H}^0(A, \mathcal{O}_A[G|_A - C])$. We can assume that $G|_{U_r}$ is trivial for each U_r and that $\{g_{rr'}\}$ are transition functions of G for $\{U_r\}$. Now α could be defined by a collection of $\{\alpha_r\}$, where each α_r is a regular function in $\overline{U}_r = U_r \cap A$ and in $\overline{U}_r \cap \overline{U}_{r'}$ $\alpha_r = \alpha_{r'} \cdot g_{rr'} \cdot \frac{\tau_{r'}}{\tau_r}$. For each r choose a regular function α_r' (in U_r) with $\alpha_r'|\overline{U}_r = \alpha_r$. Let $\hat{\alpha}_r' = q^* \alpha_r'$ and $\tilde{\alpha}_{r_0} = \phi^* \hat{\alpha}_r' \cdot u_r$, $\tilde{\alpha}_{r_1} = \phi^* \hat{\alpha}_r'$. It is clear that $\tilde{\alpha}_{r0}$ (resp. $\tilde{\alpha}_{r1}$) is a regular function in U_{r_0} (resp. U_{r_1}) and that $\tilde{\alpha}_{r_0}$, $\tilde{\alpha}_{r_1}$ don't depend on the choice of α_r' (we use that u_r is regular in U_{r_0} and that $u_r|_{\overline{\hat{B}}} = 0$).

Let $G' = \phi^* q^*(G) - [\phi^{-1}(B_0)]$ and $\{h_{r_\epsilon, r'_{\epsilon'}; \epsilon, \epsilon'} = 0, 1\}$ be transition functions of G' for the covering $\{U_{r_0}, U_{r_1}\}$ of $\tilde{Z}$. It is easy to check directly that in $U_{r\epsilon} \cap U_{r'\epsilon'} \tilde{\alpha}_{r\epsilon} = \tilde{\alpha}_{r'\epsilon'} \circ h_{r\epsilon, r'\epsilon!}$. Thus the collection $\{\tilde{\alpha}_{r\epsilon}\}$ defines an element of $\mathcal{H}^0(\tilde{Z}, \mathcal{O}_{\tilde{Z}}[G_r'])$. Denote this element by $\tilde{\alpha}$.

Now let $\alpha^{(1)}, \ldots, \alpha^{(M)}$ be a basis for $\mathcal{H}^0(A, \mathcal{O}_A[G|_A - C])$. For each $\alpha^{(i)}$ we define $\tilde{\alpha}^{(i)} \in \mathcal{H}^0(\tilde{Z}, \mathcal{O}_{\tilde{Z}}[G'])$ as above (that is, the same way as $\tilde{\alpha}$ was defined for α).

Now let $\psi \in \mathcal{H}^0(Z, \mathcal{O}_Z[G])$, $\psi = \{\psi_r\}$, where each ψ_r is regular in U_r and in $U_r \cap U_{r'}$, $\psi_r = \psi_{r'} \circ g_{rr'}$. Set $\hat{\psi}_r = q^* \psi_r$ and $\tilde{\psi}_{r_0} = \phi^* \hat{\psi}_{r}$, $\tilde{\psi}_{r_1} = \phi^* \psi_r \circ v_r$, where $v_r = \frac{t}{\hat{\tau}_r}$, $\tilde{\psi}_{r_0}$ is considered a regular function in U_{r_0} and $\tilde{\psi}_{r_1}$ as a regular function in U_{r_1} (v_r is well defined and regular in U_{r_1}).

It is easy to check directly that in $U_{r\epsilon} \cap U_{r'\epsilon'}$ ($\epsilon, \epsilon' = 0, 1$), $\tilde{\psi}_{r\epsilon} = \tilde{\psi}_{r'\epsilon'} \circ h_{r\epsilon, r'_{\epsilon'}}$. So $\{\tilde{\psi}_{r\epsilon}\}$ defines an element in $\mathcal{H}^0(\tilde{Z}, \mathcal{O}_{\tilde{Z}}[G'])$ which we denote by $\tilde{\psi}$.

Choose a basis $\psi^{(1)}, \ldots, \psi^{(N)}$ for $\mathcal{H}^0(Z, \mathcal{O}_Z[G])$ and let $\tilde{\psi}^{(1)}, \ldots, \tilde{\psi}^{(N)}$ be corresponding elements in $\mathcal{H}^0(\tilde{Z}, \mathcal{O}_{\tilde{Z}}[G'])$.

Let us now make the following *Assumption* (A_2).

Elements of $\mathcal{H}^0(A, \mathcal{O}_A[G|_A - C])$ don't have common zeros on A (In other words, $\alpha^{(1)}, \ldots, \alpha^{(M)}$ don't have common zeros on A).

Using this assumption, it is easy to check that the map $f : \tilde{Z} \to \mathbf{C}P^{N+M-1}$ defined by:

$$x \to (\tilde{\psi}^{(1)}(x) : \ldots : \tilde{\psi}^{(N)}(x) : \tilde{\alpha}^{(1)}(x) : \ldots : \tilde{\alpha}^{(M)}(x))$$

is a regular map and that the following is true:

(i) $\forall\, t \in \mathbf{C}^1 - 0$ the map $\pi_N f|_{\phi^{-1}p^{-1}(t)} : \phi^{-1}p^{-1}(t) \to \mathbf{C}P^{N-1}$ where π_N is the projection to $\mathbf{C}P^{N-1}$ corresponding to the first N coordinates, coincides with the given embedding $Z \times t \to \mathbf{C}P^{N-1}$ corresponding to G,

(ii) $\pi_N f|_{\overline{B}_0} : \overline{B}_0 \to \mathbf{C}P^{N-1}$ coincides with $B_0 \to \mathbf{C}P^{N-1}$ corresponding to $G|_B$,

(iii) $f|_{\tilde{B}_0 \tilde{B}_0 \cap \overline{A}_0} : \tilde{B}_0 - \tilde{B}_0 \cap \overline{A}_0 \to \mathbf{C}P^{N+M-1}$ is a biregular map;

(iv) $\pi'_M f|_{A_0} : A_0 \to \mathbf{C}P^{M-1}$, where π'_M is the projection to $\mathbf{C}P^{M-1}$ corresponding to the last M coordinates, coincides with the regular map $A \to \mathbf{C}P^{M-1}$ defined by $\mathcal{H}^0(A, \mathcal{O}_A[G|_A - C])$, that is, by: $x \to (\alpha^{(1)}(x) : \ldots : \alpha^{(M)}(x))$.

Let $Z' = f\phi^{-1}(Z_0)$, $A' = f(\overline{A}_0)$, $B' = f(\tilde{B}_0)$, and $g' : Z' \to \mathbf{C}P^{N+M-1}$ the given embedding. We say that (Z', A', B', g') is obtained from (Z, A, B, g) by the construction which we call *"construction (D)"*.

It is clear that if (Z', A', B', g) satisfy to the Assumptions (A_1) and (A_2) then we can repeat the *construction* (D) and get (Z'', A'', B'', g''). Continuing we can get a sequence $\{Z^{(j)}, A^{(j)}, B^{(j)}g^{(j)}\}$.

The main point here is that at each step j we have a family of projective varieties $Z^{(j)}(t), t \in \mathbf{C}^1$, with embeddings $f_t^{(j)} : Z^{(j)}(t) \to \mathbf{C}P^{N_j + M_j - 1}$, such that $\forall\, t \in \mathbf{C}^1 - 0$, $Z^{(j)}(t) = Z^{(j-1)}$, and $\pi_{N_j} f_t^{(j)} : Z^{(j)}(t) \to \mathbf{C}P^{N_j - 1}$, where π_{N_j}, the projection to $\mathbf{C}P^{N_\alpha - 1}$ corresponding to the first N_j coordinates, coincides with $g^{(j-1)} : Z^{(j-1)} \to \mathbf{C}P^{N_{j-1} + M_{j-1}^{-1}}$, $Z^{(j)}(0) = Z^{(j)}$ and $f_0^{(j)} = g^{(j)}$.

DEFINITION 1. *Let X, Y be two complex projective algebraic varieties and $g : X \to \mathbf{C}P^L$, $h : Y \to \mathbf{C}P^M$ their projective embeddings. We say that (X, g) is a projective degeneration of (Y, h) if there exists a family of projective varieties $\{Y(t), t \in \mathbf{C}^1\}$ with embeddings $h_t : Y(t) \to \mathbf{C}P^N$, $N \geq M, L$ and projections $\pi_L : \mathbf{C}P^N \to \mathbf{C}P^L$, $\pi'_M : \mathbf{C}P^N \to \mathbf{C}P^M$ such that $Y(0) = X$, and $\forall\, t \in \mathbf{C}^1 - 0$, $Y(t) = Y$, $\pi_L|_{h_0(Y(0))}$ and $\pi'_M|_{h_t(Y(t))}$, $t \in \mathbf{C}^1 - 0$, are regular maps, $\pi_L h_0 : Y(0) \to P^L$ coincides with g, and $\forall\, t \in \mathbf{C}^1 - 0, \pi'_M h_t : Y(t) \to \mathbf{C}P^M$ coincides with h.*

We can say that after the construction (D) the pair (Z', g') is a projective degeneration of (Z, g).

§3. Degenerations of Veronese surfaces to unions of planes.

Now we apply the construction (D) to a Veronese surface V_n. Let $Z^{(0)} = V_n$, $A^{(0)} = V_n$, $B^{(0)}$ be a straight line on $V_n (\simeq CP^2)$, $g^{(0)} : V_n \to P^N$ be the Veronese embedding of V_n. It is possible to check that we can apply the construction (D), $n-1$ times. We obtain a sequence $(Z^{(j)}, A^{(j)}, B^{(j)}, g^{(j)})$, $j = 0, 1, \ldots, n-1$, where for $j = 1, \ldots, n-1$, the pair $(Z^{(j)}, g^{(j)})$ is a projective degeneration of $(Z^{(j-1)}, g^{(j-1)})$. A direct verification shows that for $j = 1, \ldots, n-1$, $Z^{(j)} = \bigcup_{k=0}^{j} T_k$, $T_0 = CP^2$, $g^{(j)}|_{T_0}$ is the Veronese embedding corresponding to monomials of degree $n-j$, and for $k = 1, \ldots, j$, T_k is the Hirzebruch (ruled) surface F_1, $g^{(j)}|_{T_k}$ corresponds to the linear system $|S_-^{(k)} + (n-j+k)F^{(k)}|$, where $S_-^{(k)}$ is the cross-section of T_k with $(S_-^{(k)})^2 = -1$, $F^{(k)}$ is a fiber of T_k.

Moreover, in $Z^{(j)}$ for $|k-\ell| \geq 2$, $T_k \cap T_\ell = \varnothing$, for $k = 0, 1, \ldots, j-2$, $T_k \cap T_{k+1}$ is a curve which is equal to $S_+^{(k)}$ on T_k ($S_+^{(k)}$ is a cross-section with $(S_+^{(k)})^2 = 1$ for $k \neq 0$ and a straight line on T_0 for $k = 0$) and to $S_-^{(k+1)}$ on T_{k+1}, and also all intersections T_k with T_{k+1} are transversal.

Observe that $A^{(j)} = T_0 \ (\subset Z^{(j)})$, $B^{(j)} = \bigcup_{k=1}^{j} T_j \ (\subset Z^{(j)})$ and $g^{(j)}$ is defined by its restrictions to $T_0, \ldots, T_j$.

Now let $\overline{Z}^{(0)} = Z^{(n-1)}$, $\overline{B}^{(0)}$ be a connected reducible curve on $\overline{Z}$ with $\overline{B}^{(0)} = \bigcup_{k=0}^{n-1} F^{(k)}$, where $F^{(0)}$ is a straight line on T_0 and for $k = 1, \ldots, n-1$ $F^{(k)}$ is a fiber on T_k, $\overline{g}^{(0)} = g^{(n-1)}$, $\overline{A}^{(0)} = \overline{Z}^{(0)}$. Applying the construction (D) to $(\overline{Z}^{(0)}, \overline{A}^{(0)}, \overline{B}^{(0)}, \overline{g}^{(0)})$ we get the corresponding $(\overline{Z}^{(1)}, \overline{A}^{(1)}, \overline{B}^{(1)}, \overline{g}^{(1)})$. It is easy to check that actually we can repeat the construction (D) to get finally a sequence $(\overline{Z}^{(j)}, \overline{A}^{(j)}, \overline{B}^{(j)}, \overline{g}^{(j)})$, $j = 0, 1, \ldots, n-1$. It is is possible to verify that for $j = 1, \ldots, n-1$, $\overline{Z}^{(j)} = \left[\bigcup_{k=0}^{n-1} \bigcup_{\ell=0}^{\min(k,j-1)} T_{k\ell}\right], \cup \left[\bigcup_{r=j}^{n-1} T_j'\right]$ where for $\ell < k$, $T_{k\ell}$ is a quadric F_0, T_j' and all $T_{\ell\ell}$, $\ell = 0, \ldots, j-1$, are CP^2, all T_r', $r = j+1, \ldots, n-1$ (for $j < n-1$) are F_1's. Moreover, for $\ell < k$, $g^{(j)}|_{T_{k\ell}}$ corresponds to the linear system $|P_{k\ell} + Q_{k\ell}|$, where $P_{k\ell}, Q_{k\ell}$ are generators of $T_{k\ell}(\simeq CP^1 \times CP^1)$, $g_{T_j'}^{(j)}$ and $g^{(j)}|_{T_{\ell\ell}}$, $\ell = 0, \ldots, j-1$, correspond to the linear systems of straight lines, and for $r = j+1, \ldots, k-1$ (if $j < n-1$) $g^{(j)}|_{T_r'}$ corresponds to the linear system $|S_-^{(r)} + (r+1-j)F^{(r)}|$ where $S_-^{(r)}$ is the cross-section with $(S_-^{(r)})^2 = -1$ and $F^{(r)}$ is a fiber on T_r'. Finally, all intersections of irreducible components in $Z^{(j)}$ are transversal and could be described by the schematic picture shown in Fig. 1 where

(1) rectangles correspond to quadrics, vertical and horizontal edges of rectangles correspond to two types of generators of quadrics;

(2) triangles correspond to projective planes with edges standing for projective lines;

(3) trapezoids correspond to ruled surfaces F_1 with the vertical edge standing for a fiber, smaller base (resp. greater base) for a cross-section with self-intersection equal -1 (resp. $+1$).

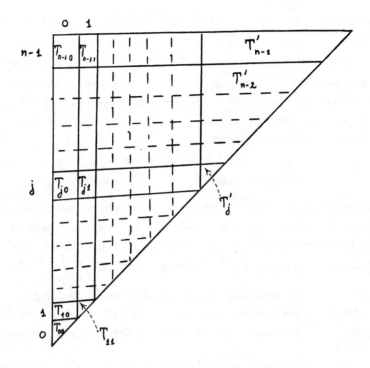

FIGURE 1

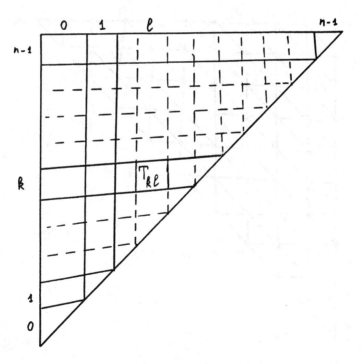

FIGURE 2

Applying the construction (D) for $(\overline{Z}^{(j)}, \overline{g}^{(j)})$ $(j = 1, \ldots, n-2)$ we take $\overline{B}^{(j)} = \bigcup_{k=0}^{n-1} \bigcup_{\ell=0}^{(\min(k,j-1)} T_{k\ell}$, $\overline{A}^{(j)} = \bigcup_{r=j}^{n-1} T'_j$.

Denoting the last step (that is, for $\overline{Z}^{(n-1)}$) T'_j by $T_{n-1\,n-1}$ and using Fig. 1, we can describe $\overline{Z}^{(n-1)}$ by the picture in Fig. 2 where (as above) rectangles $(T_{k\ell}, \ell < k)$ correspond to quadrics $CP^1 \times CP^1$ and triangles $(T_{\ell\ell}, \ell = 0, \ldots, n-1)$ to projective planes CP^2. Clearly $\overline{Z}^{(n-1)} = \bigcup_{k=0}^{n-1} \bigcup_{\ell \le k} T_{k\ell}$. Denote by $E_{k\ell}$ the line bundle on $T_{k\ell}$ corresponding to

$$\begin{cases} (CP^1 \times pt) + (pt \times CP^1) & \text{for } \ell < k \\ \text{projective line} & \text{for } \ell = k. \end{cases}$$

Let $\overline{G}$ be the line bundle on $\overline{Z}^{(n-1)}$ corresponding to a hyperplane section (for the embedding $\overline{g}^{(n-1)}$). It is easy to show that $\overline{G}$ is the only line bundle on $\overline{Z}^{(n-1)}$ such that it's restriction on any $T_{k\ell}$ is equal to $E_{k\ell}$. It follows from that that any projective embedding $\overline{g}$ of $\overline{Z}^{(n-1)}$ with $\deg(\overline{g}(T_{k\ell})) = \begin{cases} 2 & \text{for } \ell < k \\ 1 & \text{for } \ell = k \end{cases}$ could be connected to $\overline{g}^{(n-1)}$ by continuous families of projective embeddings.

Each quadric $T_{k\ell}$ could be degenerated to a union of two planes, say $T_{k\ell 1}$ and $T_{k\ell 2}$, so that $\overline{Z}^{(n-1)}$ will degenerate to a union of planes Y_n which is schematically described by Fig. 3.

Here, as above, triangles correspond to projective planes CP^2 and their edges to projective lines. It is convenient to consider Fig. 3 as being obtained from

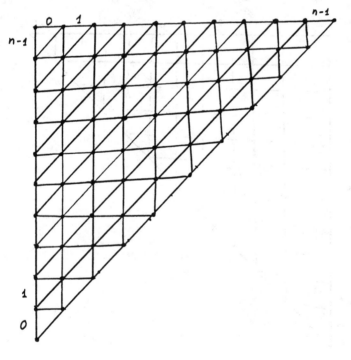

FIGURE 3

Fig. 2 by rectangles breaking apart along a diagonal (which we can associate with the diagonal of the corresponding $CP^1 \times CP^1$).

There is a unique line bundle H on Y_n such that it's restriction to any plane $T_{k\ell\epsilon}, \epsilon = 1, 2$, corresponds to a straight line. Now let $\varphi : \Gamma \to \Delta_\alpha (\Delta_\alpha = \{z \in C, |z| < \alpha)$ be the map describing the degeneration of $Z^{(n-1)}$ to Y_n, that is for certain $t_0 \in \Delta_\alpha - 0$, $\varphi^{-1}(t_0) = \overline{Z}^{(n-1)}$ and $\varphi^{-1}(0) = Y_n$. It is possible to show (taking α sufficiently small) that there exists a projective embedding $\mu : \Gamma \to P^\Lambda$ such that $\mu|_{\varphi^{-1}(0)}$ coincides with a projective embedding h_n of Y_n corresponding to H_n and such that on $\varphi^{-1}(t_0)$, μ embeds quadrics as degree two surfaces and planes as degree one surfaces. Set $\overline{g} = \mu|_{\varphi^{-1}(t_0)} = \mu|_{\overline{Z}^{n-1)}}$. Because $\overline{g}$ is connected to $\overline{g}^{(n-1)}$ by continuous families of projective embeddings (see above) we can think of the pair (Y_n, h_n) as a degeneration of $(\overline{Z}^{(n-1)}, \overline{g}^{(n-1)})$. Finally, we get a chain of degenerations from $(V_n, g^{(0)})$ to $Y_n, h_n)$, that is from a Veronese surface to a union of planes.

Taking generic projections to CP^2 for all steps we get a chain of continuous families of maps $f_t^{(r)} : Y_t^{(r)} \to CP^2$, $t \in \Delta_\alpha$, $r = 0, 1, \ldots, m$, such that $Y_0^{(m)} = Y_n$, and $\exists t_0 \in \Delta_\alpha - 0$ with $Y_{t_0}^{(0)} = V_n$, $f_{t_0}^{(0)}$ a generic projection corresponding to $g^{(0)}$, for $r = 1, \ldots, m$, $f_{t_0}^{(r)} = f_0^{(r-1)}$, $f_0^{(m)}$ a generic projection corresponding to h_n, and for $\forall r = 0, 1, \ldots, m$ all $f_t^{(r)}$ with $t \in \Delta_\alpha - 0$ topologically the same. In particular we can get the branch curve on CP^2 of $f_{t_0}^{(0)} : V_n \to CP^2$ by "a regeneration" of the branch curve of $f_0^{(m)} : Y_n \to CP^2$.

§4. Degenerations of surfaces $F_0(m, n)$ to unions of planes.

For $F_0(m, n)$ we use sequences of constructions (D) almost as in §3. We shall also use notations similar to that of §3. Let $Z^{(0)} = F_0(m, n)$, $g^{(0)}$ be a projective embedding corresponding to $|m\ell_1 + n\ell_2|$, $A^{(0)} = Z^{(0)}$, $B^{(0)} = \ell_1$. Starting with $(Z^{(0)}, A^{(0)}, B^{(0)}, g^{(0)})$ we can perform the construction (D), $m - 1$ times and get a pair $(Z^{(m-1)}, g^{(m-1)}$, where $Z^{(m-1)} = \bigcup_{k=0}^{m-1} T_k$, each T_k is a copy of F_0, $g^{(m-1)}$ is such that for $k = 0, \ldots, m-1, g^{(m-1)}|_{T_k}$ corresponds to $|\ell_1 + n\ell_2|$, all intersections of T_k's on $Z^{(m-1)}$ are transversal and could be described by Fig. 4, where each rectangle stands for a copy of F_0, vertical edges of rectangles correspond to ℓ_2's and horizontal edges to ℓ_1's.

Now let $\overline{Z}^{(0)} = Z^{(m-1)}$, $\overline{A}^{(0)} = \overline{Z}^{(0)}$, $\overline{g}^{(0)} = g^{(m-1)}$ and $\overline{B}^{(0)}$ be a connected curve on $Z^{(n-1)}$ with $\overline{B}^{(0)} = \bigcup_{k=0}^{m-1} \ell_2^{(k)}$ $\ell_2^{(k)}$ a generator ℓ_2 on T_k. Starting with $(\overline{Z}^{(0)}, \overline{A}^{(0)}, \overline{B}^{(0)}, \overline{g}^{(0)})$ we perform the construction (D), $n - 1$ times and come to a pair $(\overline{Z}^{(n-1)}, \overline{g}^{(n-1)})$, where $\overline{Z}^{(n-1)} = \bigcup_{k=0}^{m-1} \bigcup_{\ell=0}^{n-1} T_{k\ell}$, each $T_{k\ell}$ is a copy of F_0, $\overline{g}^{(n-1)}$ is such that for $k, \ell = 0, \ldots, m-1, \overline{g}^{m-1}|_{T_{k\ell}}$ corresponds to $\ell_1 + \ell_2$, all intersections of $T_{k\ell}$'s in $\overline{Z}^{(n-1)}$ are transversal and could be described by Fig. 5, where, as above, rectangles correspond to copies of F_0, their vertical edges, to ℓ_2's and their horizontal edges to ℓ_1's.

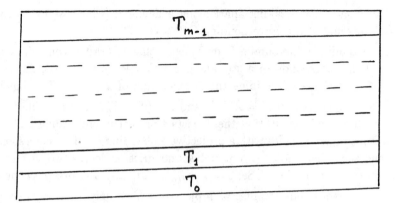

FIGURE 4

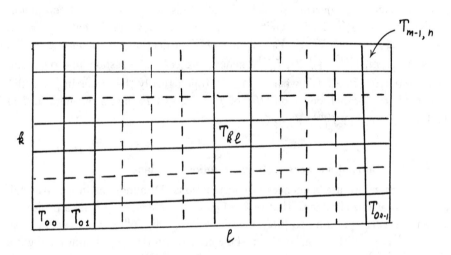

FIGURE 5

Now arguing as in §3, we degenerate each $T_{k\ell}$ to a union of projective planes $T_{k\ell 1}$, $T_{k\ell 2}$ and $\overline{Z}^{(n-1)}$ to a union of planes $Y_{m,n}$ which is schematically described Fig. 6, where the triangles correspond to copies of CP^2 and their edges to projective lines on CP^2's.

There exists a unique line bundle $H_{m,n}$ on $Y_{m,n}$ such that for any irreducible component $T_{k\ell\epsilon}(\epsilon = 1, 2)$ of $Y_{m,n}$, $H_{m,n}|T_{k\ell\epsilon}$ corresponds to a projective line. Let $h_{m,n}$ be a projective embedding of $Y_{m,n}$ corresponding to $H_{m,n}$. As in §3 we can degenerate $(\overline{Z}^{(n-1)}, \overline{g}^{(n-1)})$ to $(Y_{m,n}h_{m,n})$. Finally we get a chain of degenerations from $(F_0(m,n), g^{(0)})$ to $(Y_{m,n}, h_{m,n})$, that is from the projective surface $F_0(m,n)$ to a union of planes.

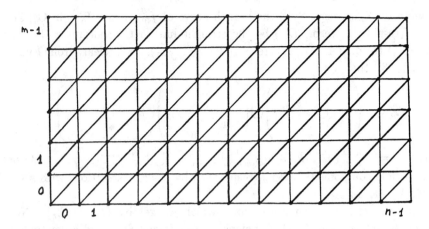

FIGURE 6

Taking generic projections to CP^2 for all steps we get (as in §3) a chain of continuous families of maps $\tilde{f}_t^{(r)} : \tilde{Y}_t^{(r)} \to CP^2$, $t \in \Delta_\alpha$, $r = 0, 1, \ldots, q$, such that $\tilde{Y}_0^{(q)} = Y_{m,n}$, and $\exists t_0 \in \Delta_\alpha - 0$ with $\tilde{Y}_{t_0}^{(0)} = F_0(m,n)$, $\tilde{f}_{t_0}^{(0)}$ is a generic projection of $F_0(m,n)$ (projectively embedded by $g^{(0)}$) to CP^2, for $r = 1, \ldots q$ $\tilde{f}_{t_0}^{(r)} = \tilde{f}_0^{(r-1)}$, $\tilde{f}_0^{(q)}$ is a generic projection corresponding to $h_{m,n}$, and for $r = 0, \ldots, q$ all $\tilde{f}_t^{(r)}$ with $t \in \Delta_\alpha - 0$ are topologically the same.

Thus we can get the branch curve on CP^2 of $\tilde{f}_{t_0}^{(0)} : F_0(m,n) \to CP^2$ by "regeneration" of the branch curve of $\tilde{f}_0^{(q)} : Y_{n,m} \to CP^2$.

§5. Braid monodromies for branch curves of $\tilde{f}_0^{(q)}$ and $f_0^{(n)}$.

Our arguments for $\tilde{f}_0^{(q)}$ and $f_0^{(n)}$ are very similar. We shall consider $\tilde{f}_0^{(q)}$ and in parenthesis show what changes must be made for $f_0^{(n)}$.

It is clear that up to continuous deformations our projective embedding of $Y_{m,n}$ (resp. Y_n) in a high-dimensional CP^N is well defined by a choice of vertices a_j, $i = 0, \ldots, n$, $j = 0, \ldots, m$, (resp. $0 \le i \le j$, $j = 0, \ldots, n$) of the simplical complex of Fig. 6 (resp. Fig. 3) in CP^N. The points a_{ij} have to be in general position in CP^N, so that the planes T_{ij0}, T_{ij1}, $i = 0, \ldots, n-1$, $j = 0, \ldots m-1$, (resp. $0 \le i \le j$, $j = 0, \ldots, n-1$) defined respectively by triples $(a_{ij}, a_{i+1j}, a_{i+1j1})$, and $(a_{ij}, a_{ij+1}, a_{i+1,j+1})$ will intersect each other according to Fig. 6 (resp. Fig. 3. We take $T_{ii0} = \varnothing$). Setting $A_{ij} = \tilde{f}_0^{(q)}(a_{ij})$ (resp. $f_0^{(m)}(a_{oj})$), we obtain points A_{ij}, $i = 0, \ldots, n, j = 0, \ldots, m$, (resp. $0 \le i \le j$, $j = 0, \ldots, n$), which define uniquely the branch curve of $\tilde{f}_0^{(q)}$ (resp. of $f_0^{(m)}$) in CP^2. We shall use an explicit description. First we replace CP^2 by $\mathbf{C}^2$ with coordinates (x, y). For explicit computations, it is convenient to choose the points A_{ij} such that $A_{ij} = (u_{ij}, v_{ij})$, where $u_{ij} = c^{i+dj}$, $v_{ij} = (u_{ij})^2$, $c, d \in \mathbb{R}$, $c \gg 1$, $d \gg 1$.

Denote by S_{mn} (resp. $S^{(n)}$) the branch curve of $\tilde{f}_0^{(q)}$ (resp. $f_0^{(n)}$) in $\mathbf{C}^2$, and let Hij (resp. V_{ij}, resp. D_{ij}) be the straight line in $\mathbf{C}^2$ containing the pair (A_{ij}, A_{i-1j}) (resp. (A_{ij}, A_{ij-1})), resp. $(A_{ij}, A_{i-1,j-1})$). It is clear (from Fig. 6 (resp. Fig. 3)) that

$$S_{mn} = \left(\bigcup_{\substack{i=1,\ldots,n \\ j=1,\ldots,m-1}} \mathcal{H}_{ij} \right) \cup \left(\bigcup_{\substack{i=1,\ldots,n-1 \\ j=1,\ldots,m}} V_{ij} \right) \cup \left(\bigcup_{\substack{i=1,\ldots,n \\ j=1,\ldots,m}} D_{ij} \right)$$

(resp. $S^{(n)} = (\bigcup_{i=1}^{n-1} \bigcup_{j=i}^{n-1} \mathcal{H}_{ij}) \cup (\bigcup_{i=1}^{n-1} \bigcup_{j=i+1}^{n} V_{ij}) \cup (\bigcup_{i=1}^{n-1} \bigcup_{j=i+1}^{n} D_{ij}))$. Order all irreducible components of S_{mn} (resp. $S^{(n)}$) by their intersections with a line $x = M$, $M \in \mathbb{R}$, $M \gg c, d$.

Denote the ordered sequence of lines which we get by $\{L_1, \ldots, L_r\}$. Now consider the set of all A_{ij}'s which belong to S_{mn} (resp. $S^{(n)}$) and order this set by x-coordinates of points. We get a sequence $\{\alpha_1, \ldots, \alpha_\nu\}$. It is clear that for each k, $1 \le k \le v$, there exist exactly two points in $\{\alpha_1, \ldots, \alpha_\nu\}$ belonging to L_k. We get two functions $f(k)$, $g(k)$ with $f(k) \le g(k)$ and $\alpha_{f(k)}$, $\alpha_{g(k)} \in L_k$.

From $x(A_{ij}) = c^{i+dj}$, $y(A_{ij}) = c^{2(i+dj)}$ it is easy to deduce the following:

(1) $\forall\, k_1, k_2$, $1 \le k_1 < k_2 \le r$ we have $g(k_1) \le g(k_2)$ and if $g(k_1) = g(k_2)$ then $f(k_1) < f(k_2)$;

(2) $\forall\, i$, $1 \le i \le \nu$ and $\forall\, k_1, k_2 \le \max_{g(k)=i} k$ we have either $L_{k_1} \cap L_{k_2} = \alpha_i$ or $x(L_{k_1} \cap L_{k_2}) < x(\alpha_i)$, $y(L_{k_1} \cap L_{k_2}) < y(\alpha_i)$,

(3) $\forall\, k_1, k_2$, $k_1, k_2 \in \{1, 2, \ldots, r\}$ with $L_{k_1} \cap L_{k_2} \notin \{\alpha_1, \ldots, \alpha_v\}$ we have that $L_{k_1} \cap L_{k_2}$ is a double point of S_{mn} (resp. $S^{(n)}$).

To obtain the braid monodromy of S_{mn} (resp. $S^{(n)}$), we use the following:

PROPOSITION 1. *Let $\{L_1, \ldots, L_r\}$ be a sequence of lines in the (x, y)-plane $\mathbf{C}^2$ defined over $\mathbb{R}$. Assume that all $L_1, \ldots, L_r$ have different and positive slopes and that $\{L_1, \ldots, L_r\}$ is ordered by the slopes of the L_i's. Let $\{\alpha_1, \ldots, \alpha_\nu\}$ by a set of real points in $\mathbf{C}^2$ ordered by their x-coordinates, which are all different, and such that each α_i, $1 \le i \le \nu$, belongs to $\bigcup_{k=1}^r L_k$ and $\forall\, k$, $1 \le k \le r$, there are exactly two points in $\{\alpha_1, \ldots, \alpha_\nu\}$ belonging to L_k, say $\alpha_{f(k)}$, $\alpha_{g(k)}$, $f(k) < g(k)$. Suppose that we have the following:*

(1) $\forall\, k_1, k_2 \in (1, \ldots, r)$, $k_1 < k_2$, *either* $g(k_1) < g(k_2)$ *or* $g(k_1) = g(k_2)$ *and* $f(k_1) < f(k_2)$;

(2) $\forall\, i \in (1, \ldots, \nu)$ *and* $\forall\, k_1, k_2 \le \max_{g(k)=i} k$ *either* $L_{k_1} \cap L_{k_2} = \alpha_i$ *or* $x(L_{k_1} \cap L_{k_2}) < x(\alpha_i)$, $y(L_{k_1} \cap L_{k_2}) < y(\alpha_i)$;

(3) $\forall\, k_1, k_2 \in (1, \ldots, r)$ *if* $L_{k_1} \cap L_{k_2} \notin \{\alpha_1, \ldots, \alpha_v\}$ *then* $L_{k_1} \cap L_{k_2}$ *is a double point of* $\bigcup_{k=1}^r L_k$.

Choose $M \gg 1$ and let $\tilde{\mathbf{C}}^1 = \{(x, y) \in \mathbf{C}^2 | x = M\}$. Set $S = \bigcup_{k=1}^r L_k$, $Q = S \cap \lambda p \tilde{\mathbf{C}}^1$, $q_k = L_k \cap \tilde{\mathbf{C}}^1$, $1 \le k \le r$, $S_i = \bigcup L_k \{k | \alpha_i \in L_k\}$, $1 \le i \le \nu$, $T_i = S_i \cap \tilde{\mathbf{C}}^1$. Connecting all points of T_i to each other by simple paths in $\tilde{\mathbf{C}}^1$ going under the real axis we get for each $i \in \{1, \ldots, \nu\}$ an embedding of braid groups $\psi_i : B[\tilde{\mathbf{C}}^1, T_i] \to B_r = B[\tilde{\mathbf{C}}^1, Q]$. Let $\Delta_i^2 = \psi_i(\Delta^2(B[\tilde{\mathbf{C}}^1, T_i]))$. For all $a, b \in \{1, \ldots, r\}$,

$a < b$, $q_a \notin T_{f(b)} \cup T_{g(b)}$, denote by $\tilde{z}_{b,a}$ a simple path in $\tilde{\mathbf{C}}^1$ which is defined as follows: $\tilde{z}_{b,a}$ starts at q_b, goes to the left under the real axis of $\tilde{\mathbf{C}}^1$ until the line $Re(y) = \min_{q_{b'} \in T_{g(b)}} q_{b'} - \epsilon$ ($\epsilon > 0$ is small) (of course all $q_{b'}$ are real), then goes up along the line $Re\,y = \min_{q_{b'} \in T_{g(b)}} q_{b'} - \epsilon$ until it intersects the real axis and after that goes to the left over the real axis to the point q_a (see Fig. 7).

Let $\tilde{Z}_{b,a}$ be the element (half-twist) of $B_r = B[\tilde{\mathbf{C}}^1, Q]$ defined by $\tilde{z}_{b,a}$ and $\tilde{Z}_b^{(2)} = \prod_{\substack{1 \le a < b \\ q_a \notin T_{f(b)} \cup T_{g(b)}}} \tilde{Z}_{b,a}^2$, where the order in the product coincides with the natural order of a's and $Z_b^{(2)} = 1$ if for each $a < b (a \in (1, \ldots, r)) q_a$ belongs to $T_{f(b)} \cup T_{g(b)}$.

Set $C_i = \prod_{\{b \mid f(b) = i\}} \tilde{Z}_b^{(2)}$, where the order in the product coincides with the (natural) order of b's, and $C_i = 1$ if there are no b's with $f(b) = i$.

Then

(a) the braid monodromy of S (corresponding to x-projection and initial point $x = M$) is given by

(*1)
$$\Delta_{(r)}^2 = \Delta^2(B[\tilde{\mathbf{C}}^1, Q]) = \prod_{i=\nu}^{1} (C_i \cdot \Delta_i^2);$$

(b) the system of good paths on the x-plane corresponding to (*1) is such that they all initially go from $x = M$ along the real axis till the point $x(\alpha_\nu) + \epsilon$ ($\epsilon > 0$ is small), then the first continues along the real axis to $x(\alpha_\nu)$ and all others go from $x(\alpha_\nu) + \epsilon$ along the lower semicircle of radius ϵ centered at $x(\alpha_\nu)$ until $x(\alpha_\nu) - \epsilon$ and after that never return to the half-plane $Re\,x \ge x(\alpha_\nu) - \epsilon$.

PROOF. $\forall r' \in (1, \ldots, r)$ set $S(r') = \bigcup_{k=1}^{r'} L_k$ and let $\{\alpha_1^{(r')}, \ldots, \alpha_{\nu(r')}^{(r')}\}$ be the subsequence of $\{\alpha_1, \ldots, \alpha_\nu\}$ consisting of all α_i's which belong to $S(r')$.

It is easy to check that all suppositions of the Proposition made for S and $\{\alpha_1, \ldots, \alpha_\nu\}$ are true also for $S(r')$ and $\{\alpha_1^{(r')}, \ldots, \alpha_{\nu(r')}^{(r')}\}$, $\forall r' \in (1, \ldots, r)$.

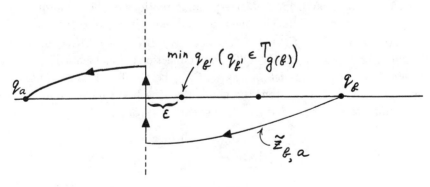

FIGURE 7

Thus we can use induction on r: Let $\hat{S} = S(r-1)$, $\hat{Q} = \hat{S} \cap \tilde{\mathbf{C}}^1$, $\{\hat{\alpha}_1, \ldots, \hat{\alpha}_{\hat{\nu}}\}$ $= \{\alpha_1^{(r-1)}, \ldots, \alpha_{\nu(r-1)}^{(r-1)}\}$ and $\hat{\Delta}_i^2$, $\hat{C}_i$ be defined for $\hat{S}$, $\{\hat{\alpha}_1, \ldots, \hat{\alpha}_{\hat{\nu}}\}$ the same way as Δ_i^2, C_i were defined for S, $\{\alpha_1, \ldots, \alpha_\nu\}$. Clearly it is enough to prove the following: If the braid monodromy for $\hat{S}$ (and x-projection) could be given by

$$(*2) \qquad\qquad \Delta_{(r-1)}^2 = \Delta^2(B[\tilde{\mathbf{C}}^1, \hat{Q}] \underset{\sim}{=} \prod_{i=\hat{\nu}}^{1} (\hat{C}_i \cdot \hat{\Delta}_i^2)$$

with the condition (b) formulated above then the braid monodromy for S (and x-projection) is given by (*1) and the condition (b) is satisfied.

We thus assume that the braid monodromy for $\hat{S}$ could be given by (*2). Let $\{\hat{\gamma}_\nu, \ldots, \hat{\gamma}_1\}$ be the corresponding ordered system of good paths on the x-plane from $x = M$ to the projections of singular points of $\hat{S}$. If $\hat{\alpha}_{\hat{\nu}} \neq \alpha_\nu$, we do not use the condition (b) for (*2). If $\hat{\alpha}_{\hat{\nu}} = \alpha_\nu$, then our inductive assumption includes also the condition (b) for $\{\hat{\gamma}_\nu, \ldots, \hat{\gamma}_1\}$ related to $x(\alpha_\nu)$. Introduce υ' as follows: $\upsilon' = \upsilon$ if $\hat{\alpha}_{\hat{\nu}} \neq \alpha_\nu$ and $\upsilon' = \upsilon - 1$ if $\hat{\alpha}_{\hat{\nu}} = \alpha_\nu$.

Let $L(0)$ be a line in $\mathbf{C}^2$ defined over $\mathbb{R}$ with a positive slope, passing through α_ν and such that it is very close to the line $x = x(\alpha_\nu)$.

We can assume that x-projections of all $L(0) \cap L_k$, $k \in \{1, \ldots, r-1\}$ are in $(x(\alpha_\nu) - \epsilon, x(\alpha_\nu)]$. We can change $\{\hat{\gamma}_{\upsilon'}, \ldots, \hat{\gamma}_1\}$ if necessary by replacing the segment $[x(\alpha_\nu) - \epsilon, x(\alpha_\nu) + \epsilon]$ by the lower-semicircle c_ν (see (b) above). For all $x_0(\in \mathbf{C})$ set $\tilde{\mathbf{C}}_{x_0} = \{(x, y) \in \mathbf{C}^2 | x = x_0\}$ and $q(x_0) = y(\tilde{\mathbf{C}}_{x_0} \cap L(0))$. When x is moving along c_ν from $x(\alpha_\nu) - \epsilon$ to $x(\alpha_\nu) + \epsilon$ the corresponding $q(x)$ is moving along a big semicircle in the y-plane (centered on the real axis) from $q(x(\alpha_\nu) - \epsilon)$ to $q(x(\alpha_\nu) + \epsilon)$.

When x moves on the real segment from $x(\alpha_\nu) + \epsilon$ to M (the initial point) $q(x)$ goes along a segment of the y-plane from $q(x(\alpha_\nu) + \epsilon)$ up to $q(M)$.

Taking $L(0)$ close enough to $x = x(\alpha_\nu)$ we can assume that $|q(x(\alpha_\nu) - \epsilon)|$, $|q(x(\alpha_\nu) + \epsilon)|$ are very big and that all motions of $q(\tilde{x})$, when $\tilde{x}$ moves along any of $\hat{\gamma}_{\upsilon'}, \ldots, \hat{\gamma}_1$ are far away from the corresponding motions of the y-coordinates of $L_k \cap \mathbf{C}_x$, $k = 1, \ldots, r-1$, which define the braid monodromy of $\hat{S}$ for $\hat{\gamma}_{\upsilon'}, \ldots, \hat{\gamma}_1$.

Let $S(0) = \hat{S} \cup L(0)$, $Q(0) = \hat{S} \cap \tilde{\mathbf{C}}$ and $\psi_0 : B[\tilde{\mathbf{C}}, \hat{q}] \to B[\tilde{\mathbf{C}}, Q(0)]$ be the embedding corresponding to $\hat{Q} \subset Q(0)$ and connections of elements of $\hat{Q}$ by real segments, for $i = \upsilon, \ldots, 1$, let $\hat{\beta}(\hat{\gamma}_i) \in B[\tilde{\mathbf{C}}, Q]$ be the braid corresponding to $\hat{\gamma}_i$ by (*2). Now it follows from the above that the braid monodromy of $S(0)$ corresponding to each $\hat{\gamma}_i$, $i = \upsilon', \ldots, 1$, is given by $\psi_0(\hat{\beta}(\hat{\gamma}_i))$. Set

$$T_\nu(0) = \begin{cases} T_{\hat{\nu}} \cup q(M) & \text{if } \hat{\alpha}_{\hat{\nu}} = \alpha_\nu \\ q(M) & \text{if } \hat{\alpha}_{\hat{\nu}} \neq \alpha_\nu \end{cases}$$

and let $\eta_0 : B[\tilde{\mathbf{C}}, T_\nu(0)] \to B[\tilde{\mathbf{C}}, Q(0)]$ be the embedding corresponding to $T_\nu(0) \subset Q(0)$ and connections by real segments of elements of $T_\nu(0)$. Actually we use the following fact here: if $\alpha_\nu \in L_{k_1} \cap L_{k_2}$, and $k_1 < r < k_2$, then $\alpha_\nu \in L_r$. This follows immediately from $g(k_1) \leq g(r) \leq g(k_2)$, and $\nu = g(k_1)$, $\nu = g(k_2)$.

Let $\Delta_\nu^2(0) = \eta_0(\Delta^2(B[\tilde{\mathbf{C}}, T_\nu(0)])$. Denote by Γ the oriented real segment from M to $x(\alpha_\nu)$. From our choice of $L(0)$ and from (2) above, it follows easily that the braid monodromy of $S(0)$ corresponding to Γ is given by $\Delta_\nu^2(0)$.

Let e be a simple path on $\mathbf{C}$ from $q(M)$ to q_r obtained from the real segment $[q(M), q_r]$ by replacement by semicircles of small segments around all such point $q \in (q(M), q_r)$ for which the line L in $\mathbf{C}^2$ passing through α_ν and q is not in general position to $\hat{S}$ outside of α_ν. Let $\omega : [0,1] \to \tilde{\mathbf{C}}$ be a parametrization of e and $L(t)$, $t \in [0,1]$, be the line in $\tilde{\mathbf{C}}^2$ passing through α_ν and $\omega(t)$. Set by $S(t) = \hat{S} \cup L(t)$, $Q(t) = S(t) \cap \tilde{\mathbf{C}}$. Considering the family $\{S(t), t \in [0,1]\}$ together with x-projection we will get from $\{\hat{\gamma}_\nu, \ldots, \hat{\gamma}_1\}$ a continuous family $\{\hat{\gamma}_\nu(t), \ldots, \hat{\gamma}_1(t)\}$ of systems of good paths on the x-plane from M to x-projections of singular points of $S(t)$. It is clear that $\forall\, t \in [0,1]$ and for $j = \upsilon, \ldots, 1$ the braid monodromy of $S(t)$ corresponding to $\hat{\gamma}_j(t)$ is obtained from the braid monodromy of $S(0)$ corresponding to $\hat{\gamma}_j$ by moving $q(M)$ to $q(t)$ along the real segment from $q(M)$ to $q(t)$.

It is clear also that $S(1) = S$ and $Q(1) = Q$. Set $\gamma_j = \hat{\gamma}_j(1)$, $j = \upsilon, \ldots, 1$. If $\alpha_{f(r)} \in \hat{S}$ then denote by Γ' the element γ_i corresponding to $x(\alpha_{f(r)})$.

It easily follows from the above that $\forall\, \gamma_j \neq \Gamma'$, Γ, the braid monodromy of S corresponding to γ_j is given by $\psi_1(\hat{\beta}(\hat{\gamma}_j))$ were $\psi_1 : B[\tilde{\mathbf{C}}^1, \hat{Q}] \to B[\tilde{\mathbf{C}}^1, Q]$ corresponds to $\hat{Q} \subset Q$ and connections of elements of $\hat{Q}$ by real segments.

From (2) above it follows that the braid monodromy of S corresponding to Γ is given by Δ_ν^2. From our definition of C_i it is easy to see that $C_\nu = 1$.

If $\alpha_{f(r)} \notin \hat{S}$ denote by Γ' a simple path from M to $x(\alpha_{f(r)})$ such that Γ' does not intersect $\bigcup_{i=1}^\upsilon \gamma_i$ outside of M and Γ' is between γ_{j_0+1} and γ_{j_0} where j_0 is defined as follows: $\hat{\beta}(\hat{\gamma}_{j_0})$ is the leftmost factor in $(\hat{C}_{i_0} \cdot \hat{\Delta}_{i_0}^2)$ from (*2) where i_0 is maximal among i with $x(\hat{\alpha}_i) < x(\alpha_{f(r)})$.

Let $x_1, \ldots, x_\varsigma$ be x-projections of intersections points of L_r with $\hat{S}$ which are different from $\alpha_{f(r)}$, $\alpha_{g(r)}$ $(\alpha_{g(r)} = \alpha_\nu)$. Set $\delta_0 = \Gamma'$ and denote by $\delta_1, \ldots, \delta_\varsigma$ a system of paths such that: (i) $\delta_\varsigma, \ldots, \delta_1, \delta_0$ is a good ordered system of paths in x-plane from M to $x_\varsigma, \ldots, x_1$, $x_{f(r)}$; (ii) each of $\delta_1, \ldots, \delta_\varsigma$ does not intersect $\bigcup_{i=\upsilon}^1 \gamma_i$ outside of M and all (iii) $\delta_\varsigma, \ldots, \delta_1$ go very close to Γ' until they come to a small neighborhood of $x_{f(r)}$. We add $\{\delta_\varphi, \ldots, \delta_1, \delta_0\}$ to $\{\gamma_\upsilon, \ldots, \gamma_1\}$ as follows: If $\gamma_{f(r)} \in \hat{S}$ we identify δ_0 with Γ' and put $\delta_\varsigma, \ldots, \delta_1$ immediately before it; and if $\alpha_{f(r)} \notin \hat{S}_r$ we put $\delta_\varsigma, \ldots, \delta_0$ between γ_{j_0+1} and γ_{j_0} defined as above. Thus we get a good ordered system of paths, defining a braid monodromy for S.

From induction and our remarks above, it follows that this system gives the following formula for braid monodromy:

$$(*3) \qquad \Delta_{(r)}^2 = \prod_{i=\nu}^{f(r)+1} (C_i \cdot \Delta_i^2) \cdot C'_{f(r)} \cdot \prod_{\ell=\varsigma}^{0} A_\ell \cdot \prod_{i=f(r)-1}^{1} (C_i \Delta_i^2)$$

where each A_ℓ is the braid corresponding to $\delta_\ell, \ell = \varsigma, \ldots, 0$, and

$$C'_{f(r)} = \begin{cases} 1 & \text{when } \alpha_{f(r)} \notin \hat{S} \\ \hat{C}_{i'} & \text{with } \hat{\alpha}_{i'} = \alpha_{f(r)}f \text{ when } \alpha_{f(r)} \in \hat{S}. \end{cases}$$

Let $\underline{z}_{b,a}$, $1 \le a < b \le r$, be a simple path from q_b to q_a in $\tilde{\mathbf{C}}^1$ which goes under the real axis, $\underline{Z}_{b,a}$ be the braid (half-twist) in $B[\mathbf{C}^1, Q]$ defined by $\underline{z}_{b,a}$. Set

$$\underline{Z}_r^{(2)} = \prod_{\substack{a < r \\ \text{with} \\ q_a \in T_{g(r)}}} \underline{Z}_{r,a}^2, \underline{\underline{Z}}_r^{(2)} = \prod_{\substack{a < r \\ \text{with} \\ q_a \in T_{f(r)}}} \underline{Z}_{r,a}^2 \ (\underline{\underline{Z}}_r^{(2)} = 1 \text{ if } \alpha_{f(r} \notin \hat{S}),$$

where the order in the products is opposite to the natural order of a's. It is clear that $\Delta_{(r)}^2 = \underline{Z}_r^{(2)} \cdot \tilde{Z}_r^{(2)} \cdot \underline{\underline{Z}}_r^{(2)} \cdot \hat{\Delta}_{r-1}^2$ and that $\underline{Z}_r^{(2)} \cdot \tilde{Z}_r^{(2)} \cdot \underline{\underline{Z}}_r^{(2)}(= (\hat{\Delta}_{r-1})^{-1} \cdot \Delta_{(r)}^2)$ commutes with any factor in $\hat{\Delta}_{(r-1)}^2$ given by (*2). Thus we can write

$$(*4) \qquad \Delta_{(r)}^2 = \hat{\Delta}_{\hat{\nu}}^2 \cdot \prod_{i=\nu-1}^{f(r)+1} (C_i \cdot \Delta_i^2) \cdot \hat{C}_{i'} \cdot \underline{Z}_r^{(2)} \cdot \tilde{Z}_r^{(2)} \cdot \underline{\underline{Z}}_r^{(2)} \cdot \hat{\Delta}_{i'}^2 \cdot \prod_{i=f(r)-1}^{1} C_i \cdot \Delta_i^2,$$

where $\hat{\Delta}_{i'}^2 = \hat{C}_{i'} = 1$ if $\alpha_{f(r)} \notin \hat{S}$, and when $\alpha_{f(r)} \in S'$, i' is defined by $\hat{\alpha}_{i'} = \alpha_{f(r)}$.

Let $a_0 = \min_{\substack{a < r \\ q_a \in T_{g(r)}}} a$, so that $T_{g(r)} = \{q_{a_0}, q_{a_0+1}, \ldots, q_r\}$.

From (1) above, it follows that $\forall a \in \{a_0, a_0 + 1, \ldots, r-1\}$ we have $f(a) < f(r)$. This means that all points $q_{a_0}, q_{a_0+1}, \ldots, q_r$ are on the right side of a domain in which all braids which are factors in $\prod_{i=\nu-1}^{f(r)+1}(C_i \cdot \Delta_i^2) \cdot \hat{C}_{i'}$ actually are acting (that is, we can represent these braids by diffeomorphisms which are identities outside of such a domain). Hence $\underline{Z}_r^{(2)}$ commutes with all factors in $\prod_{i=\nu-1}^{f(r)+1}(C_i \cdot \Delta_i^2)\hat{C}_{i'}$. If $\hat{\nu} < \nu$ then $\underline{Z}_r^{(2)} = 1$ and $\Delta_\nu^2 = 1$. If $\hat{\nu} = \nu$ it is evident that $\hat{\Delta}_{\hat{\nu}}^2 \underline{Z}_r^{(2)} = \Delta_\nu^2$. Thus we get from (*4)

$$(*5) \quad \Delta_{(r)}^2 = \prod_{i=\nu}^{f(r)+1} (C_i \cdot \Delta_i^2) \cdot \hat{C}_{i'} \cdot \tilde{Z}_r^{(2)} \cdot \underline{\underline{Z}}_r^{(2)} \cdot \hat{\Delta}_{i'}^2 \cdot \prod_{i=f(r)-1}^{1} (C_i \cdot \Delta_i^2) \text{ (we use } C_\nu = 1).$$

Comparing (*5) with (*3), we see that $\prod_{\ell=\varsigma}^0 A_\ell = \tilde{Z}_r^{(2)} \cdot \underline{\underline{Z}}_r^{(2)} \cdot \hat{\Delta}_{i'}^2$. From $\delta_0 = \Gamma'$ and $\psi_1(\hat{\beta}(\Gamma')) = \hat{\Delta}_{i'}^2$, when $\alpha_{f(r)} \in \hat{S}$, it follows that $A_0 = \hat{\Delta}_{i'}^2 \circ A_0'$ where A_0' is conjugate to $\underline{\underline{Z}}_r^{(2)}$ in the subgroup of $B[\tilde{\mathbf{C}}^1, Q]$ generated by all $\underline{Z}_{r,a}^2$ $a = 1, \ldots, r-1$.

Denote by F the subgroup of $B[\mathbf{C}^1, Q]$ generated by all $\tilde{Z}_{r,a}^2, q_a \notin T_{f(r)} \cup T_{g(r)}$, $\underline{Z}_{r,b}^2$, $q_b \in T_{g(r)} - q_r$ and $\underline{\underline{Z}}_r^{(2)}$. It is clear that F is a free group. A simple geometrical analysis shows that actually A_0' is conjugated to $\underline{\underline{Z}}_r^{(2)}$ in F and that each $A_\ell, \ell = \varsigma, \ldots, 1$, is an element of F which is conjugated to one of $\tilde{Z}_{r,a}^2, q_a \notin T_{f(r)}$. Because $A_\varsigma \cdot \ldots \cdot A_1 \cdot A_0' = \tilde{Z}_r^{(2)} \cdot \underline{\underline{Z}}_r^{(2)} = \prod_{q_a \notin T_{f(r)} \cup T_{g(r)}} \tilde{Z}_{r,a}^2 \cdot \underline{\underline{Z}}_r^{(2)}$ we can use a modification of E. Artin's Theorem (see [1]) and get that

$$(*6) \qquad A_\varsigma \cdot \ldots \cdot A_1 \cdot A_0' \underset{\sim}{=} \prod_{q_a \notin T_{f(r)} \cup T_{g(r)}} \tilde{Z}_{r,a}^2 \cdot \underline{Z}_r^{(2)}.$$

It is clear that $\hat{\Delta}_{i'}^2$ commutes with A_0' and with all $A_\varsigma, \ldots, A_1$. That means that we can replace A_0' in $(*6)'$ by $A_0 = \hat{\Delta}_{i'}^2 \cdot A_0'$ and apply the same sequence of elementary transformations, which transformed $A_\varsigma \ldots A_1 \cdot A_0'$ in the right side of $(*6)'$, to $A_\varsigma \cdot \ldots \cdot A_1 \cdot A_0$. This will give

$$(*6)' \qquad A_\varsigma \cdot \ldots \cdot A_1 \cdot A_0 \underset{\sim}{=} \prod_{a \notin T_{f(r)} \cup T_{g(r)}} \tilde{Z}_{r,a}^2 \cdot \hat{\Delta}_{i'}^2 \cdot \underline{Z}_r^{(2)} \underset{\sim}{=} \tilde{Z}_r^{(2)} \cdot (\hat{\Delta}_{i'} \underline{Z}_r^{(2)}).$$

Because $(\hat{\Delta}_{i'}^2 \cdot \underline{Z}_r^{(2)}) = \Delta_{f(r)}^2$ and $C_{f(r)} = \hat{C}_{i'} \cdot \tilde{Z}_r^{(2)}$ we get from $(*3)$ and $(*6)$ that the braid monodromy for S (and x-projection) could be given by the following formula:

$$\Delta_{(r)}^2 = \prod_{\nu=1}^{1} (C_i \Delta_i^2)$$

The condition (b) (see p.10) is also satisfied. □

§6. Regeneration to unions of quadrics.

In this section we show how to transform the formula $(*1)$ from §5 when we want to obtain braid monodromies for branch curves of generic projections (to CP^2) of the surface $\overline{Z}^{(n-1)}$ from §4 (resp. $Z^{(n-1)}$ from §3) which is a regeneration of $Y_{n,m}$ (resp. Y_n) (see §3 and 4) to a union of quadrics (resp. quadrics and plains). As in §5, we consider mainly the case of $Y_{n,m}$ and indicate in parenthesis the changes necessary for Y_n.

We use the explicit construction of §5 (see p.8). For regeneration we have to replace each pair $\{T_{ij0}, T_{ij1}\}$ (when $T_{ij0} \neq \varnothing$) by a quadric Q_{ij} such that the new configuration will correspond to Fig. 5 (resp. Fig. 2).

Consider $f_0^{(q)}$ (resp. $f_0^{(n)}$) as a composition $\tilde{\varphi}_0(q) \circ \Pi$ (resp. $\varphi_0^{(n)} \circ \Pi$) where Π is a generic projection to CP^3 and $\tilde{\varphi}_0^{(q)}$ (resp. $\varphi_0^{(n)}$) corresponds to a generic projection from CP^3 to CP^2. Take a (generic) affine part $\mathbf{C}^3$ of CP^3 with coordinates (x, y, z). Set $A_{ij}' = \Pi(a_{ij})$. We can choose a_{ij} so that if $A_{ij}' = (u_{ij}, v_{ij}, w_{ij})$ then (u_{ij}, v_{ij}) are as above (see p.8) (in particular $v_{ij} = (u_{ij})^2$) and $w_{ij} = (u_{ij})^3$.

Set $T_{ij0}' = \Pi(T_{ij0})$, $T_{ij1}' = \Pi(T_{ij1})$. Let P_{ij0}, P_{ij1} be linear forms in CP^3, which define T_{ij0}' and T_{ij1}', and $\overline{P}_{ij0}, \overline{P}_{ij1}$ linear forms in CP^3 which vanish respectively at triples: $(A_{ij}', A_{i+1j}', A_{ij+1}')$ and $(A_{ij+1}', A_{i+1j}', A_{i+1j+1}')$. Define $K_{ij}'(\lambda) = P_{ij0}P_{ij1} + \lambda \overline{P}_{ij0}\overline{P}_{ij1}$, $|\lambda| < \epsilon$. Let $Y_{n,m}' = \Pi(Y_{n,m})$, (resp. $Y_n' = \Pi(Y_n)$). Replacing each pair T_{ij0}', T_{ij1}' (when $T_{ij0} \neq \varnothing$) by the surface $Q_{ij}(\lambda) = \{K_{ij}(\lambda) = 0\}$ we get a family $\{Y_{n,m}'(\lambda))\}$ (resp. $Y_n'(\lambda)$ with $Y_{n,m}'(0) = Y_{n,m}'$ (resp. $Y_n'(0) = Y_n'$).

Denote by $\{Y_{n,m}(\lambda)\}$ (resp. $Y_n(\lambda)\}$) an (abstract) family obtained from $\{Y_{n,m}'(\lambda)\}$ (resp. $Y_n'(\lambda)$) by removing all intersections of quadrics (resp. quadrics and planes) which are not present in Fig. 5 (resp. Fig. 2). It is easy to show that there exists a family of embeddings of $Y_{n,m}(\lambda)$ (resp. $Y_n(\lambda)$) into a high-dimensional space CP^N such that the given embeddings $Y'_{n,m}(\lambda) \to CP^3$ (resp. $Y'_n(\lambda) \to CP^3$) will correspond to generic projection Π from CP^N to CP^3. We can take $\{Y_{n,m}(\lambda)\}$ (resp. $Y_n(\lambda)$) with projective embeddings to CP^N as a representative for the regeneration $\overline{Z}^{(n-1)}$ (of §3) from $Y_{n,m}$ (resp. $\overline{Z}^{(n-1)}$ (of §3) from Y_n).

Let $\tilde{f}_\lambda : Y_{n,m}(\lambda) \to CP^2$ (resp. $f_\lambda : Y_n(\lambda) \to CP^2$) be the composition of Π with a generic projection from CP^3 to CP^2 such that $\tilde{f}_0 = \tilde{f}^{(q)}$ (resp. $f_0 = f_0^{(n)}$).

Denote by $S_{nm}(\lambda)$ (resp. $S^{(n)}(\lambda)$) the branch curve in CP^2 corresponding to $\tilde{f}_\lambda$ (resp. f_λ). It is clear that $S_{nm}(\lambda)$ (resp. $S^{(n)}(\lambda)$) is obtained from S_{nm} (resp. $S^{(n)}$ (see §5, p.8) by replacement of each D_{ij} (see p.8) by the branch curve $\hat{D}_{ij}(\lambda)$ (in CP^2) of the map $Q'_{i-1,j-1}(\lambda) \to CP^2$ corresponding to the projection $CP^3 \to CP^2$ which we are using (in our C^3 it is given by $(x,y,z) \to (x,y)$).

Explicit computations show that each $\hat{D}_{ij}(\lambda)$, ($\lambda \in \mathbb{R}$), is defined over $\mathbb{R}$, that it's real part in the (x,y)-plane $\mathbb{R}^2 \subset C^2$ is a hyperbola, which for small $|x|$ is very close to the real part of D_{ij} and that branch points of $\hat{D}_{ij}(\lambda)$ for the x-projection are very close to the points $A_{i-1,j-1}$ and A_{ij}. It is clear that this hyperbola must be tangent to the lines $\mathcal{H}_{ij}$, V_{ij} and $V_{i-1,j}$, $\mathcal{H}_{i-1,j}$.

Using this information we see that real parts of $S_{nm}(\lambda)$ (resp. $S^{(n)}(\lambda)$) in small neighborhoods of the points A_{ij} could be described by the following figures:

(1)   For $i = 0, j = 0$

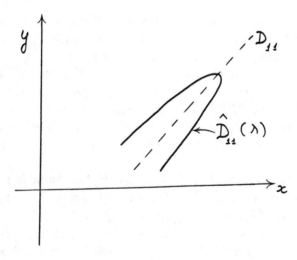

FIGURE 8.1

(2)   for $i = 0,\ j = 1, \ldots, m-1$ (resp. $j = 1, \ldots, n-1$)

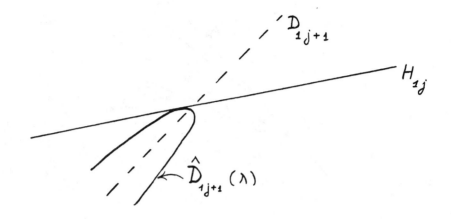

FIGURE 8.2

(3)   for $i = 1, \ldots, n-1,\ j = 0$

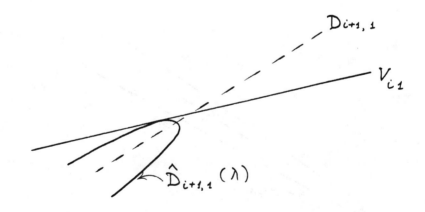

FIGURE 8.3

(4)   for $i = 1, \ldots, n-1, \ j = m \ (\text{ resp. } j = n)$

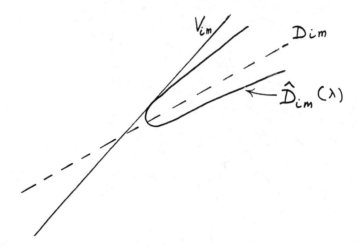

FIGURE 8.4

(5)   for $i = n, j = m \ (\text{resp. } j = n)$

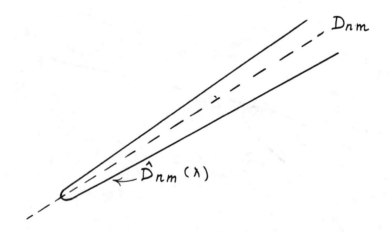

FIGURE 8.5

(6)   for $i = n, \ j = 1, \ldots, m-1$

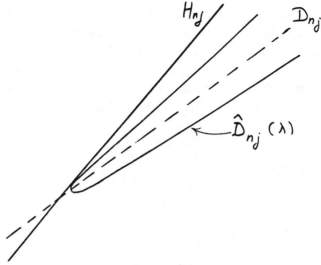

FIGURE 8.6

(7)   for $i \in (1,\ldots,n-1)$, $j \in (1,\ldots,m-1)$ (resp. $i \in (1,\ldots,n-1), j > i$)

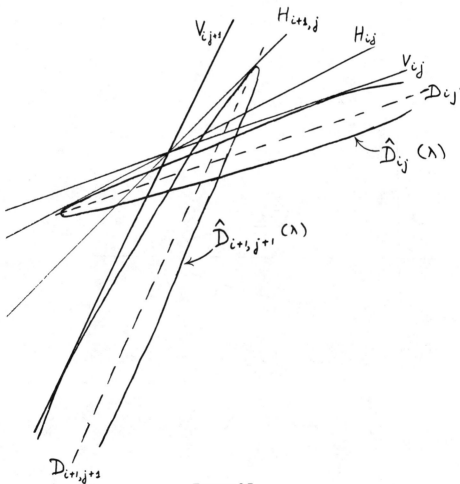

FIGURE 8.7

Consider each of the configurations of Fig. 8.1–8.7 as global (that means, in particular, that all conics on these Figures are now parabolae). Returning to complex numbers and setting $\tilde{\mathbf{C}}^1 = \{(x,y) \in \tilde{\mathbf{C}}^2 | x = M\}$, $(M > 1)$, we can describe the intersection set $K(i) = \{q_0, \ldots, \}$ of $\mathbf{C}^1 \cap C_i$ for the curve C_i (union of lines and conics) given by Fig. 8.i, $i = 1, \ldots 7$, by the following figures (in the complex y-plane corresponding to $\tilde{\mathbf{C}}^2$ (these figures also contain information related to our further arguments)

(Fig. 9.1 corresp. to Fig. 8.1)

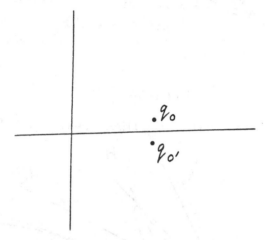

Fig. 9.1 (here $K(1) = \{q_0, q_{0'}\}$).

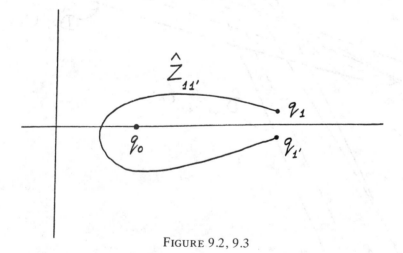

FIGURE 9.2, 9.3

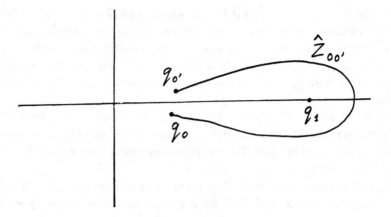

FIGURE 9.4, 9.6

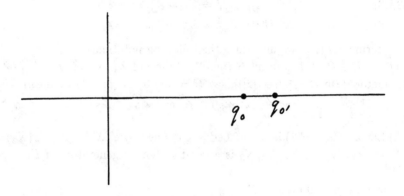

FIGURE 9.5

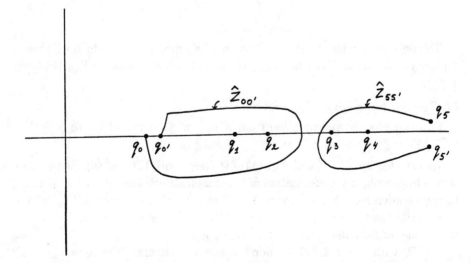

Fig. 9.7. $K(7) = \{q_0, q_0', q_1, q_2, q_3, q_4, q_{5'}, q_5\}$

Denote by $B(i) = B[\tilde{\mathbf{C}}^1, K(i)]$. For any q_r, q_s (of a given $K(i)$) denote by Z_{rs} the element (half-twist) of $B(i)$ corresponding to a straight line segment connecting q_r with q_s, if this segment does not contain other points of $K(i)$. By $\overline{Z}_{rs}$ (resp. $\underline{Z}_{rs}$) $(\in B(i))$ we denote the half-twist corresponding to a simple path connecting q_r with q_s and lying above (resp. below) the real axis (if such a path exists).

For $K(7)$ and $q_r \in \mathbb{R}$, $q_s = q_5$ (resp. $q_s = q_{5'}$) we denote by $\overline{Z}_{rs}$ (resp. $\underline{Z}_{rs}$) the half-twist corresponding to a simple path going from q_r over (resp. under) the real axis (very close to it) to $Re\, q_5$ and then continuing as a straight line segment from $Re\, q_5$ to q_5 (resp. q_5').

Explicit computations show that braid monodromies for Fig. 8.1–8.7 could be given by the right side of the following formula: ((Bi) corresponds to Fig. 8.i):

(B1,B5) $\Delta^2(B(i)) \cdot Z_{00'}^{-1} = Z_{00'}$, $i = 1, 5$

(B2,B3) $\Delta^2(B(i)) \cdot Z_{11'}^{-1} = \hat{Z}_{11'} \cdot Z_{01}^4$, $i = 2, 3$

(B4,B6) $\Delta^2(B(i)) \cdot Z_{00'}^{-1} = Z_{0'1}^4 \cdot \hat{Z}_{00'}$, $i = 4, 6$.

To formulate (B7) we introduce the following notations:

1) For $R \in B(7)$ denote by $\dot{R}$ (or $R^\bullet$) the braid $Z_{45'}^2 \cdot Z_{0'1}^{-2} \cdot R \cdot Z_{0'1}^2 \cdot Z_{45'}^{-2}$. If R is factored (in $B(7)^+$ (see [2])), say $R = \Pi u_j$ by $\dot{R}$ (or $R^\bullet$) we mean $\Pi \dot{u}_j$.

2) Set $Z_{5',5;2,1,0',0}^2 \underset{\sim}{=} Z_{5'2}^2 \cdot Z_{5'1}^2 \cdot Z_{5'0}^2 \cdot Z_{5'0}^2 \cdot Z_{52}^2 \cdot Z_{51}^2 \cdot Z_{50'}^2 \cdot Z_{5'0}^2$;
$Z_{0'0;4,3}^2 \underset{\sim}{=} Z_{0'4}^2 \cdot Z_{0'3}^2 \cdot Z_{04}^2 \cdot Z_{03}^2$.

3) Let $\varphi : B_4 \to B(7)$ be an embedding defined by $\varphi(X_1) = Z_{12}$, $\varphi(X_2) = Z_{23}$, $\varphi(X_3) = Z_{34}$, where X_1, X_2, X_3 is a good system of generators of B_4. Set $\tilde{B} = \varphi(B_4)$.

Now we can formulate (B7).

$$\Delta^2(B(7)) \cdot Z_{00'}^{-1} \cdot Z_{55'}^{-1} =$$

(B7)
$$= Z_{45'}^4 \cdot \hat{Z}_{55'} \cdot \overline{Z}_{35}^4 \cdot \underline{\dot{Z}}_{5',5;2,1,0,0'}^2 \cdot (\Delta^2(\tilde{B}))^\bullet \cdot \underline{\dot{Z}}_{0',0;4,3}^2 \cdot$$
$$\cdot \overline{Z}_{0'2}^4 \cdot \hat{Z}_{00'} \cdot \underline{Z}_{01}^4$$

The righthand sides of (B1)–(B7) give us the expressions by which we have to replace factors Δ_j^2 in (*1) when $Y_{n,m}$ (resp. Y_n) is regenerated to $Y_{n,m}(\lambda)$ (resp. $Y_n(\lambda)$).

§7. The last two steps of regeneration..

Consider now the regeneration from $\overline{Z}^{(n-1)}$ of §4 (resp. of §3) to $Z^{(m-1)} = \bigcup_{k=0}^{m-1} T_k$ of §4 (resp. to $Z^{(n-1)} = \bigcup_{k=0}^{n-1} T_k$ of §3).

We can identify $\overline{Z}^{(n-1)}$ with $Y_{nm}(\lambda)$ of §6 (resp. with $Y_n(\lambda)$ of §6). So the question is to describe a transformation of braid monodromy formulae corresponding to regeneration from $Y_{nm}(\lambda)$ (resp. $Y_n(\lambda)$) to $Z^{(m-1)}$ of §4 (resp. $Z^{(n-1)}$ of §3).

Actually the main problem is to transform local expressions $R_i, i = 1, \ldots, 7$ (right sides of formulae (Bi), $i = 1, \ldots, 7$, of §6).

For R_i with $i = 1, 2, 5, 6$ we don't make any changes (the lines $V_{k\ell}$ don't participate there (see Fig. 8.1, 8.2, 8.6)).

In ([2],§4) we explained how to make such replacements. The local pictures given by Fig. 9.3 and 9.4 (see p.16) after regeneration look as follows:

Fig. 10.3 (splitting of the point q_0 in Fig. 9.3)

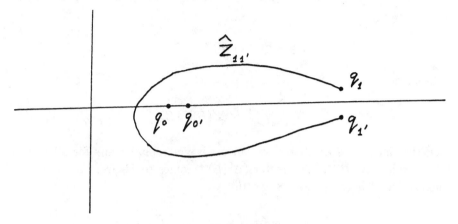

FIGURE 10.3

Fig. 10.4 (splitting of q_1 in Fig. 9.4)

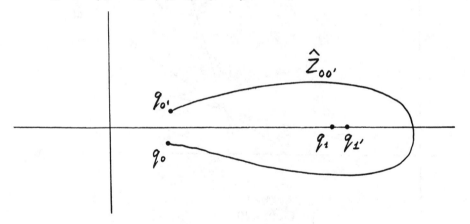

FIGURE 10.4

Denote the half-twist corresponding to the straight line segment connecting q_k with q_ℓ ($\forall k, \ell$) by $Z_{k\ell}$. From ([2],§4) it follows that after regeneration (which we are considering) R_3 has to be replaced by

(B3)(1) $$\hat{Z}_{11'} \cdot Z_{01}^3 \cdot Z_{0'1}^3 \cdot (Z_{00'} Z_{0'1} Z_{00'}^{-1})^3$$

and R_4 by

(B4)(1) $$(Z_{11'}^{-1} Z_{0'1} Z_{11'})^3 \cdot Z_{0'1}^3 \cdot Z_{0'1'}^3 \cdot \hat{Z}_{0'0}$$

Consider now R_7. The local picture given by Fig.9.7 after regeneration becomes the following:

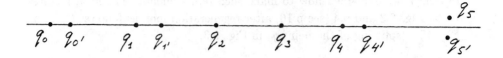

FIGURE 10.7

Recall that

$$R_7 = Z_{45'}^4 \cdot \hat{Z}_{55'} \cdot \overline{Z}_{35}^4 \cdot \dot{\underline{Z}}_{5'5;2,1,0,0'}^2 \cdots (\Delta^2(\tilde{B}))^\bullet \cdot \dot{\underline{Z}}_{0,0';3,4}^2 \cdot \overline{Z}_{0'2}^4 \cdot \hat{Z}_{00'} \cdot \hat{Z}_{01}^4$$

Using the same rules as above, we can easily get replacements for all factors of R_1 besides $(\Delta^2(\tilde{B}))^\bullet$. The factor $(\Delta^2(\tilde{B}))^\bullet$ comes to R_7 from the following configuration of lines (see Fig. 8.7, p.16):

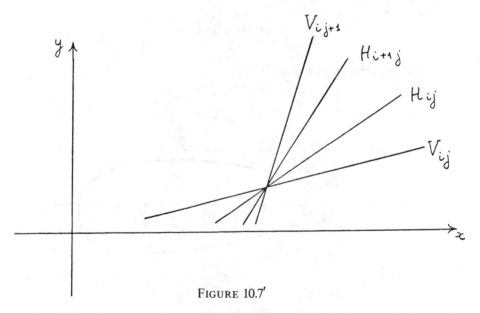

FIGURE 10.7'

For the local analysis of regeneration we can use explicit computations with the monoidal transformations involved in the construction of §2. It is easy to show that in some affine coordinates (X, Y, Z, T), the local family of surfaces (defined over $\mathbb{R}$) corresponding to "regeneration" of $\Delta^2(\tilde{B})$ could be given by the formulae:

(1) $\begin{cases} XT = u \\ XZ = 0 \end{cases}$ (u a parameter.)

To make it consistent with the order of (the real parts of) lines given by Fig. 10.7' we can use a change of affine coordinates (defined over $\mathbb{R}$), for instance the following:

$$T = \tau$$
$$X = z - \tau$$
$$Y = y + z + \tau$$
$$Z = x - y + z$$

So in the new coordinates, the local family will be given by the formulae

(2)
$$\begin{cases} \tau(z - \tau) = u \\ (y + z + \tau)(x - y + z) = 0 \quad (u \quad \text{a parameter}). \end{cases}$$

Fix $u = u_0 \neq 0$.

Simple computations show that the equation of the (x, y)-projection of the singular curve of the surface given by (2) with $u = u_0$ is the following:

(3)
$$(x - 2y)(2x - 3y) = -u_0 .$$

This curve corresponds to the regeneration of $\mathcal{H}_{ij} \cup \mathcal{H}_{i+1j}$ of Fig. 10.7'.

The branch curve for (x, y)-projection of our surface (corresp. $u = u_0$) consists of two pairs of lines:

(4)'
$$y = \pm\sqrt{8\,u_0}$$

(corresponds to the regeneration of V_{ij} of Fig. 10.7')

and

(4)''
$$y - x = \pm 2\sqrt{u_0} .$$

(corresponds to the regeneration of V_{ij+1} of Fig. 10.7').

From (3), (4)', (4)'' it follows that the real picture corresponding to the regeneration of Fig. 10.7' is the following ($u_0 \in \mathbb{R}, u_0 > 0$).

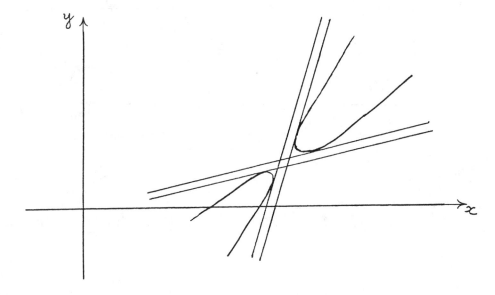

FIGURE 10.7''

From Fig. 10.7″ we get an expression for the replacement of $\Delta^2(\tilde{B})$ corresponding to our regeneration. To describe it we introduce the following notation:

We denote by B' the image of B_6 in $B_{10} = B[\tilde{\mathbf{C}}^1, \{q_0, q_{0'}, q_1, q_{1'}, q_2, q_3, q_4, q_{4'}, q_5, q_{5'}]$ corresponding to the subset $\{q_1, q_{1'}, q_2, q_3, q_4, q_{4'}\}$ and straight line segment connections of the points of this subset.

Let $\tilde{z}_{ij}, i, j = 1, 2$, $\alpha^{(1)}$ and $\alpha^{(2)}$ be braids (half-twists) corresponding to paths with the same notations as in Fig. 10.7″, (a), (b), (c).

The braid monodromy obtained by explicit computations corresponding to (Fig. 10.7″) is given by the right side of the following formula:

$$\Delta^2(B') \cdot Z_{11'}^{-2} Z_{44'}^{-2} = Z_{1'2}^4 \cdot Z_{34}^4 \cdot \alpha^{(1)} \cdot \tilde{z}_{11}^2 \cdot \tilde{z}_{21}^2 \cdot \tilde{z}_{12}^2 \cdot \tilde{z}_{22}^2 \cdot \alpha^{(2)} \cdot \underline{Z}_{12}^4 \cdot \overline{Z}_{34'}^4$$

In the last step of the regeneration, the points q_2 and q_3 split in pairs q_2, q_2' and $q_3, q_{3'}$ (see Fig. 11.7).

Let D_2, D_3 be small disks containing respectively $q_2, q_{2'}$ and $q_3, q_{3'}$.

We can assume that paths $\alpha^{(1)}$ and $\alpha^{(2)}$ (of Fig. 10.7‴) (b), (c)) intersect ∂D_2 and ∂D_3 in the same points a_2 and a_3. For $i = 2, 3$ let e_i and $e_{i'}$ be straight line segments connecting a_i with q_i and $q_{i'}$ as in Fig. 11.7. Introduce the following paths:

$$\alpha_1^{(j)} = e_{2'} \cup \overline{\alpha^{(j)} - \alpha^{(j)} \cap (D_2 \cup D_3)} \cup e_3$$
$$\alpha_2^{(j)} = e_2 \cup \overline{\alpha^{(j)} - \alpha^{(j)} \cap (D_2 \cup D_3)} \cup e_{3'}$$

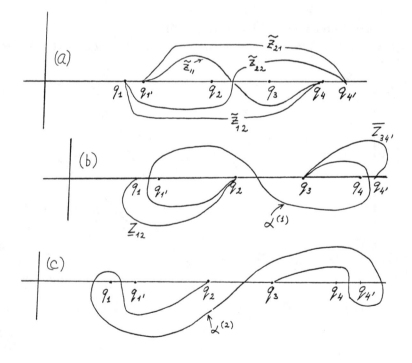

FIGURE 10.7‴

for $j = 1, 2$. Denote the braids (half-twists) corresponding to $\alpha_i^{(j)}$ by $\alpha_i^{(j)}$ again. Let $\tilde{\alpha}_1^{(1)} = \alpha_1^{(1)} \cdot \alpha_1^{(2)}$, $\tilde{\alpha}^{(2)} = \alpha_1^{(2)} \cdot \alpha_2^{(2)}$. Denote by B'' the image of B_8 in $B_{12} = B[\tilde{C}^1, q_{i'}, i = 0, \ldots, 5]$ corresponding to the subset $\{q_i, q_{i'}, i = 1, 2, 3, 4\}$ and straight line segment connections of the points in this subset.

From the construction of §2, it follows that the local family corresponding to the regeneration of the family given by (2) above could be defined by equations

(6)
$$\tau(z - \tau) = u$$
$$(y + z + \tau)(x - y + z) = v \quad (u \text{ and } v \text{ parameters}).$$

As above, we fix a real non-zero u_0 and then take real v_0 with v_0 with $v_0 \neq 0$, $|v_0| \ll |u_0|$. The local regenerated surface is defined by

(7)
$$\tau(z - \tau) = u_0$$
$$(y + z + \tau)(x - y + z) = v_0$$

Denote by C the projective closure of the branch curve for the surface defined by (7) and (x, y)-projection. Explicit computations show that C has four nodes at infinity and that the braid monodromy for these nodes (and $\tilde{C}^1$) is given by braids $Z_{11'}^2$, $Z_{22'}^2$, $Z_{33'}^2$, $Z_{44'}^2$. Thus the product of braids corresponding to the braid monodromy (and $\tilde{C}^1$) for the affine part of C is equal to $\Delta^2(B'') \cdot \prod_{i=1}^4 Z_{ii'}^{-2}$.

From (5) and ([2]§4) it now follows that the braid monodromy for the affine part of C is given by the righthand side of the following formula

$$\Delta^2(B'') \cdot \prod_{i=1}^4 Z_{ii'}^{-2} =$$

$$= \prod_{i=0}^2 (Z_{22'}^i (Z_{1'2}) Z_{22'}^{-i})^3 \cdot \prod_{i=0}^2 [Z_{33'}^i (Z_{3'4}) Z_{33'}^{-i}]^3 \cdot$$

$$\cdot \tilde{\alpha}_{p,q}^{(1)} \cdot \tilde{z}_{11}^2 \cdot \tilde{z}_{21}^2 \cdot \tilde{z}_{12}^2 \cdot \tilde{z}_{22}^2 \cdot \tilde{\alpha}_{r,s}^{(2)} \cdot \prod_{i=0}^2 [Z_{22'}^i (\underline{Z}_{12}) Z_{22'}^{-i}]^3 \cdot$$

$$\cdot \prod_{i=0}^2 [Z_{33'}^i (\overline{Z}_{3'4'}) Z_{33'}^{-i}]^3$$

for some integers p, q, r, s, where $\tilde{\alpha}_{p,q}^{(j)} = Z_{22'}^{-p} Z_{33'}^{-q} (\tilde{\alpha}^{(j)}) Z_{33'}^{q} Z_{22'}^{p}$. Denote the factored expression on the righthand side of (8) by $\tilde{F}$.

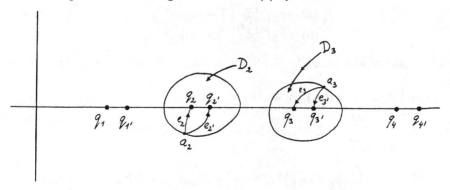

FIGURE 11.7

It follows from ([2], §1, Lemma 2) that we can take $p = 0, r = 0$ in (8). Denote by G the subgroup of B'' consisting of braids β with a representative $\tilde{\beta}$ such that for $i = 1, 4$ $\tilde{\beta}(q_i) = q_i, \tilde{\beta}(q_{i'}) = q_{i'}$. Let $B_4 = B[\tilde{\mathbf{C}}^1, \{q_2, q_{2'}, q_3, q_{3'}\}]$ and $\psi : G \to B_4$ be the canonical surjection (corresponding to "the forgetting of" $q_1, q_{1'}, q_4, q_{4'}$). It is easy to see that all factors of $\tilde{F}$ belong to G and $\psi(\tilde{\alpha}^{(1)}) = \psi(\alpha^{(2)})$. Set $\alpha = \psi(\tilde{\alpha}^{(1)})$ and apply ψ to the both parts of (8) to get:

$$(9) \qquad \Delta^2(B_4) \cdot Z_{22'}^{-2} Z_{33'}^{-2} = Z_{22'} \cdot Z_{33'} \cdot Z_{33'}^{-q}(\alpha) Z_{33'}^q \cdot Z_{33'}^{-s}(\alpha) Z_{33'}^s \cdot Z_{22'} \cdot Z_{33'}$$

Using ([2], §1, Lemma 2), it is easy to check that

$$(10) \qquad \Delta^2(B_4) = Z_{22'}^4 \cdot Z_{33'}^4 \cdot Z_{33'}^{-q}(\alpha) Z_{33'}^q \cdot Z_{33'}^{-q}(\alpha) Z_{33'}^q$$

Comparing (9) with (10) we get that $Z_{33'}^{-s}(\alpha) Z_{33'}^s = Z_{33'}^{-q}(\alpha) Z_{33'}^q$ or

$$(11) \qquad \alpha = Z_{33'}^{s-q}(\alpha) Z_{33'}^{-(s-q)}$$

Because $\alpha^{-1} Z_{33'} \alpha = Z_{11'}$ we get from (11): $1 = Z_{11'}^{s-q} Z_{33'}^{-(s-q)}$. Thus $Z_{11'}^{s-q} = Z_{33'}^{s-q}$ and so $s - q = 0$, $s = q$. We now see that in $\tilde{F}$ we must have $s = q$.

We need the following

PROPOSITION 7.1 (INVARIANCE PROPERTY FOR $\tilde{F}$). *For any integers k, ℓ we have:*

$$(12) \qquad (Z_{11'} Z_{44'})^k (Z_{22'} Z_{33'})^\ell [\tilde{F}] (Z_{22'} Z_{33'})^{-\ell} (Z_{11'} Z_{44'})^{-k} \underset{\sim}{=} \tilde{F}.$$

PROOF. Denote by $\rho = \prod_{i=1}^4 Z_{ii'}$;

$$F_1 \underset{\sim}{=} \prod_{i=0}^2 [Z_{22'}^i (Z_{1'2}) Z_{22'}^{-i}]^3 \cdot \prod_{i=0}^2 [Z_{33'}^i (Z_{3'4}) Z_{33'}^{-i}]^3 \cdot \tilde{\alpha}_{0,q}^{(1)} \cdot \tilde{z}_{11}^2;$$

$$F_2 \underset{\sim}{=} \tilde{z}_{22}^2 \cdot \tilde{\alpha}_{0,q}^{(2)} \cdot \prod_{i=0}^2 [Z_{22'}^i (\underline{Z}_{12}) Z_{22'}^{-i}]^3 \cdot \prod_{i=0}^2 [Z_{3'4}^i (\overline{Z}_{3'4}) Z_{33'}^{-i}]^3;$$

$$F_2' \underset{\sim}{=} \tilde{\alpha}_{0,q}^{(2)} \cdot \prod_{i=0}^2 [Z_{22'}^i (\underline{Z}_{12}) Z_{22'}^{-i}]^3 \cdot \prod_{i=0}^2 [Z_{33'}^i (\overline{Z}_{3'4}) Z_{33'}^{-i}]^3.$$

By direct computation it is possible to check that:

$$(13) \qquad \begin{array}{l} \text{(a)} \;\; (F_2')^{-1} \tilde{z}_{22} F_2' = \rho \tilde{z}_{11} \rho^{-1}; \\ \text{(b)} \;\; (F_2')^{-1} \tilde{\alpha}_{0,q}^{(2)} F_2' = \rho \tilde{\alpha}_{0,q}^{(1)} \rho^{-1}; \end{array}$$

(For (b) we use also that $(Z_{22'} Z_{33'})[\tilde{\alpha}^{(1)}](Z_{22'} Z_{33'})^{-1} = \tilde{\alpha}^{(1)}$ (see [2], §1, Lemma 2).

Using ([2], §1, Lemma 1''''), we can see also that

$$(14) \qquad \prod_{i=0}^2 [Z_{22'}^i (\underline{Z}_{12}) Z_{22'}^{-i}]^3 \underset{\sim}{=} \rho \prod_{i=0}^2 [Z_{22'}^i (Z_{1'2}) Z_{22'}^{-i}]^3 \rho^{-1};$$

$$\prod_{i=0}^2 [Z_{33'}^i (\overline{Z}_{3'4}) Z_{33'}^{-i}]^3 \underset{\sim}{=} \rho \prod_{i=0}^2 [Z_{33'}^i (Z_{3'4}) Z_{33'}^{-i}]^3 \rho^{-1}.$$

Now we can write (using (13)((a), (b)) and (14)):

$$F_2 = \tilde{z}_{22}^2 \cdot F_2' = F_2' \cdot \rho\tilde{z}_{11}^2\rho^{-1} =$$

(15)
$$\prod_{i=0}^{2}[Z_{22'}^i(\underline{Z}_{12})Z_{22'}^{-i}]^3 \cdot \prod_{i=0}^{2}[Z_{33'}^i(\overline{Z}_{3'4'})Z_{33'}^{-i}] \cdot \rho\tilde{\alpha}_{0,q}^{(1)}\rho^{-1} \cdot \rho\tilde{z}_{11}^2\rho^{-1} =$$

$$= \rho F_1\rho^{-1}, \text{ and } \rho F_2\rho^{-1} = \rho^2 F_1\rho^{-2}.$$

Because $\tilde{F} = \Delta^2(B'')\rho^{-2}$ and $\tilde{z}_{12}, \tilde{z}_{21}$ commute we have

(16)
$$\tilde{F} = F_1 \cdot \tilde{z}_{21}^2 \cdot \tilde{z}_{12}^2 \cdot F_2 = \tilde{z}_{21}^2 \cdot \tilde{z}_{12}^2 \cdot F_2 \cdot \rho^2 F_1\rho^{-2} =$$
$$= F_2 \cdot F_2^{-1}(\tilde{z}_{21}^2 \cdot \tilde{z}_{12}^2)F_2 \cdot \rho^2 F_1\rho^{-2} =$$
$$= \rho F_1\rho^{-1} \cdot F_2^{-1}\tilde{z}_{12}^2 F_2 \cdot F_2^{-1}\tilde{z}_{21}^2 F_2 \cdot \rho F_2\rho^{-1}.$$

Direct computations show that

(17)
$$F_2^{-1}\tilde{z}_{21}F_2 = \rho\tilde{z}_{12}\rho^{-1}; \ F_2^{-1}\tilde{z}_{12}F_2 = \rho\tilde{z}_{21}\rho^{-1}$$

Combining (16) and (17) we get

(18)
$$\tilde{F} = \rho\tilde{F}\rho^{-1}.$$

This proves the Proposition in the case $k = \ell$. This means that now we can assume that $k = 0$. In that case, the Proposition follows from ([2], §1, Lemmas 1'''' and 2). □

REMARK. *It is also possible to prove Proposition 7.1 by considering the following family of surfaces:*

$$\begin{cases} \tau(z - \tau) = u_0 e^{-ik\varphi} \\ (y + z + \tau)(x - y + z) = v_0 e^{-i\ell\varphi}, \ \ 0 \leq \varphi \leq 2\pi. \end{cases}$$

§8. The final formulae.

After final regeneration, each point q_i in $\tilde{\mathbf{C}}^1$ splits in a pair $q_i, q_{i'}$. Let D_i be a small disk in $\tilde{\mathbf{C}}^1$ containing $q_i, q_{i'}$ and let $\overline{p}_i$ be a positive half-twist in D_i which is the identity on ∂D_i. Denote also by $\overline{p}_i$ the braid represented by $\overline{p}_i$. Using local diffeomorphisms with support in D_i's we can assume that we have a (diffeomorphic) model for $\tilde{\mathbf{C}}^1$ such that each pair $q_i, q_{i'}$ is real and $q_i < q_{i'}$.

Our local models (see p.16) for $\mathbf{C}^1$ can also be choosen so that all pairs q_i, $q_{i'}$ are real and $q_i < q_{i'}$. More precisely we use for Fig. 9.1, 9.4, 9.6 negative 90° rotation in D_0, for Fig. 9.2, 9.3 positive 90° rotation in D_1 and for Fig. 9.7 positive 90° rotation in D_5.

Set $\rho_i = Z_{ii'}$ and $K = \{q_{i,i'}, i = 0, \ldots, 5\}$ when we are in a local situation. Now combining the formula (B7) of §6 with (8) of §7, we obtain that after final degeneration the formula (B7) (of §6) becomes the following: (in appropriate (diffeomorphic) model of $\tilde{\mathbf{C}}^1$ obtained from the standard $\tilde{\mathbf{C}}^1$ by diffeomorphisms with supports in D_i's):

$$\Delta^2(B[\tilde{\mathbf{C}}^1, K]) \cdot \rho_0^{-1} \rho_5^{-1} \cdot \prod_{j=1}^{4} \rho_j^{-1} =$$

$$= \prod_{i=0}^{2} [\rho_4^i(\underline{Z}_{4',5'})\rho_4^{-i}]^3 \cdot \hat{Z}_{55'} \cdot \prod_{i=0}^{2} [\rho_3^i \overline{Z}_{3'5}\rho_3^{-i}]^3 \cdot \dot{\underline{Z}}_{\{5',5\};\{q_i,q_{i'},i=0,1,2\}}^2 \cdot$$

(1) $\quad \cdot \left\{ \prod_{i=0}^{2} [\rho_2^i(Z_{1'2})\rho_2^{-i}]^3 \cdot \prod_{i=0}^{2} [\rho_3^i(Z_{3'4})\rho_3^{-i}]^3 \cdot \tilde{\alpha}^{(1)} \cdot \right.$

$$\left. \cdot \tilde{z}_{11}^2 \cdot \tilde{z}_{21}^2 \cdot \tilde{z}_{12}^2 \cdot \tilde{z}_{22}^2 \cdot \tilde{\alpha}^{(2)} \cdot \prod_{i=0}^{2} [\rho_2^i(\underline{Z}_{12}\rho_2^{-i}]^3 \cdot \prod_{i=0}^{2} [\rho_3^i \overline{Z}_{3'4'}\rho_3^{-i}]^3 \right\}^{\bullet} \cdot$$

$$\cdot \dot{\underline{Z}}_{\{0',0\};\{q_i,q_{i'},i=4,3\}}^2 \cdot \prod_{i=0}^{2} [\rho_2^i \overline{Z}_{0'2}\rho_2^{-i}]^3 \cdot \hat{Z}_{00'} \cdot \prod_{i=0}^{2} [\rho_1^i(\underline{Z}_{01})\rho_1^{-i}]^3,$$

where

   (i) $\overline{Z}_{ij}$ (resp. $\underline{Z}_{ij}$) is a half-twist defined by a simple path over (resp. under) the real axis connecting q_i with q_j;

   (ii) for any subsets A, B (of K), $Z_{A;B}^2 = \prod_{a \in A}(\prod_{b \in B} \underline{Z}_{a,b}^2)$ where in $\prod_{a \in A}$ and $\prod_{b \in B}$ the order of factors is opposite to the natural order of a's and b's,

   (iii) for any braid R we define $R^{\bullet}$ or $\dot{R}$ as

$$Z_{5;\{3,3',4,4'\}}^2 (Z_{0';\{1,1',2,2'\}}^2)^{-1}(R) Z_{0';\{1,1',2,2'\}}^2 \cdot (Z_{5;\{3,3',4,4'\}}^2)^{-1}$$

   (iv) for any factored product $A_1 \cdot \ldots \cdot A_\ell$ the factored product $(A_1 \cdot \ldots \cdot A_\ell)^{\bullet}$ is defined as $A_1^{\bullet} \cdot \ldots \cdot A_\ell^{\bullet}$,

   (v) and $\alpha^{(1)}, \alpha^{(2)}, \tilde{z}_{ij}, i, j = 1, 2$ are the same as in §7.

Denote the right side of formula (1) by H_7.

PROPOSITION 8.1 (INVARIANCE PROPERTY OF (1)). *For any integers ℓ_0, ℓ_1, ℓ_2 we have*

(2) $\qquad H_7 \underset{\sim}{=} ((\rho_0\rho_5)^{\ell_0}(\rho_1\rho_4)^{\ell_1}(\rho_2\rho_3)^{\ell_2}) H_7((\rho_0\rho_5)^{\ell_0}(\rho_1\rho_4)^{\ell_1}(\rho_2\rho_3)^{\ell_2})^{-1}$

LEMMA 8.1 (DUE TO K. CHAKIRIS). *Let G be a group, $A_1, \ldots, A_r \in G$ and C be the (factored) product $A_1 \cdot A_2 \cdot \ldots \cdot A_r$. Then $\forall$ integers m $C = \prod_{i=1}^{r}(C^{-m}A_kC^m)$.*

PROOF OF LEMMA 8.1. It is enough to consider the case $m = 1$. Set $C_j \underset{\sim}{=} \prod_{i=j+1}^{r} A_i, j = 0, \ldots, r-1, C_r = Id, C_j' \underset{\sim}{=} \prod_{k=1}^{j}(C^{-1}A_kC), j = 1, \ldots, r, C_0' = Id$. The Lemma is equivalent to the following statement: $C \underset{\sim}{=} C_r \cdot C_r'$.

We prove this statement using induction. Namely we claim that for $\ell = 0, \ldots, r$

(3)$_\ell$ $\qquad\qquad\qquad\qquad C \underset{\sim}{=} C_\ell \cdot C_\ell'$

$(3)_0$ is evidently true. Assume that $(3)_j$ is true, that is, $C \underset{\sim}{=} C_j \cdot C'_j$. Then we perform the following transformation:

$$C \underset{\sim}{=} C_j \cdot C'_j \underset{\sim}{=} A_{j+1} \cdot A_{j+2} \cdot \ldots \cdot A_r \cdot \prod_{k=1}^{j} C^{-1} A_k C \underset{\sim}{=}$$

$$\underset{\sim}{=} A_{j+2} \cdot \ldots \cdot A_r \cdot \prod_{k=1}^{j} (C^{-1} A_k C) \cdot$$

$$\cdot [(A_{j+2} \ldots A_r \cdot \prod_{k=1}^{j} (C^{-1} A_k C))^{-1} (A_{j+1})(A_{j+2} \ldots A_r \cdot \prod_{k=1}^{j} (C^{-1} A_k C))] \underset{\sim}{=}$$

$$\underset{\sim}{=} A_{j+2} \cdot \ldots \cdot A_r \cdot \prod_{k=1}^{j} (C^{-1} A_k C) \cdot [(C_j C'_j)^{-1} (A_{j+1})(C_j C'_j)] \underset{\sim}{=}$$

$$= A_{j+2} \cdot \ldots \cdot A_r \cdot \prod_{k=1}^{j} (C^{-1} A_k C) \cdot (C^{-1} A_{j+1} C) \underset{\sim}{=}$$

$$= C_{j+1} \cdot C'_{j+1}.$$

Thus $(3)_{j+1}$ is true.   □

Because $\mathcal{H}_7 = \Delta^2(B[\tilde{\mathbf{C}}^1, K]) \cdot \rho_0^{-1} \rho_5^{-1} \prod_{i=1}^{4} \rho_i^{-1}$ we get from Lemma 8.1 that the Proposition is true when $\ell_1 = \ell_0$, $\ell_2 = \ell_0$. Thus we can assume $\ell_0 = 0$. In that case we prove the Proposition using Proposition 7.1 of §7, together with the fact that $\dot{\rho}_i = \rho_i$, $\forall i$, for the expression $\{\ldots\}^\bullet$ of $(\mathcal{H}7)$, and using ([2]§1 and §4) for other factors of $\mathcal{H}7$.   □

It is clear that $\mathcal{H}_7$ is the expression which we have to finally put in $(*1)$ of §5 instead of Δ_j^2 when we "regenerate" a "6-point" (that is an intersection of six lines).

Denote by $\mathcal{H}_i$, $i = 1, \ldots 6$, the local braid monodromy expressions after final regenerations corresponding to the cases given by Fig. 8.i, $i = 1, \ldots, 6$ of §6. These expressions are (we use §6 and §7):

(4)

$$\mathcal{H}_1 = Z_{00'};$$

$$\mathcal{H}_2 = \hat{Z}_{11'} \cdot \prod_{i=0}^{2} [\rho_0^i (Z_{0'1} \rho_0^{-i}]^3;$$

$$\mathcal{H}_3 = \hat{Z}_{11'} \cdot \prod_{i=0}^{2} [\rho_0^i (Z_{0'1}) \rho_0^{-i}]^3;$$

$$\mathcal{H}_4 = \prod_{i=0}^{2} [\rho_1^i (Z_{0'1}) \rho_1^{-i}]^3 \cdot \hat{Z}_{0'};$$

$$\mathcal{H}_5 = Z_{00'};$$

$$\mathcal{H}_6 = \prod_{i=0}^{2} [\rho_1^i (Z_{0'1}) \rho_1^{-i}]^3 \cdot \hat{Z}_{00'}.$$

Invariance properties of the $\mathcal{H}_i$'s, $i = 1, \ldots, 6$ are clear from ([2],§1).

As in §5 we denote the sequence $\{A_{ij}\}$ by $\{\alpha_1, \ldots, \alpha_\nu\}$. For $j = 1, \ldots, \nu$ let $\tau(j) \in (1, \ldots, 7)$ be a number such that $\mathcal{H}_{\tau(j)}$ is the local expression of the braid monodromy after the final regeneration corresponding to Δ_j^2 of (1) §5. From (2, §4, 2.), it is clear how to replace the C_j's (of (*1) §5) after final regeneration. Denote the corresponding expressions by $\underline{C}_j$, $j = 1, \ldots, \nu$.

Let Q (resp. $\underline{T}_j$) be the set obtained from Q (resp. T_j, see p.9) of §5 when each q_i splits into $q_i, q_{i'}$, $\underline{B}(i)$ be the braid group corresponding to $\mathcal{H}_i$, $i = 1, \ldots, 7$, and $\underline{\beta}_j : \underline{B}(\tau(j)) \to B[\tilde{C}_1, \underline{Q}]$ be an embedding of braid groups corresponding to $\underline{T}_j \subset \underline{Q}$ and connections of points of $\underline{T}_j$ by simple paths which go below the real axis. Let $\underline{\mathcal{H}}_{(j)} = \underline{\beta}_j(\mathcal{H}_{\tau(j)})$.

The global braid monodromy formula after final regeneration now looks as follows:

$$(5) \qquad \Delta^2 = \prod_{j=\nu}^{1} (\underline{C}_j \cdot (e_j^{-1} \underline{\mathcal{H}}_{(j)} e_j))$$

where each e_j is an element of the subgroup $\mathcal{H}$ of $B[\tilde{C}^1, \underline{Q}]$ generated by all $\bar{p}_i$'s $i = 1, \ldots, r$ (see above).

We have to put e_j's in the formula (5) because each $\underline{\beta}_j$ is defined uniquely up to conjugation by an element of $\mathcal{H}$.

Our final result is given by the following Theorem.

THEOREM 8.2. *There exists a (diffeomorphic) model of $\tilde{C}^1$ such that the braid monodromy for the branch curve of a generic projection to $\mathbf{CP}^2$ of $F_0(m,n)$ and V_n (see §3 and §4) is given by the following formula*

$$(6) \qquad \Delta^2 = \prod_{j=\nu}^{1} (\underline{C}_j \cdot \underline{\mathcal{H}}_{(j)})$$

(for definitions of $\underline{C}_j$ see §5, Proposition 1 and discussion following. For definition of $\underline{\mathcal{H}}_j$ see §8, discussion following (4).)

PROOF. We have to show that we can make all the $e_j = 1$ in (5). We use the following

LEMMA 8.2. *There exists $A \in \mathcal{H}$ such that $\forall j \in (1, \ldots, \nu)$*

$$(7) \qquad \underline{\mathcal{H}}_{(j)} = A^{-1} e_j^{-1} \underline{\mathcal{H}}_{(j)} e_j A$$

PROOF. We proceed by induction. Assume that for certain $k \in (1, \ldots, \nu)$ we proved already that there exists $A(k) \in \mathcal{H}$ with the following property: $\forall j \in (1, \ldots, k)$

$$\underline{\mathcal{H}}_{(j)} = (A(k))^{-1} e_j^{-1} (\underline{\mathcal{H}}_{(j)}) e_j A(k)$$

Consider now $j = k + 1$. If $\tau(k+1) \neq 7$, then $\forall I e \in \mathcal{H}$ $e^{-1} \underline{\mathcal{H}}_{(j)} e = \underline{\mathcal{H}}_{(j)}$, so we have nothing to prove. Assume $\tau(k+1) = 7$. Let $\underline{T}_{k+1} = \{q_{ia}, a = 0, \ldots, 5, q_{i_0} < q_{i_1} < \ldots < q_{i_5}\}$. It is clear that there exists a sequence of integers $\{m_0, \ldots, m_5\}$ such that

$$(A(k))^{-1} e_{k+1}^{-1} \underline{\mathcal{H}}_{(k+1)} e_{k+1} A(k) = \left(\prod_{a=0}^{5} \bar{p}_{i_a}^{m_a} \right)^{-1} \underline{\mathcal{H}}_{(k+1)} \left(\prod_{a=0}^{5} \bar{p}_{i_a}^{m_a} \right).$$

From Proposition 8.1 above (invariance property of formula (1)), it follows that we can take $m_0 = m_1 = m_2 = 0$. Thus we get

$$(8) \qquad \underline{\mathcal{Y}}_{(k+1)} \widetilde{=} (\prod_{a=3}^{5} \bar{\rho}_{i_a}^{m_a}) A(k)^{-1} e_{k+1}^{-1} (\underline{\mathcal{Y}}_{(k+1)}) e_{k+1} A(k) (\prod_{a=3}^{5} \bar{\rho}_{i_a}^{m_a})^{-1}$$

Set $A(k+1) = A(k)(\prod_{i=3}^{5} \rho_{i_a}^{m_a})^{-1}$. It is clear that $\forall j < k+1$ the points $q_{i_3}, q_{i_4}, q_{i_5}$ don't belong to the set $\underline{T}_j$. That means that for all $j < k+1$ $(\prod_{i=3}^{5} \rho_{i_a}^{m_a}) \underline{\mathcal{Y}}_{(j)} (\prod_{i=3}^{5} \rho_{i_a}^{m_a})^{-1} = \underline{\mathcal{Y}}_{(j)}$ and we get using (8) that

$$A_{(k+1)}^{-1} e_j^{-1} \underline{\mathcal{Y}}_{(j)} e_j A_{(k+1)} = \underline{\mathcal{Y}}_{(j)} \; \forall \, j \in (1, 2, \ldots, k+1).$$

This finished the induction and the proof of the Lemma.

Let $\tilde{A}$ be a diffeomorphism of $\tilde{\mathbf{C}}^1$ representing the braid A of Lemma 8.2. Applying $\tilde{A}$ to $\tilde{\mathbf{C}}^1$ we obtain a new (diffeomorphic) model of $\mathbf{C}^1$ in which formula (5) for the braid monodromy becomes

$$\Delta^2 = \prod_{j=\nu}^{1} (\underline{C}_j \cdot (A^{-1} e_j^{-1} C_j^{-1} (\underline{\mathcal{Y}}_{(j)}) e_j A)$$

which by Lemma 8.2 is equivalent to

$$\Delta^2 = \prod_{j=\nu}^{1} \underline{C}_j \underline{\mathcal{Y}}_{(j)}.$$

$\square$

REMARK. *All explicit curves which we used were defined over* $\mathbb{R}$. *It follows from that in all our arguments we can change roles of upper and low-half planes in* $\tilde{\mathbf{C}}^1$ *and x-axis. Such a change requires also to replace by opposite the order of factors in all our expressions for braid monodromy. Thus we see that our final expression*

$$\Delta^2 = \prod_{j=\nu}^{1} \underline{C}_j \underline{H}_{(j)}$$

is equivalent to

$$\Delta^2 = \prod_{j=1}^{\nu} \overline{H}_{(j)} \overline{C}_j$$

where $\overline{H}_{(j)}$ *(resp.* $\overline{C}_j$*) is obtained from* $\underline{H}_{(j)}$ *(resp.* C_j*) as follows: each factor in* $\underline{H}_{(j)}$ *(resp.* C_j*) is replaced by the complex conjugate braid (in* $\tilde{\mathbf{C}}^1$*) and the order of factors is changed to the opposite.*

344   B. MOISHEZON

REFERENCES

[1] Moishezon, B. G., *Stable branch curves and braid monodromies*, in Lecture
    Notes in Math., vol. **862** (1981), pp.107–192.
[2] Moishezon, B. G., *Algebraic surfaces and the arithmetic of Braids, I*, pp.
    199–269, in "Arithmetic and Geometry, papers dedicated to I.R. Shafarevich",
    Birkhäuser, 1983.
[3] Persson, U., *Chern invariants of surfaces of general type*, Comp. Math.
    (1981).
[4] Moishezon, B.G., Teicher, M., *Simply-connected algebraic surfaces of posi-
    tive index* (to appear).

COLUMBIA UNIVERSITY, NEW YORK, NY 10027

[8] J. M. Montesinos, *Una nota a un teorema de Alexander*, Revista Mat. Hisp.-
    Amer. **32** (1972), 167–187.
[9] J. M. Montesinos, *A representation of closed orientable 3-manifolds as 3-fold
    branched coverings of S^3*, Bull. Amer. Math. Soc. **80** (1974), 845–846.
[10] J. M.Montesinos, *Three-manifolds as 3-fold branched covers of S^3*, Quar-
    terly J.Math. Oxford (2) **27** (1976), 85–94.
[11] J. M. Montesinos, *4-manifolds, 3-fold covering spaces and ribbons*, Trans.
    Amer. Math. Soc. **245** (1978), 453–467.
[12] J. M. Montesinos, *Heegaard diagrams for closed 4-manifolds*, Geometric
    Topology, Academic Press (1979), 219–237.
[13] J. M. Montesinos, *Lectures on 3-fold simple coverings and 3-manifolds*, (A.
    Stone proceedings, 1982).

FACULTAD DE CIENCIAS, UNIVERSIDAD DE ZARAGOZA

Contemporary Mathematics
Volume 44, 1985

A NOTE ON MOVES AND ON IRREGULAR COVERINGS OF S^4

José María Montesinos*

1. H. Hilden and the author have shown [3], [4], [9], [10] (cfr. [5]), that every closed, oriented 3-manifold M is a 3-fold simple covering of S^3 branched over a knot K (see [13] as a general reference). Following Fox, we can think of K as a *colored knot*, by providing the bridges of K with three colors (say G = green, R = red, B = blue), so that the representation $\omega : \pi_1(S^3 - K) \to S_3$ which sends green, red, blue meridians to (12), (13), (23) of the symmetric group S_3 of three numbers, is the monodromy map of the covering $M \to S^3$ branched over K. The move C (or its inverse C^{-1}) of Figure 1 does not change the covering manifold [2], [5], [7], [8]. We have asked [13] *if there is a set of moves which do not change the covering manifold and such that if two colored links have the same covering they are related by a finite sequence of those moves.*

In particular, it looked likely that this set of moves would consist of $C^{\pm 1}$ (compare [7], [2]). We see in this note that this is not the case.

2. Given two simple coverings $p_1, p_2 : M \to S^3$ branched over links L_1, L_2, resp., we call them *cobordant* if there is a simple covering $P : M \times [-1, 1] \to S^3 \times [-1, 1]$ which is equal to p_1 in $M \times \{-1\}$ and to p_2 in $M \times \{1\}$ and is branched over a PL immersed 2-manifold with boundary equal to the union of the branching sets of p_1 and p_2.

THEOREM. *Let $p_1, p_2 : M \to S^3$ be 3-fold simple coverings such that it is possible to pass from the branching set L_1 of p_1 to the branching set L_2 of p_2 by a sequence of m moves $C^{\pm 1}$. Then p_1 and p_2 are cobordant and the branching set of the cobordism is a PL embedded 2-manifold with m cusp singularities (i.e. cones over the trefoil).*

PROOF. It is sufficient to assume that $p_1, p_2 : M \to S^3$ are related by just one move (say C) which takes place in a 3-ball $B^3 \subset S^3$. Delete from $(S^3, L_1) \times [-1, 0] \cup (S^3, L_2) \times [0, 1]$ the part of $L_1 \times [-1, 0] \cup L_2 \times [0, 1]$ lying in the 4-ball $B^3 \times [-\frac{1}{2}, \frac{1}{2}]$ and put in their place the cone whose vertex is $(0, 0)$ and whose link is the trefoil $\partial(B^3 \times [-\frac{1}{2}, \frac{1}{2}]) \cap (L_1 \times [-1, 0] \cup L_2 \times [0, 1])$ (see Figure 2). Thus we get a PL embedded 2-manifold F in $S^3 \times [-1, 1]$ with a cusp point in $(0, 0)$.

*Supported by "Comisión Asesora de Investigación científica y técnica"

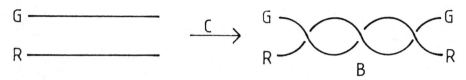

FIGURE 1

There is a natural 3-fold simple covering of $S^3 \times [-1,1]$ branched over F. What is lying over the complement of $B^3 \times [-\frac{1}{2}, \frac{1}{2}]$ is $M \times [-1,1]$ minus some 4-ball $C^3 \times [-\frac{1}{2}, \frac{1}{2}]$. What lies over $B^3 \times [-\frac{1}{2}, \frac{1}{2}]$ is the cone over the simple 3-fold covering of S^3 branched over the trefoil. Since this manifold is S^3, that cone fills in the $C^3 \times [-\frac{1}{2}, \frac{1}{2}]$, providing as covering space $M \times [-1,1]$.

Note. The above theorem was probably known to Hirsch (cfr. [6],Satz 3).

COROLLARY. *There are 3-fold simple coverings* $p_1, p_2 : 4\#S^1 \times S^2 \to S^3$ *such that the branching sets are not related by any sequence of the moves* $C^{\pm 1}$.

PROOF. The 4-dimensional torus T^4 has a handle presentation $T^4 = H^0 \cup 4H^1 \cup 6H^2 \cup 4H^3 \cup H^4$ with one 0-handle, four 1-handles, six 2-handles, four 3-handles and one 4-handle. According to [11], the manifolds $U = H^0 \cup 4H^1 \cup 6H^2$ and $V = T^4 - \mathrm{Int}\, U^4 \cong 4\ S^1 \times B^3$ are 3-fold simple coverings of D^4 branched over

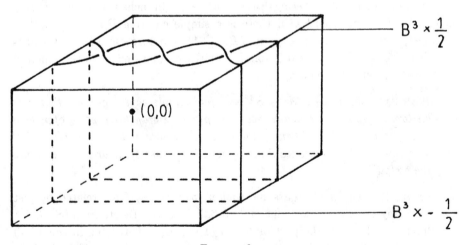

FIGURE 2

properly embedded locally flat 2-manifolds. The restrictions to ∂U and ∂V of these coverings are simple 3-fold coverings $p_1, p_2 : 4\#S^1 \times S^2 \to S^3$. If p_1 and p_2 were cobordant, then T^4 would be a 3-fold simple covering of S^4 branched over a (no locally flat) 2-manifold, which is not the case by the theorem of Edmonds [1]. Hence, by the above theorem, it is not possible to connect the colored branching sets of p_1 and p_2 by any sequence of the moves $C^{\pm 1}$.

3. Notes.

1. In Figure 3 there is a Heegaard diagram of T^4 [12]. The 3-manifold represented by the surgeries of the same figure is $\partial U = 4\#S^1 \times S^2$. Using the methods of [11], it is possible to get the branching set B of $p_1 : \partial U \to S^3$. The

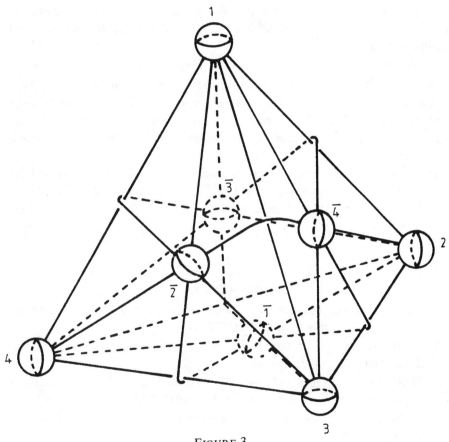

FIGURE 3

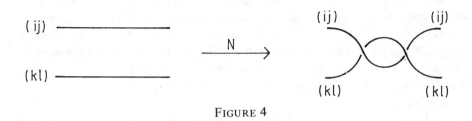

FIGURE 4

branching set of $p_2 : \partial V \to S^3$ is the trivial link of 6 components. Note that $4\#S^1 \times S^2$ is a 2-fold cover of S^3 but that the colored link B is not separable (compare [2], [13]).

2. The colored link B provides a new move independent of $C^{\pm 1}$.

3. It is still possible that the following question has an affirmative answer.

Question. *Is it true that the set of moves $S = \{C^{\pm 1}, N^{\pm 1}\}$ (Figure 4) is such that the branching sets of any two simple 4-coverings $p_1, p_2 : n\#S^1 \times S^2 \to S^3$ (coming from 3-fold simple coverings by addition of a trivial sheet) are related by a finite sequence of the moves S?*

If this was true, then an easy modification of the above theorem and [11] would show that *every closed, oriented, PL 4-manifold would be a 4-simple covering of S^4 branched over an immersed PL 2-manifold with only cusp and node singularities.*

REFERENCES

[1] A. L. Edmonds, *The degree of a branched covering of a sphere*, Geometric Topology, Academic Press (1979), 337–343.

[2] R. H. Fox, *A note on branched cyclic coverings of spheres*, Revista Mat. Hisp.-Amer. **32** (1972), 158–166.

[3] H. M. Hilden, *Every closed orientable 3-manifold is a 3-fold branched covering space of S^3*, Bull. Amer. Math. Soc. **80** (1974), 1243–4.

[4] H. M. Hilden, *Three-fold branched coverings of S^3*, Amer. J. Math. **98** (1976), 989–997.

[5] U. Hirsch, *Über offene Abbildungen auf die 3-sphäre*, Math. Z. **140** (1974), 203–230.

[6] U. Hirsch, *Bordismus verzweigter Überlagerungen von niedrig-dimensionalen sphären*, Manus. Math. **29** (1979), 1–10.

[7] J. M. Montesinos, *Sobre la Conjectura de Poincare y los recubridores ramificados sobre un nudo*, Tesis doctoral, Madrid 1971.

[8] J. M. Montesinos, *Una nota a un teorema de Alexander*, Revista Mat. Hisp.-Amer. **32** (1972), 167–187.

[9] J. M. Montesinos, *A representation of closed orientable 3-manifolds as 3-fold branched coverings of S^3*, Bull. Amer. Math. Soc. **80** (1974), 845–846.

[10] J. M.Montesinos, *Three-manifolds as 3-fold branched covers of S^3*, Quarterly J.Math. Oxford *(2)* **27** (1976), 85–94.

[11] J. M. Montesinos, *4-manifolds, 3-fold covering spaces and ribbons*, Trans. Amer. Math. Soc. **245** (1978), 453–467.

[12] J. M. Montesinos, *Heegaard diagrams for closed 4-manifolds*, Geometric Topology, Academic Press (1979), 219–237.

[13] J. M. Montesinos, *Lectures on 3-fold simple coverings and 3-manifolds*, (A. Stone proceedings, 1982).

FACULTAD DE CIENCIAS, UNIVERSIDAD DE ZARAGOZA

ABCDEFGHIJ – AMS – 898765